200 More Puzzling Physics Problems

With Hints and Solutions

Like its predecessor, *200 Puzzling Physics Problems*, this book is aimed at strengthening students' grasp of the laws of physics by applying them to situations that are practical and to problems that yield more easily to intuitive insight than to brute-force methods and complex mathematics. The problems are chosen almost exclusively from classical (i.e. non-quantum) physics, but are no easier for that. They are intriguingly posed in accessible non-technical language, and require readers to select an appropriate analysis framework and decide which branches of physics are involved. The general level of sophistication needed is that of the exceptional school student, the good undergraduate or the competent graduate student. Some physics professors may find some of the more difficult questions challenging. By contrast, the mathematical demands are relatively minimal, and seldom go beyond elementary calculus. This further book of physics problems should prove not only instructive and challenging, but also enjoyable.

PÉTER GNÄDIG graduated as a physicist from Eötvös University (ELTE) in Budapest in 1971 and received his PhD degree in theoretical particle physics there in 1980. He worked as a researcher (in high-energy physics) and a lecturer in the Department of Atomic Physics at ELTE until he retired in 2010. Between 1985 and 2004 he was one of the leaders of the Hungarian Olympic team that took part in the International Physics Olympiad. Since 1989 he has been the physics editor for the *Mathematical and Physical Journal for Secondary Schools* (Budapest). He is one of the authors of *200 Puzzling Physics Problems*.

GYULA HONYEK graduated as a physicist from Eötvös University (ELTE) in 1975 and finished his PhD studies there in 1977, after which he stayed on as a researcher and lecturer in the Department of General Physics. In 1984, following a two-year postgraduate course, he was awarded a teacher's degree in physics. In 1985 he transferred to the teacher training school at ELTE, and held a post as mentor and teacher at Radnóti Grammar School (Budapest) until his retirement in 2011. Between 1986 and 2011 he was one of the leaders of the Hungarian Olympic team that took part in the International Physics Olympiad. He is also one of the authors of *200 Puzzling Physics Problems*.

MÁTÉ VIGH graduated as a physicist from Eötvös University (ELTE) in 2010, and will shortly be proceeding to his PhD. He took part in the International Physics Olympiad (IPhO) in 2003 and 2004 as a contestant, being awarded, respectively, bronze and silver medals. Since 2012 he has been one of the leaders and trainers of the Hungarian teams preparing for and participating in the IPhO.

KEN RILEY read mathematics at the University of Cambridge and proceeded to a PhD there in theoretical and experimental nuclear physics. As well as following a research career in particle physics, he was the Senior Tutor at Clare College, Cambridge, and a University Lecturer at the Cavendish Laboratory, where he taught physics and mathematics for over 40 years. Among his other publications are the widely established mathematics textbook *Mathematical Methods for Physics and Engineering* (now in its third edition), and the physics problems books *Problems for Physics Students* and *200 Puzzling Physics Problems*, all published by Cambridge University Press.

200 More Puzzling Physics Problems

With Hints and Solutions

Péter Gnädig

Eötvös Loránd University, Budapest

Gyula Honyek

Radnóti Grammar School, Budapest

Máté Vigh

Eötvös Loránd University, Budapest

Ken Riley

Consultant Editor for translation and scientific editing

Clare College, Cambridge

Shaftesbury Road, Cambridge CB2 8EA, United Kingdom

One Liberty Plaza, 20th Floor, New York, NY 10006, USA

477 Williamstown Road, Port Melbourne, VIC 3207, Australia

314–321, 3rd Floor, Plot 3, Splendor Forum, Jasola District Centre, New Delhi – 110025, India

103 Penang Road, #05–06/07, Visioncrest Commercial, Singapore 238467

Cambridge University Press is part of Cambridge University Press & Assessment,
a department of the University of Cambridge.

We share the University's mission to contribute to society through the pursuit of
education, learning and research at the highest international levels of excellence.

www.cambridge.org
Information on this title: www.cambridge.org/9781107103856

© P. Gnädig, G. Honyek and M. Vigh 2016

First published 2016

A catalogue record for this publication is available from the British Library

Library of Congress Cataloging-in-Publication data
Names: Gnädig, Péter, author.
Title: 200 more puzzling physics problems with hints and solutions / Péter
Gnädig, Eötvös Loránd University.
Other titles: Two hundred more puzzling problems in physics with hints
and solutions
Description: Cambridge, United Kingdom; New York, NY: Cambridge University
Press, 2016. | ©2016 | Includes bibliographical references and index.
Identifiers: LCCN 2015037120| ISBN 9781107103856 (Hardback : alk. paper) |
ISBN 1107103851 (Hardback : alk. paper) | ISBN 9781107503823 (pbk. : alk. paper) |
ISBN 1107503825 (pbk. : alk. paper)
Subjects: LCSH: Physics–Problems, exercises, etc.
Classification: LCC QC32 .G523 2016 | DDC 530.076–dc23 LC record available
at http://lccn.loc.gov/2015037120

ISBN   978-1-107-10385-6   Hardback
ISBN   978-1-107-50382-3   Paperback

Dedicated to

Frederick Károlyházy

from whom we have learned so much

Contents

Preface

As was said in the preface to the predecessor of this book (*200 Puzzling Physics Problems*, Cambridge University Press, 2001), an understanding of the laws of physics is best acquired by applying them to practical problems. Many of the corresponding solutions, however, require routine, but perhaps long and boring, calculations, which tend to deter even the most curiosity-driven students of the subject. This book, like its antecedent, aims to show that not all physics problems are like that, and that a bit of careful thought, a little ingenuity and a flash of insight can go a long way.

Although we have aimed to place as many problems as possible in settings that will be familiar to, and easily understood by, most people (not just physics students and their professors), some have had to be somewhat artificially constructed in order to bring the physics involved to the fore. However, that said, many of these contrived situations can be set up in a laboratory, and theory can be tested against experiment. Even so, some 'test areas', especially those in outer space, and some apparatus, in particular a copious supply of infinitely long rods, were beyond the resources available to us!

Nevertheless, we hope that you will be intrigued by questions such as:

- How do you maximise the gravitational effect of a lump of plasticine?
- How does a spoked wheel appear in a photo-finish picture?
- What happens when a suspended Slinky is suddenly released?
- How can your square-on reflection in a plane mirror show your closed eye, but not your open one?
- How long is it before Santa Claus is discharged?
- Does an electromagnetic field carry angular momentum?
- What is the path of a ball rolled onto a rotating turntable?
- How much charge flows when a magnet is dropped through a metal loop?
- Where should you park your car to avoid a frosted windscreen?

- How large is the force between end-to-end solenoids?
- How do you bring about 'the parting of the waters' in a Florence flask?
- When does an Euler strut go phut?
- What is the pressure produced by a neutron in a box?
- How long is an 'infinite electronic chain'? Does it matter?
- How much harder is it to steer a car with flat tyres?
- What is the maximum speed with which a comet could hit the Earth?
- How do you get into shape for a 'free-wheel bike race'?
- Why do icebergs floating in the open ocean last so long?

These, those on the book cover and some 170 others are problems that can be solved elegantly by an appropriate choice of variables or coordinates, an unusual way of thinking, or some cunning idea or analogy. Of course, when such a eureka moment arrives, and the solution is then found with a minimum of effort, the reader will, quite justifiably, have every reason to be pleased with him- or herself.

However, it needs to be said that inspiration of the kind needed to produce such insights is most unlikely to come to anybody who does not have a sound knowledge and understanding of the basic laws of physics. The vast majority of the problems are based on classical physics, of the kind taught in sixth forms and the early years of university. And even the 'modern' physics questions demand little beyond elementary relativity – there is not a single quantum mechanical wavefunction in sight! This is readily understandable, as real intuition in the field of quantum physics – which often comes in the form of counter-intuition – usually merits a Nobel prize.

Although essentially correct solutions to the problems are clearly the principal goal, we should add that success is not measured by this alone. Whatever help you, the reader, may seek, and whatever stage you may reach in the solution to a problem, we hope that it will bring you both enlightenment and satisfaction, as well as increase your capacity to think in novel ways.

The 200, hopefully interesting, problems contained in this book have been collected by the authors over the course of many years. Some were invented by us, and the rest are, for the most part, taken from the Hungarian *Mathematical and Physical Journal for Secondary Schools*, covering a span of more than 100 years, or from other Hungarian physics contests. We have selected a few very challenging questions from the Boston Area Undergraduate Physics Competition (BAUPC). We have also been guided by the suggestions and remarks of our colleagues. It is impossible to determine the original authors of most of the physics problems appearing in the international 'ideas market'. Nevertheless, some of the inventors of the most puzzling problems deserve our special thanks. They include Zsolt Bihary, András Bodor, László Holics, Jaan Kalda, Frederick Károlyházy[†], Gyula

Radnai, Géza Tichy, István Varga[†] and Károly Vladár. We thank them and the other people, known and unknown, who have authored, elaborated and improved upon 'puzzling' physics problems.

Péter Gnädig, Gyula Honyek, Máté Vigh

How to use this book

This book has a two-fold objective: to teach and train students, and to intrigue and entertain everybody who likes physics and puzzles. The first chapter contains the 200 problems. If the reader's main objective is to use the book as a basis for studying physics, we recommend that the problems are tackled in the order they appear, though this does not necessarily represent their order of difficulty. Of course, if the reader is using the book for fun or as a challenge, he or she can freely 'cherry-pick' the questions.

It is an essential part of a physicist's skill to be able to recognise the type of problem in front of them, and to identify which areas of science and mathematics will need to be called upon. Almost needless to say, some of the problems could not be unambiguously assigned to, say, mechanics or gravitation or electromagnetism. Nature's secrets are not revealed according to the chapter titles of a textbook, but rather draw on ideas from various areas, and usually in a complex manner. However, after the present section, *for information*, we have included a list of topics, and the numbers of the problems that more or less belong to those topics. Some problems are listed under more than one heading. A list of symbols and numerical values for the principal physical constants, as well as several tables of material and astronomical data, not all of which will be needed, are provided in the Appendix at the end of the book. The Appendix also contains many standard mathematical results, drawn from the areas of vector algebra, conic sections, trigonometry, calculus and solid geometry.

The majority of the problems are not easy; some of them are definitely difficult. You, the reader, are naturally encouraged to try to solve them on your own and, obviously, if you do you will get the greatest satisfaction. If you are unable to achieve this, you should not give up, but turn to the relevant page of the *hints* in the short second chapter. In most cases this will help, though it will not give the complete solution, and the details will still have to be worked out. Once you have done this and want to check your result (or if you have given up and just want to see the *solution*) the final chapter should be consulted.

If a particular problem relates to another one elsewhere in this book, you will find a page reference (rather than a problem number reference) in the relevant hint or solution. Sometimes this may only be a few pages away, but on other occasions it is far removed. In a few places, reference is made to this book's predecessor (*200 Puzzling Physics Problems*, Cambridge University Press, 2001), but the hint or solution can be followed without *having* to consult this external source. Problems whose solutions require especially difficult reasoning or more demanding mathematical calculations are marked by one or two *asterisks* – and one problem has earned itself a *three-star* rating!

There are some problems whose solutions raise questions that are beyond the scope of this book. Points or issues worth further consideration, as well as outlines of possible alternative solution methods, are indicated in *Notes* at the end of the relevant main solution, but answers are not usually given.

Thematic order of problems

Kinematics: 1, 2, 3, 4*, 5, 7*, 8, 9, 10*, 11*, 12, 13*, 14**, 92.

Dynamics: 6*, 15, 16, 17*, 18, 19*, 20, 21*, 22**, 23**, 24*, 25, 27*, 37*, 38**, 39**, 40*, 45*, 130, 164*, 165*, 166*, 167**, 177, 186*.

Gravitation: 28*, 29**, 30, 31, 32, 33, 34, 35**, 36**, 37*, 38**, 39**, 40*.

Mechanical energy: 7*, 8, 23**, 25, 26, 36**, 38**, 41, 48*, 64*, 65*, 69, 71, 80*, 84*.

Collisions: 35**, 36**, 42*, 43, 44**, 45*, 52, 196*.

Rigid-body mechanics: 38**, 46, 47*, 48*, 49, 50**, 51*, 52, 53*, 54**, 55**, 56*, 57**.

Elasticity: 24*, 58, 59*, 60*, 61**, 62**, 63*, 64*, 65*, 66, 67**.

Statics: 68*, 69, 70, 71, 72, 73, 74**, 76*, 78**, 79*, 80*.

Ropes and chains: 75, 76*, 77, 78**, 79*, 80*, 81**, 82*, 83**.

Liquids and gases: 11*, 85, 86, 87, 88, 89*, 90, 91*, 92, 93**, 95*, 142*.

Surface tension: 67**, 91*, 94*, 96, 97*, 98*, 99*, 142*.

Thermodynamics: 32, 44**, 84*, 93**, 100*, 101*, 102*, 103*, 104, 105*, 106**, 107**, 108, 109*, 198*.

Phase transitions: 93**, 110, 111, 112, 113*, 114*, 115, 116*.

Optics: 107**, 117**, 118**, 119, 120*, 121*, 122*, 123*, 124*, 125, 126*, 127*, 128**.

Electrostatics: 129*, 130, 131*, 132*, 133*, 134, 135, 136*, 137*, 138**, 139*, 140**, 141*, 142*, 143, 144, 145**, 146*, 147*, 148**, 149**, 150*, 151, 152*, 153*, 154**, 160**, 172*.

Magnetostatics: 155, 156*, 157*, 158*, 159**, 160**, 162*, 163*, 164*, 165*, 166*, 167**, 172*, 197.

Electric circuits: 106**, 161***, 168, 169*, 170, 171**, 172*, 173*, 174*, 175, 176*, 183*, 184*, 185**.

---

* Asterisks indicate problems that require more difficult reasoning or somewhat more advanced mathematics. Many problems warrant two asterisks and one warrants three.

Problems

P1   The trajectories of two bodies moving with non-relativistic constant speeds are parallel in a particular inertial reference frame.

a) Is it possible to choose another inertial frame of reference in which the two trajectories cross each other?

b) If such a frame can be found, and the bodies are started with suitable initial conditions, then it could be arranged that they reach the crossing point at the same time. How can this be consistent with the parallel trajectories observed in the first frame of reference?

P2   Ann is sitting on the edge of a carousel that has a radius of 6 m and is rotating steadily. Bob is standing still on the ground at a point that is 12 m from the centre of the carousel. At a particular instant, Bob observes Ann moving directly towards him with a speed of 1 m s^{-1}. With what speed does Ann observe Bob to be moving at that same moment?

P3   A cart is moving on a straight road with constant velocity v. A boy, standing in an adjoining meadow, spots the cart and hopes to get a ride on it. In which direction should he run to catch the cart? Solve the problem generally: denote the speed of the cart by v, the maximal speed of the boy by u, and take the initial positions of the cart and boy to be as shown in the figure.

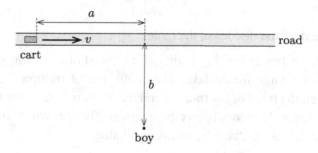

P4* A group of Alaskan gold prospectors reach a wide straight river that flows with uniform speed v. What immediately catches all their eyes is a huge gold nugget lying on the further bank, directly across the river. The laws governing prospecting in Alaska state that the first person to reach any particular place has the right to establish a mine there; speed is of the essence!

Joe, one of the prospectors, has a canoe, which he can paddle in still water at the same speed u as he can hike along a river bank. What course of action should he take if u/v is (*a*) smaller than or (*b*) larger than a certain critical value? Assume that Joe first paddles across the river (in a straight line) and then, if necessary, hikes along the bank to reach the nugget.

P5 The top surface of a horizontal laboratory table is a square of side $3d = 3$ m. Running centrally across the table, and parallel to one of its sides, is a conveyor belt consisting of an endless rubber band of width $d = 1$ m, which moves with a constant velocity $V = 3$ m s^{-1}. The height of the belt's upper surface exactly matches that of the static part of the table.

A small, flat disc is placed at the middle of one of the edges of the table (at the point A shown in the figure), and the disc is hit so that it starts sliding with velocity $v_0 = 4$ m s^{-1} at right angles to the belt. The friction between the disc and the static part of the table is negligible, while the coefficient of kinetic friction between the disc and the rubber band is $\mu = 0.5$.

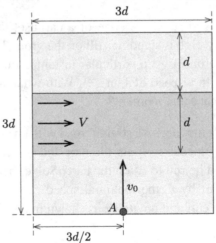

Where does the disc leave the table?

P6* A boy is running north, with a speed of $v = 5$ m s^{-1}, on the smooth ice cover of a large frozen lake. The coefficient of friction (both kinetic and static) between the tread of his trainers and the ice is $\mu = 0.1$. For the sake of simplicity, assume that the normal force he exerts on the ice, which in reality changes with time, can be substituted by its average value.

a) What is the minimal time that he needs to change direction, so that he is running east with the same speed v?

b) Find the boy's trajectory during the turn in this optimal case.

P7* A simple pendulum is released from rest with its string horizontal. What kind of curve is the locus of the end of its acceleration vector?

P8 A simple pendulum is released from rest with its string horizontal. Which of the two arcs, AP and PB as defined in the figure, will its bob cover in a shorter time?

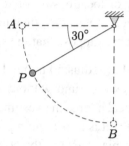

P9 The trajectory of a projectile with initial speed v_0 is parabolic in a vacuum (e.g. on the Moon). How far is the focus of this parabola from the launch point? What initial angle of elevation of the projectile is needed if the focus is to be at the same altitude as the launch point?

P10* Point-like objects are thrown with an initial speed of v_0 in various directions from the top of a tower of height h. If the air resistance is negligible, what is the maximum distance from the foot of the tower that they can reach?

P11* At the top of a long incline that makes an angle θ with the horizontal, there is a cylindrical vessel containing water to a depth H. A hole is to be drilled in the wall of the cylinder, so as to produce a water jet that lands a distance d down the incline. How far, h, from the bottom of the vessel should the hole be drilled in order to make d as large as possible? What is this maximum value of d?

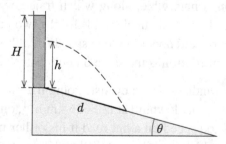

P12 Because of the finite exposure needed, in a side-on photograph of the front wheel of a moving bicycle, the spokes seem blurred. However, there will be some apparently sharp points in the picture. Where are these sharp points? For the sake of simplicity, suppose that the bicycle spokes are radial.

P13* Investigate the form of the image of a spoked bicycle wheel as recorded by a photo-finish camera. Such cameras use very narrow strip photography, electronically capturing a vertical cross-section of the sequence of events only on the finish line; every part of each body is shown as it appeared at the moment it crossed the finish line. The horizontal axis of the image represents time; anything stationary on the finish line appears as a horizontal streak. In a conventional photograph, the image shows a variety of locations at a fixed moment in time; strip photography swaps the time and space dimensions, showing a fixed location at a variety of times. For the sake of simplicity, suppose that the spokes of the bicycle are radial.

P14** A cartwheel of radius 50 m has 12 spokes, assumed to be of negligible width. It rolls along level ground without slipping, and the speed of its axle is 15 m s^{-1}. Use a graphical approach to estimate the minimal speed a crossbow bolt, 20 cm long, must have if it is to pass unimpeded between the spokes of the wheel? Neglect any vertical displacement of the bolt.

P15 A small bob can slide downwards from point A to point B along either of the two different curved surfaces shown in cross-section in the figure. These possible trajectories are circular arcs in a vertical plane, and they lie symmetrically about the straight line joining A to B. During either motion the bob does not leave the curve.

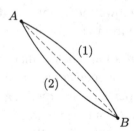

a) If friction is negligible, along which trajectory does the bob reach point B more quickly? How do the final speeds for the two paths compare?

b) What can be said about the final speeds if, although friction is not negligible, the coefficient of friction is the same on both paths?

P16 On a windless day, a cyclist 'going flat out' can ride uphill at a speed of $v_1 = 12 \text{ km h}^{-1}$, and downhill at $v_2 = 36 \text{ km h}^{-1}$, on the same inclined road. What is the cyclist's top speed on a flat road if his or her maximal effort is independent of the speed at which the bike is travelling?

The rather imprecise term 'effort' could be interpreted scientifically to mean either *a)* the magnitude of the *force* exerted on the pedals by the rider (which is then transmitted to the wheels via the crank arms, sprockets and chain), or

b) the rider's mechanical *power*.

Solve the problem for both interpretations.

P17* Ann and Bob arrange a 'free-wheel bike race' on a very long slope that makes an angle θ with respect to the horizontal. They have the same type of bicycle, and neither of them pedals during the race. The total mass of Ann and her bike is $m_A = 60$ kg, whereas the corresponding figure for Bob is $m_B = 110$ kg. Because he is overweight (and 'out of shape'), the air drag on Bob is one-and-a-half times larger than that on Ann when they have equal speeds. Which one is going to coast further on the horizontal road at the bottom of the slope?

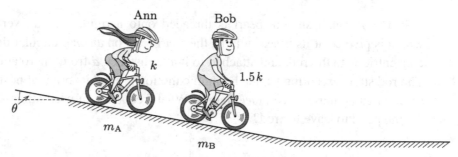

Assume that their decelerations are due to air drag (proportional to the square of the speed), friction at the bearings hub and rolling friction. The latter two effects should be treated as a sort of kinetic friction, with an 'effective frictional coefficient' of μ.

P18 A small feather with vanishingly small mass is attached to one end of a riding crop by a flimsy thread. The crop is then rotated steadily about an axis passing through its other end and perpendicular to it. What is the trajectory of the feather?

P19* A small pearl moving in deep water experiences a viscous retarding force that is proportional to its speed (Stokes' law). If a pearl is released from rest under the water, then it soon reaches its terminal velocity v_1, and continues sinking with this velocity.

In an experiment, such a pearl is released horizontally with an initial speed v_2.

a) What is the minimal speed of the pearl during the subsequent motion?

b) In which direction should the pearl be projected, with the same initial speed v_2 ($< v_1$), in order that its speed increases monotonically during its descent?

P20 Two spherical bodies, with masses m and M, are joined together by a light thread that passes over a table-mounted pulley of negligible mass. Initially they are held in the positions shown in the figure; then, at a given moment, both of them are released. Mass M is many times – say, one thousand times – larger than mass m. The friction between the smaller ball and the surface of the table is negligible. Will the lighter ball be lifted from the table-top immediately after the release?

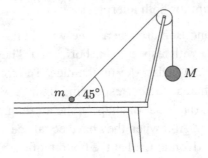

P21* A small smooth pearl is threaded onto a rigid, smooth, vertical rod, which is pivoted at its base. Initially, the pearl rests on a small circular disc that is concentric with the rod, and attached to it at a distance d from the rotational axis. The rod starts executing simple harmonic motion around its original position with small angular amplitude θ_0 (*see* figure). What frequency of oscillation is required for the pearl to leave the rod?

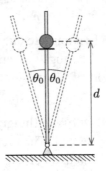

P22** The plane of a flat, rigid board of length $L = 6$ m makes an angle of $\alpha = 10°$ with the horizontal, and a small rectangular block is situated at the top of this incline. The board starts vibrating with simple harmonic motion in the direction parallel to a line of steepest descent; the amplitude of the motion is $A = 1$ mm, and its angular frequency is $\omega = 500$ rad s^{-1}. The coefficients of kinetic and static friction between the block and the board are both $\mu = 0.4$. Estimate how long the block, which does not topple over during its motion, takes to reach the bottom of the incline.

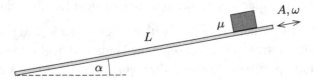

P23**   The cord of a swinging simple pendulum passes through a small hole in a ceiling and into a loft above. There, a scientist's assistant holds the loose end of the cord and pulls it up very slowly (*see* figure). Does the linear amplitude (the largest horizontal excursion) of the pendulum change? If so, how?

P24*   A mountaineer (a former circus artist) has to spend the night on the (vertical) side of a high mountain. So, as shown in the figure, he clamps himself to four carabiners fixed to the rock face using four extraordinarily flexible springs. The masses of the springs and their unstretched lengths are negligible, and their spring constants are $k_1 = 150$ N m^{-1}, $k_2 = 250$ N m^{-1}, $k_3 = 300$ N m^{-1} and $k_4 = 400$ N m^{-1}. The mountaineer can be considered – for the sake of simplicity – as a point-like body with a mass $m = 70$ kg.

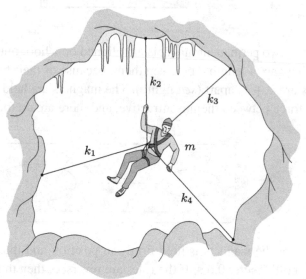

What is the mountaineer's period of oscillation if he is displaced from his equilibrium position and then released?

P25   A small rubber eraser is placed at one edge of a quarter-circle-shaped track of radius R that lies in a vertical plane and has its axis of symmetry vertical (*see* figure); it is then released. The coefficient of friction between the eraser and the surface of the track is $\mu = 0.6$. Will the eraser reach the lowest point of the track?

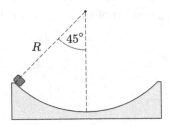

P26   A box of mass 1 kg is placed on an incline on which it does not spontaneously start to slide. It is pulled up, and then down, the incline very slowly in such
a way that the traction force is always parallel to the slope (*see* figure). The total
work done is 10 J. What is the maximum height h of the incline? Assume that the
coefficients of static and kinetic friction are equal.

P27*   Two permanent magnets are aligned on a horizontal, very slippery table-
top with a gap of length d between them; because of their finite sizes, their centres
of mass are $d + d_0$ apart (*see* figure). The magnets are held in such a position that
the net force between them is attractive, and there are no torques generated.

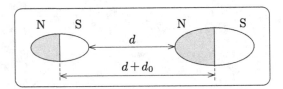

If one of the magnets is held firmly in position and the other is released, then
the two collide after 0.6 s. If the roles are reversed, then the time interval between
the release and the collision is 0.8 s. How long would it take the two magnets to
collide, if both were released simultaneously?

P28*   A U-shaped tube contains liquid that initially is in equilibrium. If a heavy
ball is placed below the left arm of the tube, how do the liquid levels in the two
arms change?

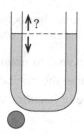

P29** We wish to produce the maximum possible gravitational acceleration at a given point in space, using a piece of plasticine[1] of uniform density and given volume. Into what shape should the plasticine be moulded?

P30 One of the planets of a star called 'Noname' has the shape of a long cylinder. The average density of the planet is the same as that of the Earth, its radius R is equal to the radius of the Earth, and the period of its rotation around its long axis is just one day.

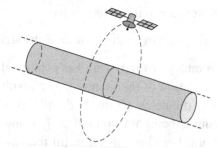

a) The first cosmic velocity $v_{c,1}$ is the speed of a satellite in a stable orbit just above the planet's surface. How large is it for this planet?

b) What is the altitude of a 'geostationary' communications satellite above the surface of this 'sausage' planet?

c) What can be said about the second cosmic velocity (the escape velocity) for this planet?

P31 The Examining Institute for Cosmic Accidents (EXINCA) sent the following short report to one of its experts:

> One of the exploration space ships belonging to the titanium-devouring little green people has found a perfectly spherical, homogeneous asteroid; it has no atmosphere, but is made of pure titanium. As part of the preparations for mining, a straight tunnel was constructed, and railway lines were laid in it. The length of the tunnel was equal to the radius of the asteroid, with both ends on the latter's surface. Unfortunately, although its braking system was on and locked, one of the mine wagons slipped into the shaft at one end of the tunnel. Initially it speeded up, but later it gradually slowed down, reversed and finally stopped exactly in

---

[1] Although this word has passed into common usage, technically it is a registered trademark for Plasticine.

the middle of the tunnel. Just before it reversed, the wagon came very close to running down the mine captain, who was standing on the track.

EXINCA asked the expert (you) to obtain numerical values for the following:

  a) how far along the tunnel the mine captain was standing,

  b) the coefficient of kinetic friction between the wheels of the mine wagon and the rails,

  c) the total time of the wagon's motion.

Assume that the volume of the tunnel is negligible compared to that of the asteroid.

P32   A space ship carrying titanium-devouring little green men has found and landed on a perfectly spherical planet of radius R. A narrow trial shaft has been bored from point A on the surface of the planet to O, its centre; this has confirmed that the whole planet is made of homogeneous (edible) titanium. In addition, according to the measurements made, the temperature inside the narrow shaft is constant, and equal to T_0. The planet has an atmosphere with molar mass M, and the atmospheric pressure at its surface is p_A.

  a) Find the air pressure at the bottom of the shaft.

After the exploratory drilling, work has continued, and the little green men have started secret excavation of the titanium, as a result of which they have formed a spherical cavity of diameter AO inside the planet, as illustrated in the figure. The excavated titanium is being transported away using expendable cargo space craft. Air from the atmosphere has moved to fill the cavity and, as a consequence, the pressure at the access point A has decreased from p_A to p'_A.

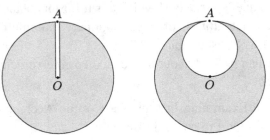

  b) Assuming that the temperature everywhere inside the cavity is the same as it was in the shaft, how has the atmospheric pressure at O changed?

P33   Mr Tompkins[2] visited Wonderland in his dream, where the laws of physics are almost the same as we know them – except that gravity deviates 'slightly' from Newton's well-known law. Awakening with a start, he remembered

---

[2] He is the eponymous main character in the physicist George Gamow's book, *Mr Tompkins in Wonderland*, first published in 1940.

that several planets orbit the only 'sun' in Wonderland, and that these planets obey the following three 'Kepler's laws':

1. Planets travel in elliptical orbits with the sun at the *centre* of the ellipse.
2. A line drawn from a planet to the sun *sweeps* ... (unfortunately, Mr Tompkins forgot how to continue).
3. The orbital periods of all the planets (independent of the sizes of their major and minor axes) are the *same*, namely one 'year'.

How does the gravitational law look in the physics textbooks of Wonderland, and what is the missing part of the statement of 'Kepler's second law'?

P34   In the absence of an atmosphere on Earth, what would be the maximal and minimal impact speeds with which a comet, which is orbiting the Sun, can strike the Earth?

P35**   Two comets with identical masses and speeds are found by astronomers to be approaching the Sun along parabolic trajectories that lie in the same plane. The comets collide at their common perihelion P (the point in their trajectories that is nearest to the Sun S), and break into many pieces that then go in all directions, but with identical initial speeds (*see* figure).

What shape is the envelope of the subsequent trajectories of the pieces of debris?

P36**   We aim to make a space probe launched from Earth leave the Solar System with the help of a single gravitational slingshot, which utilises the relative movement and gravity of one of the planets that orbit the Sun. In astronomical units (the mean Sun–Earth distance = 1 AU), how far from the Sun would the 'ideal' planet be if the initial launch speed of the probe is to be kept to a minimum?

When solving the problem, make the following approximations:

- The orbits of the planets are circles all lying in the same plane.
- Near a planet, it is sufficient to take into account only the gravity due to that planet.
- Far from all planets, only the Sun's gravity is relevant.

Does a real planet exist at or near the optimal orbit?

P37* Because of the effects of air drag, abandoned satellites, at the end of their useful lives, lose energy in the upper layers of the atmosphere, before finally burning up when they reach the denser lower layers. It can be shown that satellites originally moving along circular trajectories will continue to travel in approximately circular orbits, with their orbital radii slowly decreasing.

A half-tonne satellite is orbiting the Earth on a roughly circular orbit when it is abandoned. The drag acting on this particular satellite can be expressed as $c\varrho v^2$, where $c = 0.23$ m^2, ϱ is the density of air at the altitude of the satellite and v is the speed of the satellite.

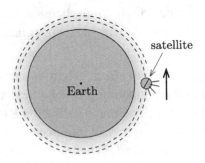

a) Does the satellite brake or accelerate as a result of the air drag? How can your answer be explained from the point of view of dynamics?

b) A simple connection can be found between the drag force and the tangential acceleration of the satellite. What is it?

c) What is the density of air at an altitude of 200 km, if in this region the orbital radius of the satellite decreases by 100 m during a single revolution?

P38** It is a well-known fact that the Moon always shows, more or less, the same face to the Earth. This curiosity is not a coincidence, but a straightforward consequence of the tidal forces acting between the Earth and the Moon. Over time, tidal forces continuously slowed the Moon's rotation about its own axis until the period of that rotation became equal to the Moon's orbiting period around the Earth. It is for the same reason that the Earth's rotation around its own axis is continually slowing, and the orbital speed of the Moon is further decreasing.

a) Estimate the ratio of the rates of decrease of the Earth's and Moon's kinetic energies.

b) During the Apollo Program (flights 11, 14 and 15), retro-reflectors (arrays of corner-cube laser mirrors) were placed on the Moon. According to the extremely accurate laser ranging measurements thereby made possible, the

Moon's linear distance from Earth is currently increasing at a rate of 3.8 cm per year. Using this measured datum, estimate the change in the length of an Earth day during a year.

c) If the Earth–Moon system continued its motion undisturbed, then, as a result of the braking effect of the tidal forces, after a sufficiently long time, the Earth would always show the same face to the Moon, i.e. the rotations and the orbital motions of these two bodies would be synchronised.[3] How many times larger than at present would an Earth day and the Earth–Moon distance be with such perfect synchrony?

Assume that the orbit of the Moon remains circular, and neglect the tidal effect of the Sun.

P39[**] An astronaut, who seems light-headed, but who is an expert in celestial mechanics, jumps away from the International Space Station (ISS) in the direction directly opposed to that of the Earth, with a speed of $v_0 = 0.1$ m s^{-1}. He has an oxygen cylinder, but no lifeline tether or auxiliary jet pack. What will be his greatest subsequent separation from the ISS? If it is finite, for how long must his oxygen supply last?

P40[*] For an interstellar space mission, set in the future, an attempt is made to partially compensate for the lack of gravity by uniformly rotating the reasonably long and very heavy cylindrical space ship, which has a diameter of $2R = 20$ m, around its symmetry axis. The period of rotation is adjusted so that the astronauts feel an Earth-like 'gravitational acceleration' of $g = 10.00$ m s^{-2} at the outer edges of the cylinder.

During the long journey, the astronauts exercise in a 5 m 'high' gym, whose 'floor' is the outer casing of the space ship; they notice that things are not quite like they are on Earth.

a) How much work is done by a (point-like!) astronaut with a mass of 80 kg, when he climbs up to the gym's ceiling on a fixed climbing pole. By how much, measured in an inertial frame of reference fixed to the space ship's centre of mass, does the astronaut's kinetic energy change as a result of the climb? How can the work done be reconciled with the change in his kinetic energy?

b) If the astronaut fell from the top of the pole, how long would he take to fall, and how far from the bottom of the climbing pole (measured along the floor) would he 'hit the ground'?

P41 The height difference between the top and bottom of a downward-moving escalator is $h = 20$ m. A mischievous boy of mass $m = 50$ kg runs up from the

---

[3] In reality, before this could happen, the Sun will expand into a red giant and engulf the current orbit of the Earth.

bottom to the top at an (average) speed, relative to the steps, that is one-and-a-half times their translational speed. Find the work done by the boy and explain how it is accounted for.

P42* On an air-hockey table, there are N identical small discs lying equally spaced around a semicircle (*see* figure); the total mass of the discs is M. Another small disc, D, of mass m, travelling in a direction perpendicular to the closing diameter of the semicircle, strikes the first of the stationary discs. By some miracle, it subsequently bounces off all of the other $N - 1$ discs in turn, after which it is travelling in a direction directly opposed to that of its initial motion. All the collisions are perfectly elastic, and friction is everywhere negligible.

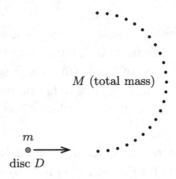

a) In the limiting case of $N \rightarrow \infty$, what is the minimal value of the mass ratio M/m for such a miracle to be possible?

b) When the mass ratio has the critical value found in part *a)*, what is the ratio of the final and initial speeds of D?

P43 Two identical balls are suspended on (vertical) threads of length ℓ so as to just touch each other and, as shown in the figure, can swing in the plane defined by their suspension points and their centres. One of the balls is drawn aside in this plane through a distance d ($\ll \ell$) and then released. Each time the balls collide, they do so inelastically, and, as measured in their centre-of-mass frame, their velocities decrease by a factor of k, the coefficient of restitution, where $0 < k < 1$.

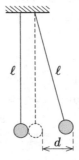

How will the balls be moving at a much later time? What will be the amplitude of the swings after a large number of collisions have taken place? Assume that the damping due to air resistance is very small.

P44** Two small balls are threaded onto a frictionless horizontal rod that protrudes from a vertical wall. The lighter ball with mass m is initially at rest, a distance L from the wall, while the much heavier second ball, of mass M, approaches the wall from a distance greater than L (*see* figure). After their elastic collision, the ball of mass m slides towards the wall, bounces back elastically, and again collides with the heavier ball. The process then repeats itself, over and over.

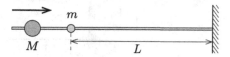

How close does the heavier ball get to the wall?

Consider the balls as point-like particles, and, where appropriate, assume that $m \ll M$.

P45* In a very dense fog, there are many tiny water drops that 'float' in the air with negligible speed. If one of the water drops, which is a little larger than the rest, begins to sink, it absorbs those smaller drops that lie in its path (*see* figure). The ever-growing drop, which can be regarded as spherical, is found to be accelerating uniformly, despite the air drag – proportional to the square of the speed and the cross-sectional area of the drop – acting upon it. What is the maximum possible value for this acceleration?

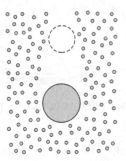

P46 A carved wooden American Indian statuette, which can be considered (to a good approximation) as a solid cylinder with a homogeneous mass distribution, has a height of $H = 6$ cm and a diameter of $d = 1$ cm (*see* figure). A thread is attached to the statuette at a point that is $h = 2$ cm above the base, and the statuette is placed in the middle of a rough horizontal table-top. The coefficient of friction between the table-top and the statuette is $\mu_{\text{kinetic}} = \mu_{\text{static}} = 1/3$.

As a challenge, contestants are required to pull horizontally on the thread and drag the statuette (in one continuous movement) to the edge of the table, without it falling over.

Is it possible to do this? If yes, how? If not, why not?

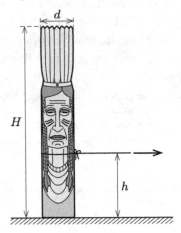

P47*   In the shop window of a toy store, a miniature sailing boat is suspended (with its deck horizontal) by two vertical thin rubber bands (*see* figure). The masses of the mast and sails are negligible compared to that of the boat's hull, the length of which, from bow to stern, is much greater than its vertical height, from keel to gunwale.

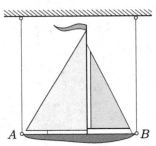

If the left-hand rubber band is cut, does point B sink or rise immediately afterwards?

P48*   Four identical, homogeneous rods are connected by four light frictionless knuckle joints[4] to form a square, which is then placed on a horizontal, smooth, polished table-top. Vertex P is pushed in the direction of the diagonal of the square that passes through P, and, as a result, acquires an initial acceleration of a_P (*see* figure).

---

[4] A knuckle joint allows the angle between the two components it joins to vary freely in one particular plane.

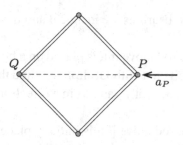

In which direction, and with how much acceleration, does the opposite vertex Q of the square start to move?

P49 A cylindrically symmetric (but not necessarily homogeneous) body is attached to two identical cords at points near its ends. The cords are partially wound in the same sense around the cylinder, and their free ends are fastened to points on a ceiling; initially, the cords are vertical and the cylinder is horizontal. A third cord is attached to and wound (in the same sense as the other two cords) around the middle of the cylinder; a heavy weight is tied to the free end of this cord (*see* figure).

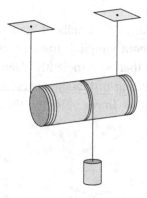

When the system is released from rest, what is the acceleration of the heavy weight?

P50** A homogeneous flat disc (such as an ice-hockey puck) is both sliding and rotating on an icy surface. Because of friction, both kinds of motion decelerate and finally stop. Which of the two motions stops the earlier, the rotation or the translation?

Assume that the disc presses uniformly on the ice, that the frictional force between two surfaces does not depend upon their relative speed, and that air drag is negligible.

P51* A uniform rod of mass m and length L is fitted at each end with a frictionless bearing in the form of a freely rotating wheel, for which the rod acts as an axle. The two bearings are identical and have negligible masses compared to that of the rod.

a) How does the rod move if it is placed on a horizontal rough surface – meaning that the bearings roll on it without slipping – and the two ends of the rod are initially given parallel velocities of v_1 and v_2 in a direction that is perpendicular to the axis of the rod?

b) How does the rod move if it is initially placed, with an angular velocity of ω_0 and zero centre-of-mass velocity, across the slope of a broad rough plane that is inclined at an angle of θ to the horizontal (*see* figure)?

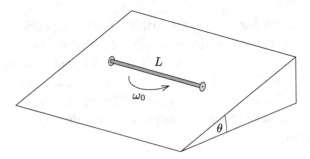

P52 Two identical billiard balls of diameter 5 cm, each moving with a speed of 3 m s^{-1}, roll, without slipping, towards each other on a horizontal rectangular U-shaped trough that is sufficiently deep that the balls are clear of its base (*see* figure). The resulting instantaneous collision is perfectly elastic, and, during it, each ball reverses its linear velocity, though their angular velocities are not affected.

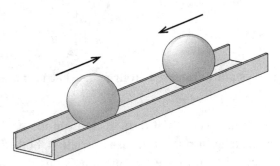

a) How wide does the trough need to be for the balls to collide twice?

b) Find the speed of the balls just before the second collision if the width of the trough is 4 cm.

P53* A billiard ball, initially at rest on a billiard table, is struck by a cue tip at the point T shown in the figure. The cue lies in the vertical plane containing T, the centre C of the ball and the ball's point of contact P with the table;

consequently, so does the line of action of the resulting impulse. Find the direction in which the cue should be aligned in order that, after the shot,

a) the ball's subsequent rotational and slipping motions terminate at the same instant, and the ball comes to a halt,

b) the ball rolls without slipping, whatever the value of the coefficient of static friction between it and the table.

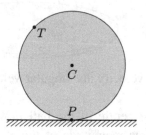

Assume that, as a result of chalking the cue tip, the coefficient of friction between it and the ball is sufficiently large that there is no slippage between them during the cue stroke.

P54**   If the line of action of the impulse in the previous problem does not lie in the vertical plane defined by the points *T*, *C* and *P*, then, just after the shot, the ball's angular velocity vector will not be perpendicular to the velocity of its centre of mass. Billiards players call this shot a *Coriolis-massé*.

Such a shot is shown in the figure, in which the line of action of the impulse meets the ball's surface (for a second time) at *T′* and the table at *A*.

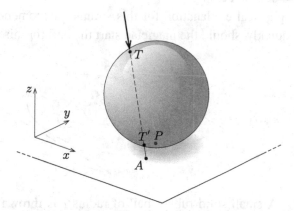

a) What kind of trajectory does the ball's centre of mass follow from just after the shot until the point at which simultaneous rolling and slipping cease?

b) In which direction, relative to the line *PA*, will the ball continue its path once it starts to roll without slipping?

Assume that, whatever the downward force acting on it, the billiard cloth does not 'become squashed', and the ball's contact with it is always a point contact.

P55** A very large flat horizontal disc with a rough surface is rotating about its vertical axis of symmetry with angular velocity Ω. A solid rubber ball of radius R is placed on the disc in such a way that it rolls without slipping, and its centre moves in a circle that has radius r_0 and is concentric with the disc.

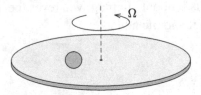

a) Find the initial velocity and angular velocity that the ball must have for this kind of motion.

b) How would the centre of the ball move if it were started from the same position with the same magnitude of initial velocity as specified in part *a*), but with the opposite direction?

P56* A large flat disc with a rough surface rotates around the axis of symmetry that is normal to its plane, and does so with constant angular velocity Ω. The plane of the disc is tilted at an angle θ relative to the horizontal. A magician places a solid rubber ball of radius R and mass m on the rotating disc and starts it off in an appropriate direction. Then, to the audience's great surprise, the centre of the ball moves uniformly in a straight line, until it reaches the rim of the rotating disc. Throughout the ball's motion, it does not slip on the disc, and the angular velocity of the disc does not change.

Find a physical explanation for this strange phenomenon. In which direction, and how quickly, should the magician start the ball for this stunt to be successful?

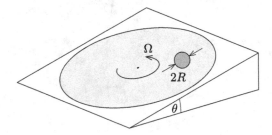

P57** A small solid rubber ball of radius r is thrown onto the inner wall of a long cylindrical tube, which has radius R and is fixed with its axis of symmetry vertical. If the ball is started off with a sufficiently large horizontal velocity v_0, then it starts to oscillate periodically in the vertical direction, while still maintaining contact with the tube. Describe the 'dance' performed by the centre of the ball.

The static friction is quite large, and so the ball never slides on the wall. Assume that the ball is sufficiently incompressible that its contact with the tube is only ever through a single point, and that air drag and rolling friction are negligible.

P58 Slinky,[5] a well-known toy, is a pre-compressed helical spring, with a negligibly small unstressed length. To a good approximation, it obeys Hooke's law, but with a small 'Young's modulus', even being significantly stretched by its own weight.

Such a spring is first hung vertically with its upper end fixed, and is then re-suspended with its two ends attached to supports at the same vertical height, but separated horizontally. The ends of the Slinky make angles of 45° with the vertical, as shown in the figure. In which case is the stretched spring longer?

P59* What shape does a Slinky take up when its two ends are fixed to points that are at the same height and separated by a moderate distance (the Slinky remains a helical spring)?

P60* A Slinky of mass m was initially resting on a table-top with its axis vertical. Its top end was then slowly raised until its lowest coil was just clear of the table. At that point, the length of the Slinky was L.

a) How much work was done during the lifting stage?

b) If the upper end of the Slinky is now released (from rest), then, curiously, the lowermost coil does not start moving until the Slinky has completely collapsed (*see* figure). What speed, v_0, does the collapsed Slinky have immediately afterwards?

c) How long does the Slinky take to completely collapse?

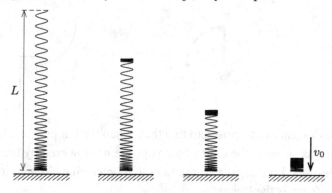

---

[5] Slinky was invented by Richard James in the early 1940s; his wife dubbed the toy 'Slinky' (meaning 'sleek and graceful') after finding the word in a dictionary, and deciding that the word aptly described the sound of a metal spring expanding and collapsing.

P61** A Slinky is placed inside a frictionless horizontal tube, with one of its ends attached to a fixed point of the tube. The fixed point is a distance r_0 from a vertical axis about which the tube rotates with uniform angular velocity ω (*see* figure). The Slinky spring is 'ideal': its unstressed length is negligible; its potential elongation is unlimited; and it obeys Hooke's law.

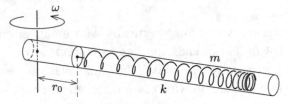

What is the length ℓ of the stretched spring, if its spring constant is k and its total mass is m?

Consider what happens in the limiting case $r_0 \to 0$.

P62** Find the shape of a Slinky inside the International Space Station (i.e. in weightless conditions) if it is rotating uniformly – like a skipping rope – with both ends of the spring twirled in unison.

P63* A tree in the editor's garden has a thin, light and elastic horizontal side branch that in early spring has very few leaves on it and provides a popular perch for wild birds. Is the end of the branch depressed more when a pigeon sits at its midpoint or when a blackbird, which only weighs one-quarter as much as the pigeon, perches on the end of the branch?

P64* A hat peg is planned by a post-modern interior designer in two versions. In both, a thin but strong elastic metal wire in the form of a quarter-circle is attached to a solid vertical post. The alternative mounting arrangements are shown in the figure.

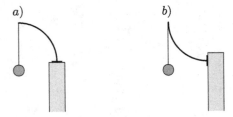

The designer is surprised to find that, when the hat pegs are loaded with the same weight (as shown), the ends of the pegs are not lowered by the same amount in the two designs. Find a simple argument that determines in which design the droop of the peg's tip is the larger.

P65* Hanging a body with a mass of 1 kg from the end of a uniform, horizontal, 1 m long rod, which is fixed rigidly at its other end, causes the loaded end to deflect by 1 cm. If the rod is stood vertically on one end (*see* figure), *estimate* the vertical load F, applied to the other end, that is required to make it buckle.

P66 A cable of given thickness, and large but finite tensile strength, is to hang vertically downwards from a ship, with the intention that it should reach the sea bed. However, when the lowered cable is 1 km long, it snaps (at the top) because of its own weight. How much cable is required to reach the sea bed, at a depth of 3 km, if filaments can be doubled up and run parallel to each other along particular lengths? Estimate the total length of original cable needed to construct a composite one that will reach to the bottom of the Mariana Trench, which lies 11 km beneath the surface.

Assume that the density of seawater is constant, and that the stretching of the cables is negligible.

P67** It is a common observation in the kitchen that, if a frankfurter sausage (a long, straight sausage in ovine gut skin, sometimes called a 'Frankfurter Rindswurst') splits during the boiling process, it always does so 'lengthways', and never 'across'. What is the reason for this?

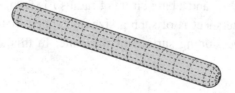 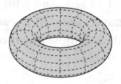

If it were possible to produce one, where, and along which direction, would a toroidal sausage split while being cooked? Assume that, in both cases, the thickness of the sausage skin is uniform.

P68* A thin, strong, but flexible steel tight-rope is installed horizontally above a wide street, and highly tensioned. An acrobat moves slowly onto and along the tight-rope. When he reaches the quarter-way point (Q in the figure), the nearer tri-section point of the rope (T) has been depressed by 5 cm from its original position. By how much is Q depressed, when the acrobat reaches point T?

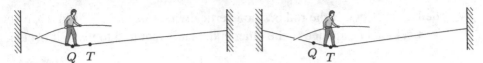

Can your result be generalised to arbitrary points Q and T?

You may assume that the weight of the rope is negligible, that its depression is always very small compared to its total length, and that the tension in it can be treated as constant.

P69   Devise a mechanical system that uses a fundamental physics principle to determine the point P on a general triangle that has the property that the sum of its distances from the three vertices is a minimum. Can your device be adapted to minimise a *weighted* sum of the distances?

P70   A sack, full of sand and of mass M, lies on a carpet, which, in turn, lies on a relatively smooth horizontal surface. The distribution of the sack's weight is not uniform, but it is known that its centre of mass (CM) lies at distances s_1 and s_2 from its extremities.

By pulling horizontally on the carpet, both it and the sack are to be moved onto an immediately adjacent rougher surface at the same height (*see* figure). How much work is required to do this if the coefficients of friction between the carpet and the original and final surfaces are μ_1 and μ_2, respectively?

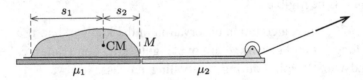

P71   A cone with height h and a base circle of radius r is formed from a sector-shaped sheet of paper. The sheet is of such a size and shape that its two straight edges almost touch on the sloping surface of the cone. In this state the cone is stress-free.

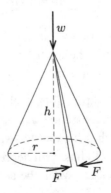

The cone is placed on a horizontal, slippery table-top, and loaded at its apex with a vertical force of magnitude w, without collapsing. The splaying of the cone is opposed by a pair of forces of magnitude F acting tangentially at the join in the base circle (*see* figure). Ignoring any frictional or bending effects in the paper, find the value of F.

P72   Frank has made an iron triangle by welding together three thin iron rods of identical cross-section. He decides to try to identify the centre of mass of the triangle, and so he lays it on a sheet of paper, and draws the lines connecting each vertex to the midpoint of the opposite side. At that moment, Lisa arrives and says that the centre of mass of the triangle is not at its centroid (its geometrical centre). She claims that Frank should connect the midpoints of the triangle's sides to form a smaller triangle, and that the centre of mass of the larger iron triangle coincides with the *incentre*[6] of the smaller one. Who, if anybody, is right?

P73   A triangle is cut out from a uniform sheet of cardboard, and placed on a level table-top standing on one of its edges. It is loosely supported on both sides so that it can move only in its own plane – but can do so freely in that plane. Is it possible to make a triangle that tumbles over from two of its edges, and has only a single equilibrium position?

P74**   We describe any particular face of a tetrahedron as 'unstable' or 'stable' according to whether or not the tetrahedron spontaneously falls over if it is placed on a level table-top with that face as its base. Is it possible to make a (homogeneous) tetrahedron that has three unstable faces and only a single stable one?

P75   The cable strung between two neighbouring electricity posts sags a little. The mass per unit length of the cable is λ, the distance between the posts is L, and the 'maximum sag' of the cable is d ($d \ll L$). What is the (approximate) tension in it?

P76*   In a gymnasium, a climbing rope and a climbing pole, both with uniform cross-sections, have the same length; they also have equal weights. Each is attached to the ceiling of the gymnasium by a small pivot, and both are pulled aside by identical horizontal forces at their lower ends. Determine whether it is the rope's or the climbing pole's lowest point that is lifted higher.

P77   The two ends of a 40 cm long chain are fixed at the same height, as shown in the figure. Find the radius of curvature of the chain a) at its lowest point and b) at the suspension points.

---

[6] The incentre of a general triangle is the (unique) point at which the internal bisectors of its angles meet; this point is the centre of the triangle's incircle.

P78** A uniform flexible rope passes over two small frictionless pulleys mounted at the same height (*see* figure). The length of rope between the pulleys is ℓ, and its 'sag' is h. In equilibrium, what is the length s of the rope segments that hang down on either side?

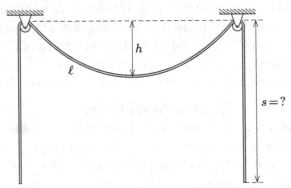

P79* One end of a necklace of small pearls is attached to the outer surface of a fixed cylinder that has radius R and a horizontal axis; the attachment point P is at the same level as the axis. The necklace is wound once round the slippery surface of the cylinder, and the free end is left to dangle (*see* figure). How long, ℓ, does this free end need to be if the rest of the necklace is to touch the cylinder surface everywhere?

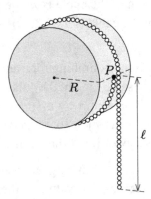

P80* A flexible scatter rug (also known as a runner), of mass M and length L, is tightly coiled into a cylinder of radius R ($\ll L$), as in the figure. If the coiled rug is released, then, in the absence of rolling friction, it spontaneously unrolls to its full length.

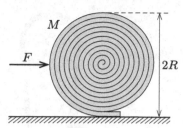

a) Explain in terms of the forces involved why this happens.

b) How large a horizontal force F, applied as in the figure, is required to prevent the unrolling of the rug?

P81** The committee of the Glacier Climbing Club has decided to introduce a new challenge for its members: they have to climb as high as they can on an artificial right-circular cone, the surface of which has been made into a very smooth, slippery 'iceberg', by letting water trickle down it in sub-zero temperatures. A lasso is to be their only piece of climbing equipment!

The lasso for novices is as shown in figure *a*) and consists of a length of rope attached by a small eyelet to a closed loop of *fixed* length. That for experts is shown in figure *b*) and is a single rope with, at one end, an eyelet, through which the other end is threaded. All the ropes are light compared to the mass of a climber, and friction between them and the ice, and within the eyelet, is negligible.

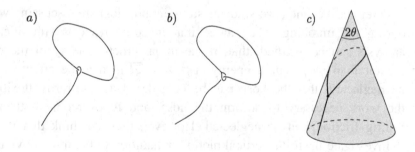

For what ranges of cone angle, 2θ, are (i) the novices and (ii) the experts able to climb up the iceberg using their lassos in the way illustrated in figure *c*)? Assume that the straight segment of a lasso follows one of the cone's lines of steepest descent.

P82* The spindle of a bicycle chain assembly is mounted horizontally, and a loop of bicycle chain is placed on the toothed wheel, as in the figure. The wheel is then rotated around its axis at a steadily increasing rate until it has achieved a high, but constant, angular velocity. What is then the shape of the chain?

P83**   A fire hose of mass M and length L is coiled into a roll of radius R ($\ll L$). The hose is sent rolling across level ground with an initial speed v_0 (and angular velocity v_0/R), while the free end of the hose is held at a fixed point on the ground (*see* Fig. 1). The hose unrolls and becomes straight.

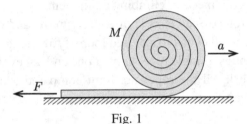

Fig. 1

Peter and Pauline (two students studying physics) are discussing what happens during the unrolling. They agree that in some respects simplification of the analysis can be justified: that, if the initial kinetic energy of the roll is much greater than its potential energy ($v_0 \gg \sqrt{gR}$), then the effect of gravity can be neglected; that the hose can be considered as arbitrarily flexible; and that the work necessary to deform the hose, and to overcome both air drag and rolling friction, can be neglected. However, they do think that it is important to investigate the roll's vertical motion in addition to the more obvious horizontal one.

The speed of the diminishing roll continually increases, and its acceleration a is clearly a vector pointing in the same direction as its velocity. On the other hand, the vector resultant of the horizontal external forces that act on the roll (the frictional force and the restraining force at the fixed end of the hose) points in the opposite direction. This strange fact can be understood by recognising that the total momentum of the moving roll (and hence of the whole system) is

$$p(x) = m(x)\, v(x) = M(1 - (x/L))\frac{v_0}{\sqrt{1 - (x/L)}} = M v_0 \sqrt{1 - (x/L)},$$

where $m(x) = M(1 - (x/L))$ is the mass of the moving part of the hose after it has travelled a distance x. Its speed $v(x)$ has been determined using the conservation of mechanical energy (but ignoring gravitational effects), and is given by $v(x) = v_0/\sqrt{1 - (x/L)}$.

Clearly, as x increases, $p(x)$ decreases – reflecting the fact that the mass of the piece in motion decreases faster than the rate at which its speed increases. The direction of the resultant force $F(x)$ acting on the system is therefore opposed to that of the motion,[7] with

$$F(x) = \frac{dp}{dt} = \frac{dp}{dx}\frac{dx}{dt} = \frac{dp(x)}{dx}v(x) = -\frac{Mv_0^2}{L}\frac{1}{2(1 - (x/L))}.$$

Peter (using the conservation of mechanical energy and mass) has determined not only the mass $m(x)$ of the roll, its radius $r(x)$ and its angular velocity $\omega(x)$, but also its moment of inertia $I(x)$. He has also found the horizontal and vertical velocities, v_x and v_y, of the roll's centre of mass (*see* Fig. 2). All of these quantities are functions of x, and therefore also of time t.

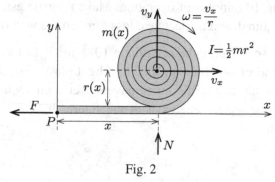

Fig. 2

Peter says that:

> I have found v_x and v_y, and multiplying them by the instantaneous mass $m(x)$, I have calculated the components p_x and p_y of the linear momentum of the roll, and from the rates of change of the latter, the external forces, F and N, can be found.

He decided to calculate the angular momentum J of this strange system, and also its rate of change, and so check whether or not the torques due to the external forces would produce this rate of change, i.e. whether or not $\tau_{\text{net}} = dJ/dt$. As both the centre of the roll and the centre of mass of the whole system are accelerating, he expressed the formulae for the angular momentum and the torques, not about these points, but with respect to the fixed point P, the stationary end of the hose.

---

[7] This calculation appears in the predecessor of this book, as the solution to 'Problem 108' of P. Gnädig, G. Honyek & K. F. Riley, *200 Puzzling Physics Problems* (Cambridge University Press, 2001). There, a further question was posed: How long does it take for the hose to completely unroll?

To his great surprise, he has found that – though the net torque of the external forces is *zero* – the angular momentum of the system does not remain constant, but *changes with time*! After he had checked his calculations for the umpteenth time, he was close to stating:

> In this system, the angular-momentum theorem, a basic law in classical mechanics, is not obeyed.

Pauline, however, thought differently! She claimed that there was an error in the method Peter had used, in that he had applied a well-known standard formula to a situation to which it did *not* apply.

Who was right?

P84* A cylindrically shaped closed container rotates uniformly around its principal symmetry axis, which lies horizontally, at a rate of 0.5 revolutions per second. Both the cylinder's inner diameter and its length measure 1 m, its inner wall is rough and it contains 100 kg of sand. Estimate the temperature rise of the sand during 10 minutes of operation. Make realistic estimates for (nearly all of) the data required, and neglect heat loss through the wall of the container.

P85 A solid iron cube of volume 10^{-3} m^3 is fastened to one end of a cord, the other end of which is attached to a light plastic bucket containing water. The cord, which has negligible mass, passes over a pulley, and the iron cube is suspended in the water, as shown in the figure. It is found that the system is in equilibrium.

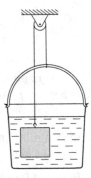

a) How many litres of water are there in the bucket?

b) What would happen if more water were poured into the bucket?

c) What would happen if some or all of the water evaporated?

P86 An air bubble with a diameter of a few centimetres was injected into the bottom of a giant *closed* vertical cylinder that had a height of 10 m and was, apart from the bubble, *completely* full of water. At the moment of injection, the water pressure at the bottom of the cylinder was 1.1 atm. The air bubble rose, and a little bit later it reached the top of the closed container. Find the pressure of the water at the bottom of the cylinder after the air bubble had risen.

For the sake of definiteness, assume that the hydrostatic pressure of 10 m of water is exactly 1 atm.

P87 There is water in a container that stands on a table. A rubber hose connects the bottom of the container with a cupboard-sized *black box* standing next to the table. If additional water is added to the container, then the water level in it *sinks*, and if some is taken out, the level rises. Whatever could the black box contain?

P88 A funnel is placed upside down on a table, as shown in the figure, and 1000 cm³ of water is poured into it. The area covered by the funnel on the table-top is 200 cm², and the height of the water is 18 cm.

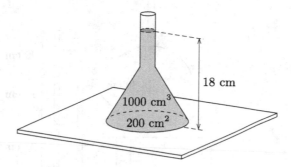

What is the minimum mass of the funnel if it is not to be lifted away from the supporting surface?

P89* A thin-walled hemispherical shell of mass m and radius R is pressed against a smooth vertical wall, and, through a small aperture at its top, is filled with a liquid (say, water) of density ϱ (*see* figure). What are the minimal magnitude and the direction of the force that has to be applied to the shell if liquid is not to escape from it? To which point on the shell should this force be applied?

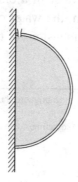

P90   The gauge pressure[8] in the tyres of quite an old car (one without power steering) is increased from 1.5 atm to 2.0 atm. What percentage change is there in the force required to rotate its steering wheel when the car is stationary?

P91*   A balloon is inflated and connected to a water manometer, as shown in the figure. The difference in height between the water columns in the two arms of the manometer is 10 cm. When a solid cylindrical iron rod, of length 50 cm and diameter 2 cm, is placed vertically and carefully on top of the balloon, will the water in the manometer overflow?

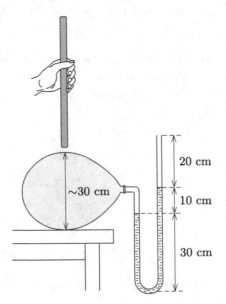

P92   The wall of a vertical measuring cylinder contains many uniformly distributed small holes. The cylinder is filled with water up to a height H, and so thin water jets are ejected horizontally through the holes. The jets do not interfere with each other, and the water level in the cylinder is continuously maintained at its initial value. What shape is the envelope of the water jets?

P93**   There is some water in a closed spherical Florence flask.[9] When the flask is turned upside down, the water collects in its long neck up to a height of about 5 cm. The internal dimensions of the flask are shown in the figure.

---

[8] The gauge pressure is not the absolute pressure, but the difference between the absolute pressure and the atmospheric pressure p_0.

[9] A Florence flask is a spherical flask (sometimes with a flattened bottom) that is mainly characterised by its long (and sometimes quite wide) neck.

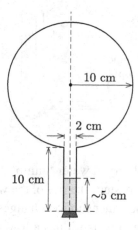

The flask is now rotated about its vertical axis of symmetry at a rate of three revolutions per second, and it is arranged that the temperature of the flask wall is everywhere the same. After a sufficiently long time, an equilibrium state is established in the flask.

Make a sketch showing the distribution of the water within the flask when this equilibrium state has been reached.

P94* A razor-blade floats on the surface of water contained in a glass. When the glass is gently shaken, the razor-blade sinks. How, if at all, does the water level change as a result?

P95* Take a disc of copper with a diameter of about 10 cm, and a thickness of approximately 0.2 mm (other metals, such as soft steel or aluminium, are also suitable for this experiment). In the centre of the disc, make a circular depression with a diameter of 15–20 mm and a depth of 2–3 mm. If such a disc is placed on a water surface, it will probably float (*see* solution on page 263). But if some water is splashed onto it, the disc will almost certainly sink, because metal is denser than water.

Now place the dry disc on the water surface and direct a strong vertical jet of water downwards into the central depression. What you will observe is paradoxical – the jet pushes the disc down, but cannot sink it![10] It will be noticed that, on the disc's surface, there is a circular 'hump' of water that is pushed outwards by the diverging thin layer of running water (*see* figure).

---

[10] This interesting phenomenon is described in an article entitled 'An unsinkable disk' by A. Luzin in *Quantum*, Sept./Oct. 1999, p. 42. Unfortunately, the explanation given there of why the disc does not sink is, in our opinion, in error.

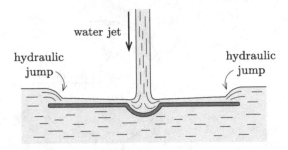

In hydraulics, the observed water 'hump' is known as a 'hydraulic jump', and is characterised by a sharp, step-like rise of the water level in an open waterway; at the jump, the nature of the water motion changes. When liquid moving at high speed discharges into a zone in which it has a lower velocity, a rather abrupt rise occurs in the liquid level: the rapidly flowing liquid is suddenly slowed and increases its depth.

A common example of a hydraulic jump is the roughly circular stationary wave that forms around a central stream of water falling on the flat bottom of a kitchen sink. The 'hump' of water near the rim of the copper disc is very similar to this easily observed phenomenon. The depression at the centre of the disc provides the disc with horizontal stability – without it, the disc would move rapidly away from under the water jet.

Now for the question! Why does the disc not sink under the pressure generated by the powerful water jet?

P96   Two spherical soap bubbles with different initial radii coagulate, and the radii of their free surfaces after the coagulation are R and r. What is the radius of curvature of the soap film that separates the two bubbles? What is the radius of this film's circular perimeter?

P97*   While cleaning his flat, a physics research student knocks over a bucket containing 5 litres of water, which then forms itself into one continuous puddle. How large, in area, does the student expect the puddle to be? The contact angle between the floor and the water is 60°.

P98*   Inside the cabin of a freely orbiting space ship, a ball of water approximately 4 cm in diameter and a nearby thin glass rod are in a state of levitation. The rod is about 8 cm in length, and has a circular cross-section and blunt ends. One end of the rod gently touches the ball. Sketch the shape the 'water drop' subsequently takes up.

P99*   Inside a space station, a freely floating, closed, spherical shell of inner diameter 8 cm has one-third of its volume occupied by water; the rest contains air at STP (standard temperature and pressure). In equilibrium, how is the water distributed, if the material of the spherical shell is

a) glass, which is 'perfectly' wetted by the water,

b) silver, for which the contact angle of water is 90°?

P100* You are given 1 kg of distilled water at 0 °C, an equal mass of boiled tap water at 100 °C, and the task of warming the distilled water to 60 °C. How would you set about it?

No additional water, hot or cold, is available, but you are provided with both insulating and conducting materials (in sheet form) and a suitable selection of tools.

What is the maximum possible temperature, in principle, to which your method could warm the distilled water?

P101* There is a *very small* hole in an otherwise totally enclosed heated furnace. Outside the furnace, the air temperature is 0 °C and the air pressure is 100 kPa. The air inside the furnace is kept at a constant temperature of 57 °C by the controlled heating system, and after a sufficiently long time its pressure also becomes stationary. Estimate the magnitude of this stationary pressure.

P102* An ideal gas, enclosed in a fixed, long, vertical cylinder with open ends, is separated from its surroundings by two identical, frictionless pistons. Between the pistons there is a fixed, rough, separating wall in which there is a small hole (*see* figure). Initially, the temperature of the enclosed gas is equal to that of the outside air, and the lower piston is held up against the separating wall. When the lower piston is released, the pistons descend slowly.

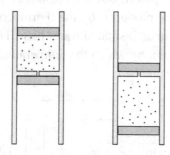

In which of the two following scenarios will the bottom piston finish in a lower position:

a) the pistons, the separating wall and the wall of the cylinder are good heat insulators,

b) the pistons, the separating wall and the wall of the cylinder are good heat conductors?

P103* It is well known to hikers and mountaineers that the air temperature decreases by 1 °C for each 100 m increase in altitude.

The ground, warmed by solar radiation, heats up the air above it, and the lowest layer of the atmosphere (the troposphere) is in a state of permanent convection. However, this mechanism is insufficient to equalise the temperatures of the various air layers, because rising air masses are not able to effectively exchange heat with their surroundings (due to a combination of their rapid motion and the low thermal conductivity of air).

There is thus a variation of temperature with height. Use simple physical considerations to explain and estimate the observed rate of air temperature decrease with altitude.

P104 A quantity, n mol say, of helium gas undergoes a thermodynamic process (neither isothermal nor isobaric) in which the molar heat capacity C of the gas can be described as a function of the absolute temperature by $C = 3RT/(4T_0)$, where R is the gas constant and T_0 is the initial temperature of the helium gas. Find the work done on the system up until the point at which the helium gas reaches its minimal volume.

P105* An ideal gas is contained within a closed bag made from material that is both easily stretched and an excellent heat insulator. The pressure outside the bag is decreased from p_1 to p_2 and the gas both expands and cools. In which case is the temperature drop in the gas greater: if the decrease in the outside pressure is slow, or if it happens suddenly?

P106** One of the junctions (A) of a thermocouple is in air at $T_A = 27\,°C$, while the other (B) is placed inside an insulated vessel containing ice at $T_B = 0\,°C$. The electrical energy produced by the thermocouple is dissipated in a resistor of resistance R placed in an insulated water bath. The masses of the water and ice are equal. Find the amount by which the temperature of the water has increased when the last of the ice has melted.

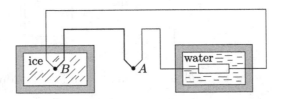

P107** By how much can a small spherical black body be warmed by sunshine, using a thin convex lens that has a focal length equal to twice its diameter? Does the result depend upon the radius of the sphere?

P108 During cloudless autumn nights, hoar frost can cover car windscreens, even if the temperature of the surrounding air does not fall below 0 °C. How can this happen? Where should a car be parked so as to avoid this phenomenon?

P109* The three vertices of a metal plate in the form of an equilateral triangle are held at constant temperatures T_1, T_2 and T_3 (by cooling or heating, as necessary). What temperature do you expect at the centre of the triangle? Prove that your answer is correct.

P110 Identical ordinary ice cubes are put into two quite large beakers, one of which contains tap water, the other a strong brine solution. The liquids have the same volume, and both of them are at room temperature. In which beaker will the ice cube melt more quickly?

P111 If 1 kg of water freezes, then 334 kJ of energy are released. How much energy is released if 1 kg of supercooled water at −10 °C freezes, and during the freezing process the temperature remains constant?

P112 Should we increase or decrease the volume of a quantity of air that contains saturated water vapour, if we wish to condense some of the vapour?

P113* Water is boiling in a narrow test tube that is open at the top. Just before the last few drops are vaporised, the tube is rapidly and hermetically sealed. The temperature at the top of the tube is then slowly increased to 200 °C, while the very bottom of it is maintained at 100 °C.

What is the pressure of the steam in the test tube?

P114* At some time in the distant future, humankind makes contact with the inhabitants of an exoplanet, on which the atmospheric pressure near the surface is the same as on the Earth, i.e. 1 atm ≈ 101 kPa. Further, the atmosphere consists of a mixture of oxygen and nitrogen gases. Because of these similarities, the planet is called Exo-Earth.

Human researchers and Exo-Earth scientists cross-check the physical and chemical data of their two atmospheres, and state that, on both planets, the boiling points of liquid nitrogen and liquid oxygen are 77.4 K and 90.2 K, respectively, at standard atmospheric pressure.

On both planets, local 'air' was isothermally compressed at a constant temperature of 77.4 K, and liquefaction set in when the pressure reached 113 kPa. However, on Earth *oxygen*, and on Exo-Earth *nitrogen*, condensed first.

a) What is the composition of the atmosphere on Exo-Earth?

b) For what atmospheric composition would the oxygen and nitrogen begin to liquefy *simultaneously* under isothermal compression at 77.4 K, and at what pressure would this happen?

P115   Two spherical flasks are almost identical. One of them has a straight neck directed upwards; the other is similar to a retort, in that its neck points downwards, as shown in the figure. The same amount of a liquid is put into each of the flasks, which are then heated from below in a controlled way that ensures that the temperatures of the liquids within them remain fixed and identical. Which flask will run out of liquid first if

a) water is used, and the constant temperature is 100 °C,

b) ether (diethyl ether, $(C_2H_5)_2O$) is used, and the constant temperature is 34.6 °C, the boiling point of ether at atmospheric pressure?

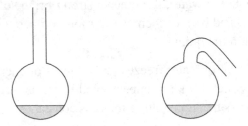

**P116*   Two narrow, straight, vertical test tubes with identical cross-sectional areas are open at the top; one of them is 20 cm long, and the other is 40 cm. We pour 1 cm^3 of eau-de-Cologne into the first test tube, and 2 cm^3 into the second. How many times longer does the latter take to evaporate than the former? Is there any change in the result if both test tubes are covered, and identical but very small holes are made, one in each cover?

P117   A point source of light is inside a solid sphere made from glass of refractive index *n*. The sphere has radius *r* and the source is a distance *d* from the sphere's centre (*see* figure). What is curious is the fact that the sphere forms a *perfect* (virtual) image of the light source, i.e. the backward continuations of all light rays starting from the source, refracting at the sphere's surface, and travelling to the right-hand side of the figure, intersect each other at a common point.

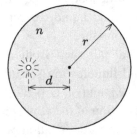

What is the distance *d*, and where is the sharp image of the light source formed?

P118** What is the shape of the blunt end of a glass rod that would focus (at a single point F within the rod) all the light rays incident upon it that are parallel to the rod's axis? Find the equation describing the shape of the surface, expressing it in terms of the (uniform) refractive index n of the glass, and the rod's focal length f, defined to be the distance between F and the point on the surface that lies on the rod's axis.

What is the corresponding solution if a beam propagating inside the glass rod, and parallel to its axis, is to be brought to a focus at an axial point outside the rod?

P119 On a spherical planet, the refractive index of the atmosphere, as a function of altitude h above the surface, varies according to the formula

$$n(h) = \frac{n_0}{1 + \varepsilon h},$$

where n_0 and ε are constants. Curiously, any laser beam, directed horizontally, but at an arbitrary altitude, follows a trajectory that circles the planet. What is the radius of the planet?

P120* Find the shape of the image that is formed by a converging thin lens with focal length f, if the object is a sphere of radius r ($< f$), and the centre of the sphere is at the focus of the lens.

P121* How many times 'brighter' is an image of the Moon when looked at through a telescope rather than with the naked eye? And what about the stars?

P122* A converging lens of focal length f is cut along a plane that contains the optical axis of the lens, and a small, black plate of thickness δ is placed between the two half-lenses. A point-like source emitting monochromatic light of wavelength λ is located on the 'optical axis', a distance p from the lens ($p > f$).

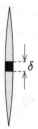

How many interference fringes can be seen on a screen placed a distance H behind the lens, with its plane perpendicular to the optical axis?

Data: $f = 10$ cm, $p = 20$ cm, $\delta = 1$ mm, $\lambda = 0.5$ μm, $H = 50$ cm.

P123* In an unusual optical diffraction grating, the distances between the neighbouring slits are not equal; they are alternately d and $3d$. The widths of the slits themselves *are* all equal and are much smaller than d. If light of wavelength λ falls at normal incidence on the grating, what kind of diffraction pattern is formed on a screen placed at a distance L from it?

P124* An optical grating is illuminated normally by a laser beam of wavelength $\lambda \ll d$, where d is the slit spacing. The grating is unusual, in that alternate slits are wider and narrower: the width of the odd-numbered slits is a, while that of the even-numbered ones is b, where $b < a$ and both of them are much smaller than d.

The special character of this grating is reflected in its diffraction pattern in a peculiar, easily noticeable, way. How? What is the diffraction pattern like (i) if $b \ll a$ and (ii) if $b \approx a$, and the screen on which the patterns are formed is a distance L behind the grating?

P125 An opaque sheet is perforated by many small holes arranged on a square grid (*see* figure), and is illuminated normally by monochromatic laser light of wavelength λ. What kind of diffraction pattern can be observed on a screen placed parallel to the sheet, and a distance L behind it, if the 'lattice constant' of the grid is d? Assume that $L \gg d \gg \lambda$.

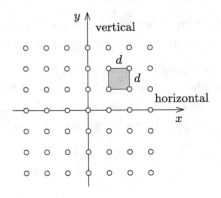

How does the diffraction pattern change if the sheet is compressed horizontally (along the x-axis in the figure) by a factor of N, with the result that the holes lie on a rectangular grid?

P126* An opaque sheet is perforated by many small holes arranged on a triangular grid (*see* figure), and is illuminated normally by monochromatic laser light of wavelength λ. What kind of diffraction pattern can be observed on a screen placed parallel to the sheet, and a distance L behind it, if the 'lattice constant' of the grid is d? Assume that $L \gg d \gg \lambda$.

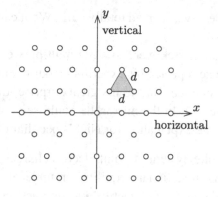

P127* *a*) Light cannot be transmitted through two coaxial polarising filters if their polarisation axes are orthogonally oriented. However, if a third filter is placed between them, some light can pass through all three. What is the maximum fraction of the incident intensity that can be so transmitted? Find the corresponding angle between the polarisation axis of the first filter and that of the one interposed – denoted by φ in figure *a*).

a)   *b*)

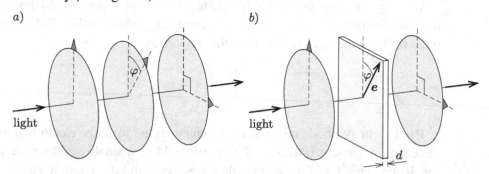

b) There is a second way in which light can be made to pass through the two orthogonal polarising filters: by placing, between and parallel to the filters, a uniform plate made from uniaxial birefringent (birefractive) material. The special feature of such a plate is that its refractive index for light polarised parallel to the direction of vector *e*, which lies in the plane of the plate (*see* figure *b*)), is n_1, whereas for light polarised perpendicular to *e* it is n_2.

What is the maximum fraction of the incident intensity that can be transmitted, if the system is illuminated (perpendicularly to the plane of the filters and the plate) by monochromatic light of wavelength λ? Find the thickness d of the birefringent plate appropriate to this case, and how we should choose the orientation of *e*?

P128** *a*) Nick, a 17-year-old grammar school pupil tidying up his room, found an old pair of spectacles used for viewing 3D movies. He tried on the specs in the bathroom, and he realised that when looking at his own image in the bathroom mirror, with one eye closed, he could see only his open eye; the other 'lens', in front of his closed eye, seemed totally dark. What can be the explanation for Nick's observation?

b) A week later, Nick went to a 3D multiplex cinema (movie theatre), and on returning home he repeated his experiment, but with the new 3D glasses he had just acquired at the cinema. But, to his great surprise, the effect was just the opposite of what he expected. In the mirror, he could see only his closed eye, and not the open one. Find the explanation for Nick's peculiar experience.

P129* A finite system of point electric charges, in vacuum and very far from anything else, is in stationary equilibrium under the influence of its own internal electric fields. What is the electrostatic interaction energy of the system? Is the equilibrium stable or unstable?

P130 Three small positively charged pearls lie one at each vertex of a triangle. Their masses are m_1, m_2 and m_3, and their charges are Q_1, Q_2 and Q_3, respectively. When the pearls are released from rest, each moves along a different straight line, the three motions taking place in a vacuum with negligible effects due to gravity.

What special condition has to be satisfied for this to happen? Find the angles of the triangle formed by the pearls at the beginning of their motion, if the charge-to-mass ratios of the three pearls are in the proportion

$$\frac{Q_1}{m_1} : \frac{Q_2}{m_2} : \frac{Q_3}{m_3} = 1 : 2 : 3.$$

P131* In a cathode ray tube, the beam emerging from the electron gun can be deflected by passing it through the electric field of a small parallel-plate capacitor. If the capacitor is uncharged, then the speed and direction of motion of the electrons are constant, and they hit the screen with an unchanged velocity.

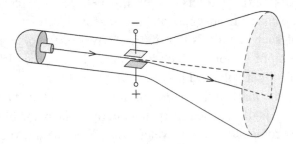

When the parallel plates of the capacitor carry a *constant charge*, the electron trajectory is bent. Is the speed with which electrons then hit the screen increased or decreased as compared to its initial value? The electron gun–capacitor and capacitor–screen distances are both much greater than the length of the capacitor's plates.

P132* A triangle is made from thin insulating rods of different lengths, and the rods are uniformly charged, i.e. the linear charge density on each rod is uniform and the same for all three rods. Find a particular point in the plane of the triangle at which the electric field strength is zero.

P133* Three very long (they can be regarded as infinitely long), thin, insulating rods are uniformly charged, i.e. the linear charge density on each rod is uniform and the same for all three rods. The rods lie in the same plane, and cross each other to make a general triangle. Where can a point charge be placed so as to be in equilibrium?

P134 Two very long, thin, insulating rods, each carrying uniform linear charge density λ, lie in perpendicular directions (in three-dimensional space), a distance d from each other (*see* figure). What is the magnitude of the force of repulsion between them?

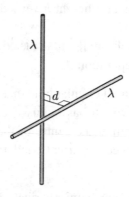

P135 Two very long, parallel, thin, insulating rods are uniformly charged with equal but opposite electric charge densities. What shape are the electric field lines?

P136* A square of side d, made from a thin insulating plate, is uniformly charged and carries a total charge of Q. A point charge q is placed on the symmetrical normal axis of the square at a distance $d/2$ from the plate. How large is the force acting on the point charge?

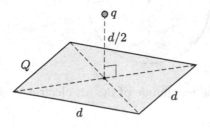

P137* A cube is made from six thin insulating square faces, each square having side d and carrying a uniformly distributed charge of Q. How large is the electrostatic force acting on each face?

P138** Two identical, large, rectangular, insulating plates lie one vertically above the other. The uniform separation between the thin plates is d, and they are both uniformly charged; the surface charge density on the upper one is $+\sigma$, and that on the lower one is $-\sigma$.

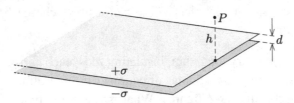

What is the magnitude and approximate direction of the electric field at the point P (*see* figure), which is at a height h vertically above the midpoint of one of the upper plate's edges?

The distance h is much smaller than the edge lengths of the insulating plates, but much larger than the separation d.

P139* There is a small circular hole in a thin-walled, uniformly charged, insulating spherical shell, which has radius R and carries a total charge Q. What is the electric field strength at the centre of the hole? Sketch the electric field lines associated with the holed spherical shell in a section that includes its axis of symmetry.

P140** Two insulating hemispherical shells (e.g. the two halves of a ping-pong ball) are placed very close to each other with their centres almost coincident, as shown in figure a). Both of them are uniformly charged, one with total electric charge Q, and the other with total charge q.

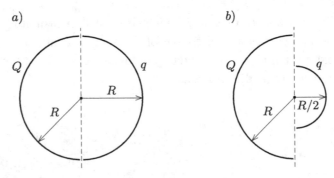

a) What is the magnitude of the net electric force exerted by the two bodies on each other?

b) Does the result change if the radius of one of the hemispherical shells is only one-half that of the other, as shown in figure *b)*?

P141* Taking the zero of electrostatic potential at infinity, how many times larger is the potential at the centre of an insulating solid cube carrying uniform charge density than that at one of its vertices?

P142* A glass capillary tube is placed vertically into mercury held in a glass container (*see* figure). A high voltage, relative to ground, is applied to the mercury. Does the meniscus in the capillary tube (the interface between the mercury and air) move, and, if it does, which way?

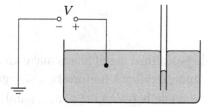

P143 Near the edges of a parallel-plate capacitor, the electric field is inhomogeneous, but the effect of this so-called fringe field is usually neglected. Would a larger or smaller value be found for the magnitude of the capacitance if the fringing effect were also taken into account?

P144 One of the plates of a parallel-plate capacitor with capacitance C carries a charge of Q_1, and the other carries Q_2. What is the voltage (potential difference) across the capacitor?

P145** Two very large, identical, rectangular, metal plates lie in the same plane, with corresponding edges parallel. The plates, which are very close to each other and connected by a wire, are initially uncharged. A point charge Q is now placed near them at the position above the plates shown in the figure. Find the resulting amounts of charge on each individual plate.

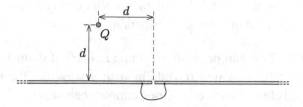

P146* Both plates of a parallel-plate capacitor, with plate separation d, are earthed, and a small pearl carrying charge Q is placed a distance δ ($\delta \ll d$) from the (imaginary) mid-plane between the plates (*see* figure). Find the force acting on the pearl.

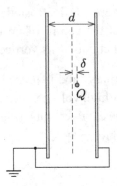

P147* Two large earthed metal plates make an angle θ with each other. A thin rod of length L, and carrying a uniformly distributed charge Q, is placed in the plane bisecting the angle between the plates, parallel to, and at a distance r from, their line of the intersection ($L \gg r$).

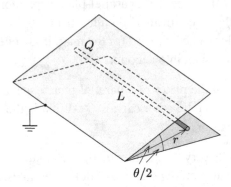

Find the force that acts upon the rod as a result of the electrostatic induction *a*) if $\theta = 180°$ and *b*) if $\theta = 18°$.

P148** A thin metal disc of radius R is charged, and carries a total electric charge of Q. Find the charge distribution on the disc.

P149** Two thin metal discs, A and B, each of diameter 5 cm, are suspended by electrically insulating threads in such a way that the discs are parallel (*see* Fig. 1) and close to each other (for example, their separation might be 2 mm). Both discs have small charges q on them. As q is small, we may neglect the associated relative displacement of the discs, and the possibility of electrical discharge. The electrostatic force between the two discs is clearly repulsive at this stage.

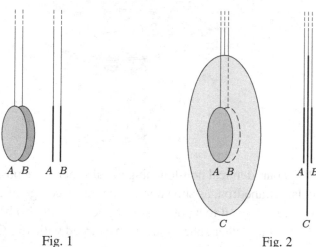

Fig. 1   Fig. 2

A third metal disc C of diameter $D > 5$ cm is carefully placed between the two original discs. The third disc is uncharged and also suspended by an electrically insulating thread. The three discs are all parallel and their centres lie along the same horizontal line. The resulting set-up is shown in Fig. 2.

Find the diameter D of the third disc that will make the net electrostatic force acting on each charged disc equal to zero.

P150* A solid metal sphere of radius R is divided into two parts by a planar cut, made in such a way that the outer surface of the smaller part of the sphere is πR^2. The cut surfaces are coated with a negligibly thin insulating layer, and the two parts are put together again, so that the original shape of the sphere is restored. Initially the sphere is electrically neutral.

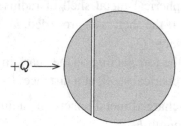

The smaller part of the sphere is now given a small positive electric charge $+Q$, while the larger part of the sphere remains neutral. Find

a) the charge distribution throughout the sphere,

b) the electrostatic interaction force between the two pieces of the sphere.

P151 A thin metal ring of radius R is charged in such a way that the electric potential at its centre is V_0. The ring is now placed horizontally above a grounded solid metal sphere of radius r, so that the centre of the ring and the top of the sphere coincide. Find the total charge induced on the sphere.

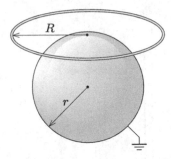

P152* Four identical non-touching metal spheres are positioned at the vertices of a regular tetrahedron, as shown in the figure. A charge of 20 nC given to one of the spheres, *A*, raises it to a potential *V*. Sphere *A* can also be raised to potential *V* if it and one of the other spheres are each charged with 15 nC.

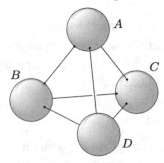

What must be the size of equal charges given to *A* and two other spheres for the potential of *A* to again be raised to *V*? And if all four spheres are used?

P153* A small pearl carrying charge *Q* is placed at a distance *d* from the centre of a thin-walled spherical metal shell of radius *R*, with $d > R$. Find the force acting on the pearl if the shell is *a*) grounded, *b*) uncharged and not grounded, and *c*) charged, with charge Q'.

What would be the force acting on the pearl, in each of the three cases, if it were placed inside the spherical shell, at a distance $d < R$ from its centre?[11]

P154** An uncharged metal sphere of radius *R* is placed in a homogeneous electric field of strength E_0.

a) What is the electric field formed around the metal sphere?

b) Find the charge distribution on the sphere.

P155 Two small, identical bar magnets are placed some distance apart, as shown in the figure.

---

[11] The essence of this final complementary question can also be found in the predecessor of this book: see 'Problem 92' in P. Gnädig, G. Honyek & K. F. Riley, *200 Puzzling Physics Problems* (Cambridge University Press, 2001). Our reason for repeating its solution here is to present in one place all of the possible variants of this question.

In which of the following cases is the work needed larger? And how many times larger?

a) The axis of the right-hand magnet is slowly rotated through 180°, so that the two magnetic moments are parallel.

b) The right-hand magnet is slowly removed 'to infinity' along the straight line connecting the magnets.

P156* A circular wire loop of radius R, and carrying an electric current I, is at rest in a horizontal plane (*see* figure).

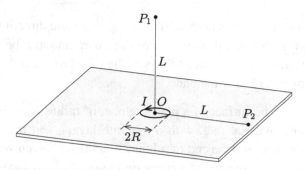

Find the magnetic field strength at a distance L ($\gg R$) from the centre O of the circular loop:

a) at a point, P_1, situated on the vertical axis passing through O,

b) at a point, P_2 in the plane of the current loop.

P157* A regular tetrahedron is made of homogeneous resistance wire of uniform cross-section. Current I is conducted into vertex A through a long, straight wire directed towards the centre O of the tetrahedron, and it is conducted away through vertex B in the same way, as illustrated in the figure. What are the magnitude and direction of the magnetic field vector at the centre of the tetrahedron?

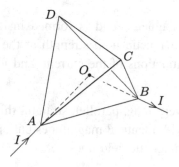

P158*   One end of a long, straight, current-carrying metal wire is electrically connected to a very large, thin, homogeneous metal plate, whose plane is perpendicular to the wire, and whose distant perimeter is earthed. One of the terminals of the battery supplying the current is also earthed.

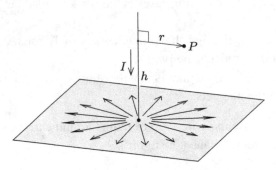

Ignoring the Earth's field, find the magnitude and direction of the magnetic field at a point P that is a distance r from the wire and at a height h above the metal plate, when a steady current I flows in the wire (*see* figure). Describe the magnetic field above and below the plate.

P159**   The surface of a plastic globe, of radius R, is uniformly covered with conducting material (e.g. with a graphite layer), and the globe is placed on an insulating support. One end of a long, straight, radial, current-carrying metal wire is electrically connected to a point on the sphere's surface. The steady current I, flowing through the surface, leaves the globe through another long, straight, radial metal wire that is perpendicular to the input wire, as shown.

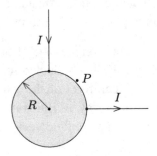

What kind of magnetic field is formed inside and outside the globe? Find, in particular, the magnetic field strength at the point P 'half-way' between the input and output junctions of the current, and just a 'whisker' above the globe's surface.

P160**   Two very long, parallel, straight, thin wires, a distance d apart, carry electric currents with identical magnitudes but opposite directions. Find the shape of the associated magnetic field lines.

P161*** A normal 8×8 chessboard is made from plates of two different metals, both of which are quite poor electrical conductors. The only other conducting
elements in its construction are two thin terminal strips, which have very good
conductivity. They are positioned one at each end of the board (but not shown in
the figure). The common thickness t of the plates is much less than the length L of
the board.

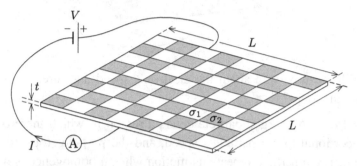

The conductivity of the light squares is σ_1, and that of the dark ones is
σ_2. Find the current flowing through the chessboard, if a steady voltage V is
applied across the terminals. Any interface resistances between the squares can be
neglected.

P162* Two long, cylindrical, unifilar, air-cored solenoids are placed end-to-
end and very close to each other, so that they have a common axis of symmetry
(*see* figure). The solenoids are identical, with cross-sectional area A and n turns
per unit length. The direct current flowing in one of them is I_1, and that in the other
is I_2. What is the magnitude of the magnetic force between them?

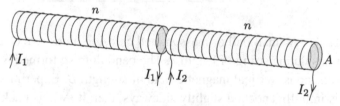

P163* In a strong magnetic field B, the two ends of a thin, flexible wire of
length ℓ are fixed at points P_1 and P_2, a distance $\ell/2$ apart. The direct current
flowing in the wire is I. What is the shape of the wire, if the magnetic field vector is
    a) perpendicular to the line segment P_1P_2,
    b) parallel to the line segment P_1P_2?

With what force does the wire pull on the anchor points in each case?

P164* Two identical circular wire loops are placed in parallel planes, with
their centres on a common normal to both planes. The direct currents they carry
are the same, both in magnitude I and in direction of flow (*see* figure). From the

midpoint O of the common normal, an electron with initial speed v_0 starts off in a direction parallel to the planes of the loops.

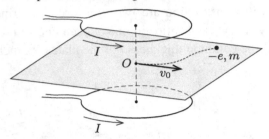

What can be said about the electron's speed and direction when it is very far from the loops?

P165* A charged particle enters a region in which there is a frictional force proportional to the particle's speed, and the particle stops 10 cm from its entry point. If the particle repeats its motion when a homogeneous magnetic field, perpendicular to the plane of its trajectory, is also present, then the particle comes to rest 6 cm from its entry point (*see* figure). How far from its entry point would the particle stop if the magnetic field were twice as large?

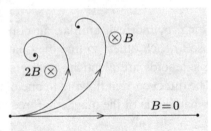

P166* A small ball of mass m, and carrying a positive charge q, is suspended by an insulating string of length ℓ. The pendulum so formed is placed, at rest, in a homogeneous, vertical magnetic field of strength B. Experiment shows that, if the ball is initially knocked slightly sideways, then it swings back and forth, with the plane of its swing slowly rotating. How long does it take for the plane to make one complete revolution?

P167** This problem investigates the motion of two electrons that are moving in a plane perpendicular to the field lines of a homogeneous magnetic field. The electrons are considered as classical point masses, affected only by electric and magnetic forces.

a) The two electrons, initially at rest, are placed a distance d apart. They are then given initial velocities of identical magnitudes v, but in opposite directions. Find the condition that d must satisfy if, in the subsequent motion, the separation is to remain constant. Find also an expression for v.

b) Show that it is possible to maintain a constant separation *d* if only one of the two electrons is given an initial velocity. What is the trajectory of the centre of mass of the system in this case? Find the minimal distance d_{min} that is necessary to realise this kind of motion. Sketch the trajectories of the electrons in that case. When will the initially moving electron first stop?

P168   The electrical resistances measured across the three pairs of terminals of the 'transparent' black box shown in the figure are

$$AB: 1\ \Omega, \qquad BC: 2\ \Omega, \qquad AC: 3\ \Omega.$$

Find the values of resistors *x*, *y* and *z*.

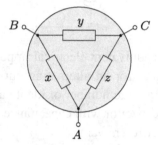

We can write the following simultaneous equations expressing the unknown resistances (in units of ohms) by means of their labels:

$$\frac{1}{x} + \frac{1}{y+z} = 1 \qquad \longrightarrow \qquad x(y+z) = x+y+z, \tag{1}$$

$$\frac{1}{y} + \frac{1}{x+z} = \frac{1}{2} \qquad \longrightarrow \qquad y(x+z) = 2(x+y+z), \tag{2}$$

$$\frac{1}{z} + \frac{1}{x+y} = \frac{1}{3} \qquad \longrightarrow \qquad z(x+y) = 3(x+y+z). \tag{3}$$

Subtracting equation (3) from the sum of (1) and (2), we get $2xy = 0$. This is *impossible*, because neither *x* nor *y* is zero!

How can this mathematical 'contradiction' be resolved?

P169*   Closed strips are made from the chain shown in the figure, which consists of 3*N* identical resistors, in two different ways:

  a) the terminals *A* and *C*, and the terminals *B* and *D*, are connected pairwise (ordinary strip),
  b) the terminals *A* and *D*, and the terminals *B* and *C*, are connected pairwise (Möbius strip).

In which of the two cases is the equivalent resistance between points *A* and *B* the larger?

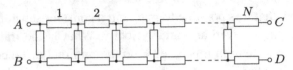

P170   Find the natural frequencies of an LC circuit with two identical inductors and two identical capacitors (a '2L2C' circuit), connected as shown in the figure.

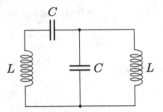

P171**   Find the equivalent electrical impedance between the terminals A and B shown in the figure, for an alternating current of frequency ω. The 'infinite' chain is made up of a large number of identical units, each consisting of a coil of inductance L and a capacitor with capacitance C. Is it possible that the equivalent impedance has two different values?

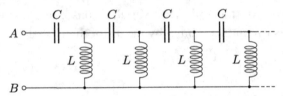

P172*   Two identical, very long, cylindrical conductors, of diameter d and negligible resistance, are placed parallel to each other with their axes separated by $D = 50d$. A battery of electromotive force (voltage) V is connected between the left-hand ends of the wires, while a resistor with resistance R is connected across their other ends (*see* figure). Find the resistance R that makes the electrical and magnetic forces between the conductors equal.

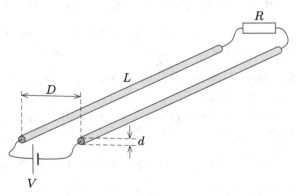

P173* In 1917, T. D. Stewart and R. C. Tolman discovered that an electric current flows in any coil wound around, and attached to, a cylinder that is rotated axially with constant angular acceleration.

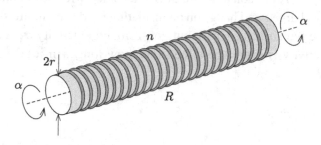

Consider a large number of rings of thin metallic wire, each with radius r and resistance R. The rings have been glued in a uniform way onto a very long evacuated glass cylinder, with n rings per unit length of the symmetry axis. The plane of each ring is perpendicular to that axis.

At some particular moment, the cylinder starts to accelerate around its symmetry axis with angular acceleration α. After a certain length of time, there is a constant magnetic field B at the centre of the cylinder. Find, in terms of the charge e and mass m of an electron, the magnitude of the field.

P174* We aim to measure the resistivity of the material of a large, thin, homogeneous square metal plate, of which only one corner is accessible. To do this, we chose points A, B, C and D on the side edges of the plate that form the corner (*see* figure). Points A and B are both $2d$ from the corner, whereas C and D are each a distance d from it. The length of the plate's sides is much greater than d, which, in turn, is much greater than the thickness t of the plate.

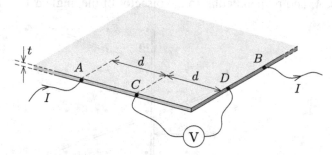

If a current I enters the plate at point A, and leaves it at B, then the reading on a voltmeter connected between C and D is V. Find an expression for the resistivity ϱ of the plate material.

P175 A metal sphere is electrically charged and hangs on an insulating cord. The sphere slowly loses its charge because the air has a small, but non-zero, conductivity σ. Assuming that the air's conductivity is everywhere the same, how long will it take for the charge on the sphere to halve?

P176* A chocolate figure of Santa Claus, wrapped in aluminium foil, is electrically charged and hangs on an insulating cord. The figure slowly loses its charge because the air has a small, but non-zero, conductivity σ. Assuming that the air's conductivity is everywhere the same, how long will it take for Santa's charge to halve?

P177 A long, straight wire of negligible resistance is bent into a V shape, its two arms making an angle α with each other, and placed horizontally in a vertical, homogeneous magnetic field of strength B. A rod of total mass m, and resistance r per unit length, is placed on the V-shaped conductor, at a distance x_0 from its vertex A, and perpendicular to the bisector of the angle α (*see* figure).

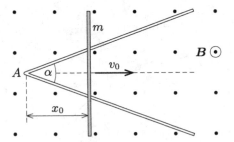

The rod is started off with an initial velocity v_0 in the direction of the bisector, and away from A. The rod is long enough not to fall off the wire during the subsequent motion, and the electrical contact between the two is good – although the friction between them is negligible.

Where does the rod ultimately stop?

P178 Inside a vertical, thin-walled, non-ferromagnetic (say, brass) tube, a quite large, strong, cylindrical permanent (say, neodymium) magnet falls very slowly. It takes a time t_1 to fall between two particular markers. If the experiment is repeated with a different non-ferromagnetic (say, copper) tube with the same length but a slightly larger diameter, then the corresponding time is t_2. How long does it take for the magnet to fall between the marks when the two tubes are fitted inside each other? The mutual inductance between the tubes is negligible.

P179 Two circular wire loops, with radii R and r ($r \ll R$), are concentric and lie in the same plane (*see* figure). The electric current in the smaller loop is increased uniformly from zero to a value of I_0 over a time interval t_0. Find the induced voltage in the larger loop.

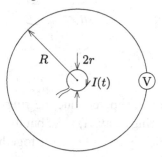

P180* A closed circular loop of radius r is made from a wire of resistance R and a diode, which can be considered ideal. The loop is held in a horizontal plane, and a long, vertical glass tube passes through its centre (*see* figure). Find the charge that flows through the diode if a small bar magnet with magnetic moment $\boldsymbol{m}$ falls through the tube.

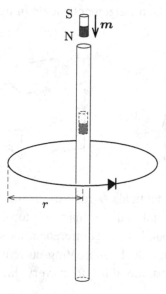

P181* Three nearly complete circular loops, with radii R, $2R$ and $4R$, and made of thin wire, are placed concentrically on a horizontal table-top, as shown in the figure. A time-varying electric current is made to flow in the middle loop. Find the voltage induced in the largest loop at the moment when the voltage between the terminals of the smallest loop is V_0.

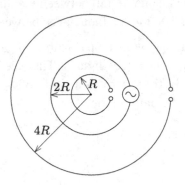

P182 Two identical superconducting rings are very far from each other. The current in one of the rings (say, A) is I_0, but there is no current in the other (B). The two rings are now slowly brought closer together. Find the current that flows in A when that in B has increased to I_1.

P183* Three metal wires of lengths $4a$, $6a$ and $6a$ are arranged along the edges of a cube of side a in three different ways, as shown in the figure. The coefficient of self-inductance of the square shown in figure a) is measured as L_1, and for the arrangement shown in figure b) it is L_2. How large, expressed in terms of L_1 and L_2, would it be for the arrangement shown in figure c)?

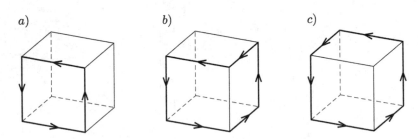

P184* Three identical, 'wide' electromagnetic coils, of negligible ohmic resistance, are wound onto a 'narrow' toroidal (doughnut-shaped) iron core, equally spaced around its circumference, as shown in the figure. The first coil is connected to an (ideal) alternating-current power supply, the second to an open switch S, and the third to a very high-resistance voltmeter. With this arrangement, the voltmeter shows a root mean square (r.m.s.) reading equal to one-half of that of the power supply.

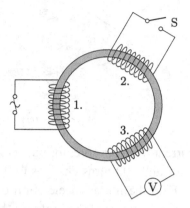

At this point, switch S is closed and short-circuits the terminals of the second coil. Assuming that the magnetic permeability of the iron core does not depend on the magnetic flux through it, determine the new reading of the voltmeter.

P185* ** Two ideal (with zero ohmic resistance) air-core toroidal coils, of identical size but with different numbers of turns, N_1 and N_2, are interlinked, as shown in the figure. The planes containing their centres and major radii are perpendicular to each other. The terminals of the coil with N_1 turns are connected to the normal household a.c. supply with r.m.s. voltage V_0, and an ideal voltmeter is connected to those of the other coil. Find the reading on the voltmeter.

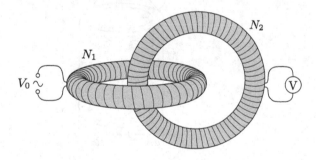

P186* * A small, electrically charged pearl, initially at rest, can move on a horizontal, frictionless plane. Not far from the pearl, there is a long, vertical solenoid, in which the electric current is first increased uniformly from zero to a given value, and then uniformly decreased back to zero. In which direction, relative to point P in the figure, will the pearl have moved by the end of the process? Or will it (still) be where it started?

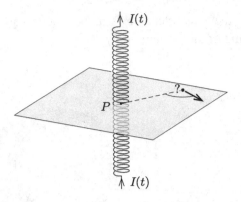

P187* The current in an air-core, toroidal coil with major radius 0.1 m, cross-sectional area 2 cm^2 and 200 turns changes uniformly (over a short interval) at a rate of 10 A s^{-1}. Find the initial acceleration of a proton, treated as a classical point mass, that starts from rest at the centre of the toroid.

P188* A cylindrical capacitor with external radius R, internal radius $R - d$ ($d \ll R$), length ℓ and mass M hangs on an insulating cord in a region where there is a homogeneous, vertical magnetic field of strength B. It can rotate freely (as a whole) around its vertical axis, but is constrained so that it cannot move horizontally. The capacitor is charged and there is a voltage difference V between its plates.

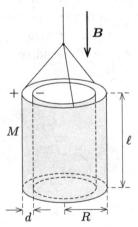

What happens, if:

a) without being mechanically disturbed, the capacitor is discharged through an internal radial wire,

b) suddenly, the magnetic field is switched off?

P189* It is well known that, in a homogeneous magnetic field $\boldsymbol{B}_0$, an observer moving with velocity $\boldsymbol{v}_0$ ($|\boldsymbol{v}_0| \ll c$) 'experiences' an electric field, given by $\boldsymbol{E}_0 = \boldsymbol{v}_0 \times \boldsymbol{B}_0$. This phenomenon is called motional electromagnetic induction.

Using only physical laws known to high-school students, investigate whether or not the inverse phenomenon exists. Does an observer 'experience' a magnetic field when moving in a homogeneous electric field?

P190** A parallel-plate capacitor, with its plates vertical, is charged in the sense shown in Fig. 1, and positioned with its lower edges above and on either side of a small horizontal compass needle. The capacitor is discharged when the tops of the plates are joined using a small conducting rod. Describe the response of the compass needle during the discharge process.

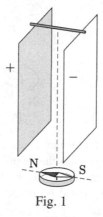

Fig. 1

This problem was used as a question in a Hungarian national physics competition, several years ago. The 'official solution' to the problem read more or less as follows:

When the tops of the plates are joined by the conducting rod, an electric current starts to flow in the rod, from left to right, with magnitude $I = -dQ/dt$. At the same time, the electric field between the plates $E = \varepsilon_0^{-1}Q/A$ changes, and J. C. Maxwell has shown that this generates the so-called displacement current. The size of the current is proportional to the rate of change of the electric field, with proportionality constant ε_0.

The displacement current density between the plates is

$$j_D = \varepsilon_0 \dot{E} = \varepsilon_0 \frac{1}{\varepsilon_0} \frac{1}{A} \frac{dQ}{dt} = -\frac{I}{A}.$$

The current in the rod (of magnitude I) and the displacement current (of magnitude $j_D A = -I$), taken together, can be considered as charge moving in a closed circuit.

According to Maxwell's equations, magnetic fields are produced not only by ordinary electric currents, but also by displacement currents. As the displacement current and the current in the rod have equal magnitudes (but opposite directions),

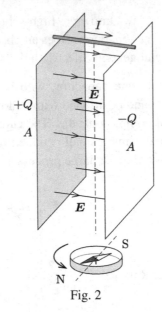

Fig. 2

and the displacement current is closer to the compass needle, its effect is larger. This is why, during the discharge process, the compass needle tends to rotate (*see* Fig. 2), with its north pole moving in an anticlockwise direction (from the position shown) when viewed from above. It will subsequently return to its original position.

Decide, whether the 'official solution' to the problem is right or wrong.

P191* A chocolate figure of Santa Claus, wrapped in aluminium foil, is electrically charged and hangs on an insulating cord. The figure slowly loses its charge because the air has a small, but non-zero, conductivity. Describe the magnetic field around Santa during the discharge process.

P192 An electron, moving with 60 % of the speed of light c, enters a homogeneous electric field that is perpendicular to its velocity. When the electron leaves the field, its velocity makes an angle of 45° with its initial direction (*see* figure).

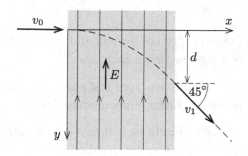

a) Find the speed v_1 of the electron after it has crossed the electric field.

b) Find the distance d shown in the figure, if the strength of the electric field is $E = 510 \text{ kV m}^{-1}$.

The rest energy, mc^2, of an electron is 510 keV.

P193 Imagine a circular evacuated tube running around the Earth's magnetic equator, in which – in principle – electrons and protons could orbit under the influence of the Earth's magnetic field.

a) Estimate the required speed for each particle, and determine the corresponding direction of circulation.

b) Express the particle energies needed for these hypothetical flights in eV units.

P194* The trajectories of charged particles, moving in a homogeneous magnetic field, can be followed by observing the trails they produce in cloud chambers (e.g. a Wilson chamber). Is it possible that, when a charged particle decays into two other charged particles, the trail segments close to the decay point (before the particles have started to slow down significantly) are arcs of circles that touch each other (as shown in the figure)?

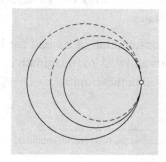

P195* The pion (π^+) is a subatomic particle with a mass 273 times larger than that of an electron. In one of its possible decay modes, it decays into a positron (e^+) and an electron-neutrino (ν_e):

$$\pi^+ \longrightarrow e^+ + \nu_e.$$

What is the minimum speed of the pion if, following its decay, the positron and the neutrino move at right angles to each other?

Assume that the neutrino has zero rest mass. For such a particle, the connection between its energy and linear momentum is $E = pc$.

P196* Investigate the elastic collision of two (ultra-relativistic) particles, moving with speeds very close to the speed of light. The linear-momentum vectors of the particles, before the collision, are shown in the figure. Determine the minimal possible angle between the directions of the particles after the collision.

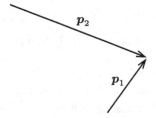

P197 An electron orbits uniformly around a circular trajectory in a homogeneous magnetic field. Is it possible for the magnitude of the magnetic field produced by the electron's motion to be larger, at the centre of the circle, than that of the field?

P198* Estimate the pressure exerted on the walls of a small cubical box, which has edges of length d, by a neutron enclosed within the box.

P199 *Positronium* is a system consisting of an electron and its antiparticle, a positron, bound together into an exotic 'atom' by their electrostatic attraction.[12] The orbit and energy levels of the two particles are similar to those of the hydrogen atom. Find – within the framework of a Bohr model – the energies of positronium's ground and excited states.

P200 The fuel consumption of a small car is 4 litre/100 km. Convert the consumption into SI units and find an (imaginative) physical interpretation for the fuel consumption expressed in these units.

---

[12] A positronium atom is unstable; the two particles annihilate each other, producing predominantly two or three gamma photons.

Hints

H1 Consider the motions of the bodies both in the original frame of reference and in another frame that moves with velocity v_0 relative to the first one. Apply the Galilean transformation formulae.

H2 Be careful, the transformation principle due to Galileo Galilei applies only to inertial reference frames. The idea that Ann simply observes Bob moving towards her with a speed of 1 m s^{-1} is false.

H3 It is easier to solve the problem in a frame of reference fixed to the cart. A vector diagram of velocities, superimposed on a geometrical one showing the initial situation, should prove helpful.

H4 It is helpful to describe Joe's motion using a reference frame moving with the river water; imagine that we are in a boat that drifts with the river. Using this frame of reference, a suitable optical analogy and the application of Fermat's principle can help us to find Joe's optimal trajectory.

H5 The description of the disc's motion in a frame of reference fixed to the table is not easy. Try using a reference frame moving with the conveyor belt.

H6 The task can be tackled in several ways. One possibility is to analyse the turning process in an appropriate projection of phase space, namely in the v_x–v_y coordinate system. Another good idea is to use an inertial frame of reference moving relative to the ice in a suitably chosen direction.

H7 From the tangential and centripetal components of the acceleration, determine the horizontal and vertical ones.

H8 Finding accurate time intervals for the two sections of the motion is very difficult. To answer the question as posed, it is easier to compare an underestimate of the time to cover arc AP with an overestimated time for arc PB.

H9   Show that the axis of symmetry of a parabola makes an angle with the tangent to the parabola at any point P that is half the angle it makes with the line connecting P to the focus.

H10   Instead of using calculus, let us try to solve the problem using a variety of techniques, e.g. by investigating the number of real roots of a quadratic equation, or by using the properties and geometric representations of relevant vectors.

H11   The answer can be obtained by using the laws of projectile motion and some straightforward, if somewhat lengthy, algebra. Find the equation that gives the height(s) h corresponding to a given point of impact, and investigate the condition under which this equation has real roots.

An alternative method is to find the directrix and focus of the parabola that describes the trajectory of the water jet, and then investigate the geometrical condition under which the water hits the incline at a given point.

H12   The sharp points on the photograph are the images of those points on the spokes that have only radial motion.

H13   Find the equation of the 'virtual' shape of the spokes in a Cartesian coordinate system.

H14   The notion that the bolt can remain between two successive spokes of the wheel for as long as it takes the wheel to rotate through an angle of $360°/12 = 30°$ is wrong. The fact is that the cartwheel does *not* rotate around a stationary axis, since it also has translational motion. An approximation to the correct answer can be found using a graphical method.

H15   In the absence of friction, the law of conservation of mechanical energy requires the two final speeds to be the same. To find the correct answers to the other questions, consider, on both trajectories, *a*) small segments with identical lengths, and determine which of the two is covered more quickly, *b*) small segments with identical slopes, and investigate on which of these a larger frictional force retards the bob.

H16   The cyclist's speed is constant in each of these three cases, and so his or her acceleration is zero. It follows that the net force acting on the cyclist (the resultant of the gravitational force, the air drag and the frictional and normal forces exerted by the road) must be zero.

H17   Because he is heavier, Bob is pulled down the slope by a larger gravitational force than the one that acts upon Ann, the forces being in the ratio of $m_B/m_A \approx 1.8$. However, the air drag acting on him is greater than that on Ann's slimmer body, with his coefficient k in the drag formula $F = kv^2$ being one-and-a-half times larger than hers. Which effect is the more important?

Write the equations of motion for the two riders and find out how their speeds at the end of the slope depend on the ratio k/m.

It is clear that their equations of motion on the horizontal road determine the total path lengths covered before each comes to a halt. The decelerations involved are *not* uniform and the equations are not straightforward. However, if we only want to know which of them coasts further, full solutions to the (differential) equations are not necessary – it is sufficient to investigate them qualitatively using suitable variables.

H18　The mass of the feather is infinitesimally small, so gravity can be ignored when considering the forces acting upon it.

H19　Examine how the difference between the instantaneous and final velocity vectors changes with time.

H20　Suppose that the smaller ball is not lifted from the table-top. Find the relation between the accelerations of the two balls, then use the fact that the larger ball is practically in free fall just after the release. What is the net force experienced by the smaller ball at that moment, and in which direction does it act?

H21　Our first thought might be that the condition for the pearl to fly off the rod is that, at the top of its trajectory, the pearl should just lift off from the disc. Starting from this (retrospectively naive) assumption, we find the inequality

$$f > \frac{1}{2\pi} \sqrt{\frac{g}{\theta_0^2 d}}.$$

But this result is false! For the correct answer, determine the vertical component of the net force acting on the pearl as a function of time, and use its time-averaged value.

H22　Most of the time, the acceleration of the board is very much larger than the acceleration the block would experience when acted upon by a force equal to the maximal value of static friction. So we can assume that the block does not adhere to the board, but slides (upwards or downwards) on it. The velocity of the block changes regularly, alternating between speeding up and slowing down. A more-or-less steady state of motion sets in, with the velocity fluctuating around a particular value v_{drift}, called the *drift* or *terminal* velocity. The block's time of arrival at the bottom of the board is determined by this drift velocity.

H23　Find the average tension in the cord, and the work done during a single swing if the length of the cord is decreased by a small amount $\Delta \ell$ during one swing.

H24   Describe, in vector notation, the equation determining the mountaineer's equilibrium position and the equation of motion when his body is displaced from equilibrium.

H25   Underestimate the work done against friction, and compare it with the initial gravitational potential energy of the eraser.

H26   Investigate the role of friction from both the static and dynamic points of view.

H27   Apply the conservation laws of energy and linear momentum. It is not necessary to explicitly find the interaction energy of the magnets. It is sufficient to know that this energy is some function of the distance between the magnets (which, in principle, can be calculated, or can be measured experimentally).

H28   Our probable first idea – that the heavy ball tends 'to pull down' the liquid above it – is *false*. In fact, just the opposite occurs, and the liquid in the left arm rises.

It is probably helpful to first analyse the extreme case in which the gravitational field due to the heavy ball is much larger than that of the Earth.

H29   Without loss of generality, we may take the point in question as the origin of our coordinate system, and the *x* direction as that of the net gravitational field. Determine the shapes of surfaces which are such that, if a small piece of plasticine of given volume is placed at *any* point upon one of them, the same additional gravitational *x* component is produced at the origin. These surfaces may be taken as 'level surfaces', differentiated only by the magnitudes of the additional components they produce; the larger the component, the closer to the origin the surface lies on average.

H30   The laws governing gravitational and electric fields are very similar; use this analogy and apply Gauss's law. *Warning:* The gravitational field near the planet has a different form from that which applies when the distance from the planet is commensurate with the length of the planet (or is much larger than it).

H31   The expert – who was not given any numerical data in the first instance – was able to provide numerical answers to all of EXINCA's queries. In his scientific report, he showed that both phases of the wagon's motion were of a simple harmonic nature.

H32   The answers to both questions can be found using brute-force methods, but consideration of gravitational potential energy in the Boltzmann distribution provides a much more elegant solution.

H33   Think about, for example, the motion of a conical pendulum.

H34 The expression 'orbiting the Sun' in the text of the problem is an indication that the total energy of the comet is negative. The impact speed of the comet is maximal if the comet approaches from a direction directly opposed to the Earth's motion, and if, at the same time, the comet is 'bound' very weakly to the Sun, i.e. its total energy is almost zero. The impact speed will be minimal if the comet arrives in the vicinity of Earth in such a way that their relative velocity is negligible.

H35 The pieces of the debris must move along a variety of elliptical orbits, but all with the same major-axis length, and all with the Sun as one of their foci. Find the possible positions of the other foci of these ellipses, and note that each point of the *envelope* can lie on only one of the ellipses.

H36 It is advisable to launch the space probe in such a way that, after 'leaving' the gravitational field of the Earth, its velocity relative to the Sun has the same direction as the velocity v_E of the Earth, because then the speeds of the Earth and the space probe are cumulative.

The conservation laws for energy and angular momentum in the presence of the Sun's gravitational field apply to the space probe when it is moving in interplanetary space. It is convenient to analyse the gravity-assisted 'swing-by manoeuvre' from a reference frame fixed to the planet.

H37 To a good approximation, the dynamical conditions for circular motion remain valid throughout the satellite's motion. Using energy considerations, it can be proved that the *decrease* in the total energy of the satellite is just the same in magnitude as the *increase* in its kinetic energy. Consequently, perhaps contrary to expectation, the air-drag loss *increases*, rather than decreases, the speed of the satellite.

The dynamical reason for this can be summarised as follows. The net force on the satellite is the vector sum of the drag force opposing the velocity and the gravitational force. As a result of the 'spiral' trajectory, this net force is *not* perpendicular to the velocity of the satellite, but has a tangential component that acts in the direction of the satellite's velocity, and so accelerates it.

H38 If we neglect angular-momentum effects due to the Sun, the Earth–Moon system can be considered as a closed system in which the total angular momentum is conserved. The total angular momentum consists of the spins of the celestial bodies rotating around their own axes, and the orbital angular momentum associated with their motions around their common centre of mass. It is useful to estimate which of these make a significant contribution to the total angular momentum, and which are negligible. For the purposes of this estimation, we can use the approximation that the Earth's Equator, the equator of the Moon and the Moon's orbit all lie in the same plane. Assume that the Moon is a homogeneous sphere.

The inertial momentum of the Earth can be taken from the table at the end of the book (*see* Appendix).

In addition to the conservation of angular momentum, another equation is needed – one that gives a connection between the orbital angular velocity of the Moon and the Earth–Moon distance. To obtain this, consider the centripetal acceleration of the Moon as it orbits in an approximate circle.

H39 We can assume that the astronaut's trajectory differs only slightly from that of the ISS. The description of the motion is best made using polar coordinates. Higher powers of any *small* dimensionless quantities that appear during the calculation can be neglected.

The problem may also be solved using Kepler's laws of planetary motion. The use of all three laws is necessary if this method is chosen.

H40 *a*) Do not forget to take into consideration the change in the space ship's kinetic energy!

b) It is convenient to use an inertial frame of reference to investigate the astronaut's motion.

H41 The naive answer, namely $mgh = 50 \times 9.81 \times 20 = 9.81$ kJ, is false.

H42 For a successful realisation of the miracle, the direction of the velocity of D must turn through an angle of π/N at each and every collision. If $m > M/N$, then D cannot scatter at an arbitrarily large angle from an initially stationary disc of mass M/N. Find an expression, in terms of m, M and N, for the maximum angle that *is* possible.

In the solution to part *b*) the following expression for the exponential function is needed:

$$e^x = \lim_{n \to \infty} \left(1 + \frac{x}{n}\right)^n.$$

H43 If the swings are small, their period is independent of their amplitude, so the balls will always collide at their initial equilibrium positions. Since air resistance is negligible, the bouncing balls will return to the lowest point of their trajectories with the same speeds as they had immediately after the previous collision. At each collision, the mechanical energy decreases, but the net linear momentum remains constant.

H44 Using the conservation laws that govern elastic collisions, it can be shown that, after each collision, the product of the relative velocity of the two balls and the distance of the collision point from the wall always has the same value. Using this and the law of conservation of mechanical energy, the required minimum distance can be found.

Another (approximate) solution to the problem can be obtained if the ball of mass m is considered as a one-dimensional 'gas', containing only a single molecule, that is being compressed adiabatically by a piston of mass M.

H45 In addition to the effects due to air drag, the motion of the increasingly heavy drop is retarded by the linear momentum changes of the smaller drops that it absorbs.

H46 If the tall narrow statuette is pulled very gently, then – because of the relatively large friction – the figurine pitches 'forwards' onto his left ear. But if the tension in the thread is too large (and the thread does not break), then the statuette tumbles over 'backwards' onto his right ear! Try to find a 'Goldilocks' tension, one that is neither too small nor too large, but just right.

H47 The tension in the remaining rubber band depends only on its current elongation, which cannot change instantaneously. Write the equations for any translational and rotational motion, and determine the condition for the upward motion of point B. It can be proved that this condition is satisfied whatever the mass distribution of the boat's hull.

H48 In principle, it is possible to formulate all the dynamical equations needed to describe the translational and rotational motions of each rod. However, those equations will contain so many auxiliary variables (e.g. the strengths and directions of the forces acting at the joints) that the required result could only be found after considerable calculation. The acceleration of Q can be derived much more easily if the work–energy theorem is applied over a very short time interval immediately following the start of the motion.

H49 At first sight it might seem that too little information has been provided. Nevertheless, pluck up courage and write the equations for translational and rotational motion.

H50 Prove that neither the translational motion nor the rotation of the disc can stop earlier than the other. To do this, consider how large the translational braking force is, even when the centre of mass is moving very slowly, and how large a torque slows the rotation, even when the angular velocity is very small.

H51 Show that the small masses of the bearings and the 'non-slipping' condition together imply that the static frictional forces that act on the bearings have negligible components perpendicular to the rod. This means, in both cases, that there is no torque acting on the rod and that its angular velocity is constant.

H52 Just after the first collision, because the balls slip on the edges of the trough, friction decelerates both their translational motions and their angular velocities. The critical point of the solution is finding the appropriate condition for the balls tỏ roll without slipping. The law of conservation of angular momentum might also be useful.

H53   In both cases, find fixed points (in space) about which the angular momentum of the ball remains constant, even during the accelerating period of the stroke.

H54   Prove that the velocity vector of the point on the ball in contact with the table has a constant direction throughout the 'slipping' motion. Part *b*) can be answered using an appropriate application of the law of conservation of angular momentum.

H55   Formulate the vector equations for the rotational motion of the ball and the motion of its centre of mass, and express the condition that the ball rolls without slipping. Using these equations, prove that, independent of the initial conditions, the ball's centre always moves uniformly around a circular trajectory, but that, in general, the centre of this circle does not coincide with the centre of the disc.

H56   It is advisable to write the equations for the translational and rotational motion of the ball in vectorial form. Investigate the condition for zero acceleration of the centre of the ball. A review of the solution to the previous problem may help.

H57   First investigate the rotation of the rubber ball about its own vertical axis, and the rotation of its centre of mass around the axis of the tube. If it could be shown that both of these rotations are uniform, then the description of the other parts of the motion would be easier. *Beware:* Any notion that the vertical motion is one with uniform acceleration is false.

H58   Compare the average tensions in the spring in the two cases.

H59   First, show that the mass of any particular, not necessarily short, piece of the Slinky is directly proportional to the length of its horizontal projection.

H60   The work done during the lifting stage increased both the gravitational potential energy and the elastic energy of the Slinky. Expressions for those changes can be found by considering the Slinky as divided into small pieces of equal mass. Parts *b*) and *c*) can be answered if it is noticed that, following the spring's release, its centre of mass is in free fall.

H61   Using the method of dimensional analysis, a partial, though not complete, solution can be found.

For a more precise analysis, it is convenient to specify any particular point on the spring by the mass m^* of that portion of the spring that lies between that point and the anchor point. If the tension in the spring, F, and the distance from the rotational axis, r, can be found as a function of m^*, then the whole problem can be solved.

To do this, write equations for the rates of change of $F(m^*)$ and $r(m^*)$ as m^* varies. From these – using an appropriate mechanical analogy – all of the details of the stretching of the Slinky (including its total elongation) can be deduced.

> *Note.* Before tackling the main problem, it is instructive to think about a simpler version of it. Instead of a spring with the continuous mass distribution present in the Slinky, consider a massless spring, with negligible unstressed length and spring constant k, that has a point-like mass m attached to its end, and is rotated in the same way as the Slinky.

H62   Find an analogy between the equation describing the shape of the rotating Slinky and the equation governing the simple harmonic motion of a point-like body. A review of the solution on page 210 could be very helpful.

H63   The displacement of the free end of a horizontal cantilevered beam that is fixed at one end, and loaded vertically at the other, is directly proportional to the loading force and to the cube of the beam's length.

H64   To make a comparison between the two end-point droops, consider the quarter-circle as divided into small segments of identical length. When the metal wire is loaded, the torque in it varies from point to point, and the small segments of the wire are tilted to varying degrees. The angle characterising the inclination of any particular segment (i.e. the difference in angle between the tangents at its two ends) is directly proportional to the torque at the position of the segment. Compare the torques acting in appropriately corresponding small segments of the two wires, and investigate the contributions these deformations make to the total droop of the end-points of the wires.

The problem can also be approached through energy considerations, using the fact that the amounts of stored elastic energy in equal-length segments of a wire are directly proportional to the square of the torques generated in them.

H65   The elastic energy stored in a deformed rod is directly proportional to its length, and inversely proportional to the square of its radius of curvature. Approximate the shape of the deformed rod by an arc of a circle in both cases, and find a connection between the stored elastic energy and the loading force.

H66   If the number of parallel cable filaments is increased, from the bottom upwards, sufficiently often, then – in principle – arbitrarily deep sea beds can be reached.

H67   The rupturing of the sausage skin is caused by the steadily rising steam pressure inside it. The resulting overpressure produces a similarly increasing tensile stress in the skin of the sausage, a stress that is anisotropic (and, in the case of the torus, also inhomogeneous). The magnitudes of the stresses can be calculated in an elementary way from the force balance of an appropriately chosen piece of the sausage.

H68   The gravitational force acting on the acrobat is balanced by the vertical components of the tension in the rope. For small angles, the approximation $\sin \theta \approx \tan \theta$ can be used.

H69   Find a mechanical system in which the energy is minimal (and so the system is in equilibrium) when the sum of the distances measured from the vertices of the triangle to a particular point is the least. Consider how the parameters of the device could be altered to determine the corresponding minimal values for a *weighted* sum of the distances.

H70   Notionally divide the area of contact between the sack and the carpet into very narrow strips perpendicular to the direction of motion, so that the load on each can be taken as uniform. The sum of the elements of work done against friction on each of these small segments of area gives the required answer. Consider whether it is important that the sack is pulled at the bottom, i.e. using the carpet. Would your analysis apply if the towing line were attached to the sack above the level of the carpet?

H71   Apply D'Alembert's principle (of virtual work) using a small (imaginary) splaying of the cone caused by the vertical load.

H72   Lisa is right. For showing that this is so, an application of the angle bisector theorem could prove useful.

H73   The triangle will tumble over if the vertical projection of its centre of mass lies outside its supporting base.

H74   It can be proved that, if one of the faces of a tetrahedron is unstable, then two of the other three faces must have larger areas than it has. It follows from this that a tetrahedron *must* have at least two stable faces, the ones with the largest and second largest areas.

H75   To estimate the tension, construct an approximate equation expressing the balance of moments about some particular point for a suitably chosen segment of the cable.

H76   The question can be resolved in an elementary way, without using the shape of the rope. Establish a relationship between the horizontal forces needed to lift each of the ends to the *same* height.

H77   To find the solution, it is not necessary to calculate the exact shape of the chain.

H78 It can be shown that the difference between the tensions in the rope at two arbitrary points on it (denoted by A and B), e.g. the centre and the right-hand pulley, depends only on the height difference between the points and the linear mass density λ of the rope. In an obvious notation:

$$F_B - F_A = \lambda g(h_B - h_A).$$

To prove this, consider the energy changes that would be involved if a small length of rope were (notionally) cut out from the neighbourhood of point A and inserted close to point B.

H79 The difference in the magnitudes of the tensions acting at two arbitrary points of the necklace is directly proportional to the height difference between the points (*see* solution on page 239). If the part of the necklace that hangs down is sufficiently long, then the equilibrium is stable. If the length of the dangling portion is decreased to below a certain critical value, then the pearl necklace drops away from the surface at the lowest point of the cylinder.

H80 Apply D'Alembert's principle of virtual work to investigate the energy changes that take place if the partially rolled rug is coiled up a bit further.

H81 The tension is constant throughout the rope of the experts' lasso. The same is true, separately, for each of the two pieces of the novices' lasso. In both cases, at equilibrium, the climber's gravitational potential energy is minimal. To determine how the rope lies on the lateral surface of the cone, it is easier if that surface is notionally spread out into a plane sheet.

H82 It can be shown that the shape of the chain remains unaltered.[13]

H83 Apply the law of conservation of mechanical energy, and, using the velocity of the roll, calculate the force as the rate of change of linear momentum.

H84 The mechanical work required to maintain the rotation of the device, which is similar to a concrete mixer, is totally transformed into heat. Try to obtain an expression for this work in terms of quantities that describe the overall situation and do not vary with time. The 'scarp angle' of the sand[14] in the mixer will be needed; estimate it on the basis of everyday experiences.

H85 Note that the downward forces acting on the bucket include not only the weight of the water inside but also the reaction to the upthrust experienced by the iron cube.

H86 Do not forget that the water is practically incompressible.

---

[13] A similar problem can be found in the predecessor of this book: see 'Problem 104' in P. Gnädig, G. Honyek & K. F. Riley, *200 Puzzling Physics Problems* (Cambridge University Press, 2001).

[14] The angle that the sloping surface of the sand makes with the horizontal.

H87   Several solutions are plausible. In one of them, an easily deformed spring is the key element.

H88   Consider notionally covering the funnel with a cylinder that has a base that coincides with the rim of the funnel, and then pouring water into it up to the level of that in the funnel. In this situation, how much force does the water exert on the funnel?

H89   Use the fact that the net force acting on the hemispherical body of water must be zero. For the determination of the application point, consider the various torques, about an appropriately chosen point, that are generated within the system.

H90   If the tyre pressure is increased, then the area of contact between the tyres and the road decreases. Consequently, the moment arm (and the torque) of the frictional force opposing the reorientation of the front wheels will be reduced.

H91   The idea that the product of the gauge pressure in the balloon and the cross-sectional area of the rod must be equal to the rod's weight is false. The force acting on the bottom of the rod is due not only to the hydrostatic force coming from the gauge pressure but also to an elastic force arising from the deformation of the balloon's rubber wall.

H92   The answer can be found with the help of the discriminant of a particular quadratic equation.

H93   At first sight, the problem seems to be a purely mechanical one, but it does, in fact, also involve thermodynamics. The stationary state that emerges 'after a sufficiently long time' involves a thermodynamic equilibrium, in which the role of water vapour is crucial.

H94   How large is the force exerted on the bottom of the glass, both before and after the razor-blade sinks?

H95   The vertical stability of the disc is provided by the pressure difference between the top and bottom of the disc. However, an explanation based on Bernoulli's law – that the pressure above the disc is lowered because of the fast horizontal outward flow of water – is false.

H96   The radius of curvature of the separating soap film can be found from the balance of pressures across it. To get the radius of the perimeter circle, consider the forces acting on a small segment of it.

H97   Two opposing effects are present: surface tension tries to shrink the puddle so that it has as small a surface area as possible; gravity would like to flatten it and make it as thin as possible. The equilibrium size of the puddle can be found by using the principle of minimum energy. Do not forget to take into account the interaction energy at the water–floor interface.

H98 Our likely first thought, that in equilibrium the water drop will coat the rod uniformly, is false. The shape of the water drop is determined by a minimum energy criterion, and the position of the rod by a balancing of the forces acting on it.

H99 The equilibrium state can be found using the principle of minimum energy, and, under weightless conditions, only surface energies need to be considered.

H100 Divide the hot water into a number of equal portions, each in its own small vessel. Then, one by one, place them in thermal contact with the originally cold distilled water.

H101 In a stationary state, the rates at which air molecules enter and leave the furnace are equal.

H102 Apply the first law of thermodynamics to the procedure whereby the gas is transferred from above the separating wall to below it.

H103 Investigate what would happen if a small volume of air in an otherwise stationary lower atmosphere were *suddenly* to rise to a slightly higher altitude. The lifted air mass would cool as a result of a fast (adiabatic) expansion, and its density would decrease. The condition for the atmosphere's stability is that the final density of the air mass under investigation should be larger than that of the air that is then around it; if it were not, the situation would be unstable and the air mass would rise still further.

H104 Consider the implication for the specific heat of the gas of reaching its minimal volume, and hence find the final temperature of the helium. Then apply the first law of thermodynamics.

H105 If the pressure decreases slowly, then the gas reaches its final state via a succession of equilibrium states, that is, by a reversible process. If the drop in the pressure is rapid, then the process is irreversible, and the net force acting on the wall of the bag during the expansion is not zero.

H106 A thermocouple is made by spot-welding one pair of ends of two wires made from different metals.

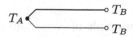

If the temperature T_A at the welded junction is different from the common temperature T_B of the free ends, then a potential difference V_{AB} is generated between those free ends, the size of the potential difference being proportional to the temperature difference (this is known as the *Seebeck effect*):

$$V_{AB} = S_{AB}(T_A - T_B),$$

where S_{AB} is the so-called Seebeck coefficient and depends upon the materials used for the wires.

In the circuit described, a further thermoelectric phenomenon, known as the *Peltier effect*, also takes place. If an electric current flows through a junction formed by two different conductors, then heat is either released or absorbed there, according to the direction of current. The rate at which heat is released or absorbed is directly proportional to the current I flowing through the junction:

$$P_A = \Pi_A I,$$

where Π_A is known as the Peltier coefficient, its numerical value depending upon the metals forming the junction. This coefficient is related to the Seebeck coefficient through the equation

$$\Pi_A = S_{AB} T_A,$$

where T_A is the temperature of the junction. Both effects need to be taken into account.

As an alternative to using the formulae given above, the problem can be solved by considering the thermocouple as a heat engine that uses room-temperature air and zero-temperature ice as its source and sink.

H107 The image of the Sun is not a point-like dot, but a small disc in the focal plane of the lens. At the black body's maximal temperature, the power of absorbed solar radiation is equal to the power of the radiation the body itself emits.

H108 In order to maintain the necessary conditions for the frost to form overnight, a balance has to be established between the outgoing heat radiated by the windscreen and the incoming heat from the (warmer) surroundings.

H109 Use the symmetry of the situation and the superposition property that follows from the linearity of the heat conduction equation (Fourier's law).

H110 The difference in melting rates is related to the fact that the density of the melted water from the ice cubes differs from the densities of the liquids in the two vessels.

H111 Apply the principle of conservation of energy.

H112 Surprisingly, water can condense from the saturated vapour in *both* cases. How is this possible?

H113   In the final, stationary state, there is a small amount of water at 100 °C at the bottom of the test tube. The wording of the question suggests that the pressure is the same all along the tube.

H114   Before the start of the liquefaction, the behaviour of the gas constituents is very close to that of an ideal gas, and so the ratio of their partial pressures is approximately equal to their molar ratio (or, equivalently, their particle number or volume ratio). During the isothermal compression, the liquefaction begins at the moment the partial pressure of one (or both) of the components reaches its saturated vapour pressure at that particular temperature.

H115   Compare the densities of ether vapour and steam with the density of air.

H116   The evaporated vapour from the eau-de-Cologne (from a physics point of view, a mixture of water and alcohol) varies in concentration continuously along the test tube, decreasing from the bottom up. The vapour escapes from the test tube as a result of *diffusion*; the speed of flow is directly proportional to the magnitude of the density gradient (change of density per unit length).

H117   If two neighbouring light rays emerge from the same point, and meet again at a different point, then, according to Fermat's principle, their optical path lengths are equal. The principle also applies to diverging rays if they are focused to a point by an image-constructing tool (say, a lens or the human eye). Using this result, it can be shown that, for any ray contributing to *perfect* image construction, its *optical path length* from the light source to the refracting point is equal to the (physical) distance between the refracting point and the image point.

As an alternative to Fermat's principle, the problem may be solved using Snell's law, applying it to the particular rays that emerge tangentially from the sphere.

H118   Apply Fermat's principle.

H119   According to Fermat's principle, light rays very close to each other traverse their trajectories in equal times.

Applying Huygens' principle also provides a means of solution. It states that every point on a constant-phase surface becomes a source of a spherical wavelet, and the new phase surface (the wavefront) is formed by the envelope of these wavelets, propagating with phase velocity c/n. Geometrical optics (which is a simplified model) describes the propagation of light in terms of 'rays' which travel perpendicularly to the wavefront.

The phenomenon of the 'circling' light can also be interpreted as a series of total internal reflections.

H120   Using the method of ray tracing, construct the image of a point on the surface of the sphere. The image of the sphere is a surface of revolution created by rotating a conic section around the optical axis.

H121   The virtual brightness of the Moon is determined by the strength of illuminance of the retina (that is, the light intensity on the photosensitive layer of the eye). When we use a telescope, more light energy enters our eye than when we look at the Moon directly, but the area of the image is also larger.

H122   The two half-lenses produce two real, point-like images of the light source. These images, acting as coherent point sources, create an interference pattern on the screen. The interference fringes can be observed only on that part of the screen where the light beams from the two image points overlap.

H123   This unusual diffraction grating can be considered as the superposition of two ordinary gratings with slit spacing $4d$, the centres of the gratings being separated by a distance d.

H124   Statements about the diffraction pattern can be made by using the *Huygens–Fresnel principle*. Accordingly, the net wave amplitude at any point on the screen can be found by summing (taking account of the appropriate phases) the amplitudes due to the elementary 'wavelets' originating at the individual slits. The square of the net amplitude is proportional to the light intensity at that point on the screen. The wavelet amplitude from a particular slit is proportional to the slit's *width*, and its phase is determined by the relevant optical path length.

H125   Interference maxima (bright spots) will be observed on the screen wherever the light waves from all the holes arrive in phase (the Huygens–Fresnel principle). Find the mathematical condition for this in terms of the holes' position vectors $r = (x, y)$, and the vector $R = (X, Y)$ giving the position on the screen of a maximum.

H126   A square grid can be transformed into one based on equilateral triangles by using *elongation* in a suitable direction, together with an appropriate scaling factor; the reverse transformation is also possible. What happens to the corresponding diffraction patterns is discussed in connection with the solution to a similar problem on page 320.

H127   *a*) A polarising filter lets through only light whose electric field is parallel to the filter's polarisation axis. The intensity of light is proportional to the square of its electric field strength.

b) It is perhaps surprising that, by choosing an appropriate plate thickness d and an optimal direction for e, up to 100 % of the incoming beam can pass through the system.

H128 *a*) The 'lenses' (in reality, the films) in the pair of 3D glasses found during the tidying up are two linear polarising filters, with polarisation axes perpendicular to each other.

b) The experience, the reverse of what happened in part *a*), indicates that the newly acquired pair of 3D glasses work with circularly polarised light. Such light can be produced from linearly polarised light (and can be analysed) with the aid of birefringent plates (as described in Note 2 on page 325), if the thickness and the orientation of the plates are suitable.

H129 Consider what happens if the whole system is magnified, with the distances between charges all increased by the same ratio. Does it remain in stationary equilibrium? What happens to the interaction energy? For the investigation of the system's stability, Gauss's law could prove useful.

H130 In the absence of external forces, the centre of mass of the system remains at rest. It is therefore convenient to choose the centre of mass as the origin of a vector coordinate system. The pearls can move in the way described if their accelerations are proportional to their instantaneous position vectors, and the proportionality factor is the same for all three bodies.

H131 The relatively strong electric field inside the capacitor influences the electrons over a short segment of their trajectory. Although the electric field outside the capacitor is weak, it exerts forces on the electrons over a longer section of their trajectory. Moreover, because the beam trajectory is bent, this weak field can also cause deceleration! What will be the net result of these two effects?

H132 You can prove, and then use, the fact that the electric field produced at some particular point *A* by a short, thin, uniformly charged rod is equal to the electric field due to a circular arc carrying the same linear charge density, if the arc is appropriately positioned with respect to *A*.

H133 The point in question must be in the plane of the rods, and inside the triangle. We are seeking a point at which the net electric field is zero, or, what is an equivalent statement, a point at which the electrostatic potential has an extremum, as a function of position.

H134 Find the force acting on small pieces of one of the rods due to the presence of the electric field of the other. It is convenient to characterise a small piece of one rod by the angle it subtends at the closest point on the other.

H135 The electric field component produced at any point *P* by one of the rods is inversely proportional to the distance of *P* from the rod. Using this fact and some geometrical considerations, it can be proved that each field line is an arc of a circle.

H136   Relate the force exerted on the point charge by a small piece of the plate to the electric flux through that same piece of plate, due to the point charge q.

H137   Start with the solution on page 340, and find the electric flux through one face of the cube due to the charge on the other five.

H138   Small pieces of the upper and lower plates that subtend the same solid angle $\Delta\Omega$ at the point P (they just cover each other when viewed from P) create electric fields at P that have the same magnitude, but opposite directions; these field components cancel each other out. It follows that the electric field produced at P by the lower plate is fully nullified by that part of the upper plate that subtends the same solid angle at P as the whole lower plate does. Thus the net electric field at P is just that generated by the 'rest' of the upper plate, i.e. the three-sided peripheral area marked off by a dashed line in the figure.

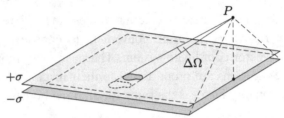

H139   Apply the method of superposition, and investigate the electric field the charged insulating spherical shell would have if the missing piece of the shell (approximately a small disc) were replaced into the hole and carried the same uniform surface charge density as the rest of the shell.

H140   *a)* First, find the force between the two hemispherical shells in the special case when their charges are equal. In this case, the electrostatic force acting on the hemispheres can be calculated as if a gas at pressure p exerted the forces on their inner surfaces. After this, the result for the general case, in which q and Q are different, can be guessed.

b) Using the principle of superposition, the result can be obtained in several different ways.

H141   Imagine the cube cut into eight identical smaller cubes, and apply both the principle of superposition and the method of dimensional analysis.

H142   On a charged conductor, the excess charge resides on its outer surface, and the electric field exerts an outward force on it normal to the surface. What is the effect of these forces on the shape of the mercury, *which has constant volume*?

H143   Using Gauss's law, find arguments that show that, for a given charge on the capacitor, the real voltage across it is smaller than that predicted by the approximate calculation.

H144 Apply the principle of superposition, and with the help of suitable super-posed charge distributions, create a capacitor that is charged 'traditionally'.

H145 Solve the problem first as if the metal plates were grounded, and then bring charges onto the two metal plates in a way consistent with the law of conservation of electric charge.

H146 The effect of the earthed plates of the capacitor can be described using an infinite number of image charges situated outside both plates. The magnitude of the net force exerted on the pearl by the image charges is finite, as the contributions due to images far from it are negligible.

H147 The problem can be solved by using the method of image charges. The linear image charges have to be placed so that the potential is zero on the planes containing the earthed plates.

H148 If a thin metal disc is electrically charged, then its charge distribution is not uniform: the surface charge density increases from the centre of the disc to its edge.

On a charged metal sphere, it is well known that the surface charge density is homogeneous, and inside the sphere the electric field is zero (the electric field contributions at any particular point, made by 'opposing' pieces of the sphere that subtend the same solid angle at the point, cancel each other out).

Prove that, if a charged metal sphere is (notionally) 'flattened' into a disc in such a way that we do not let the charges move (radially), then the surface of the disc remains an equipotential. This procedure provides the required solution to the problem.

H149 When the middle disc is just a little larger than the two charged discs, the forces acting on the latter are directed outwards. But if the middle disc is very large, then it attracts the two other discs, in the same way that an uncharged infinite conducting plate attracts an isolated charge. This means that there is a critical size for the middle disc that makes the net force acting on each charged disc zero.

The essence of the solution is that the three discs can be considered as a single disc (because they are very close to each other) with the surface charge distribution as calculated in the solution on page 362.

H150 In equilibrium, the charge distribution is such that the energy of the electrostatic field is minimal. Could a significant potential difference exist between the two parts of the sphere?

H151 At every point on the grounded metal sphere, the value of the electric potential must be zero. This net potential is produced by two sets of charges, one on the ring, and the other on the surface of the sphere. Find a point inside the sphere at which the contributions to the potential of the charges on both the ring and the sphere's surface can be calculated easily!

H152   If the distance between the spheres were much greater than their radius, then their electric fields could be calculated by approximating the charged spheres by point charges. However, in the given situation (as shown in the figure included in the problem), this is not the case: the sizes of the spheres and their separations are comparable, so the spheres cause electrostatic induction in each other. The exact surface charge distribution and electric field can be calculated only with very sophisticated mathematics, but (fortunately) this is not necessary! Instead, we apply the principle of superposition.

H153   Apply the method of spherical image charges. The basis of this method is that the electric field produced by two point charges, of opposite signs and different absolute values ($+Q_1$ and $-Q_2$), a given distance apart, has a zero-potential surface consisting of all points that satisfy

$$k_e \frac{Q_1}{r_1} - k_e \frac{Q_2}{r_2} = 0,$$

where r_1 and r_2 are the distances from the two charges of any particular point (*see* figure), and $k_e = 1/(4\pi\varepsilon_0)$ is the constant in Coulomb's law. Rearranging the equation, we get:

$$Q_1/Q_2 = r_1/r_2.$$

According to Apollonius's theorem, the locus of such points is a sphere (Apollonius's sphere) whose centre lies on the (extended) line joining the two charges.

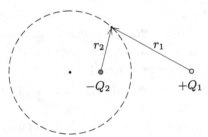

As the metal surface is an equipotential, the effects of the surface charges on the spherical shell that are produced by the electrostatic induction can be substituted by those produced by imaginary image point charges with appropriate magnitudes and positions. But how?

H154   Start with part *b*) of the solution on page 373, and investigate the change of positions and magnitudes of the image charges if the external point charge responsible for the electric field around the sphere is gradually moved away from the metal sphere to infinity! To determine the surface charge distribution on the sphere, replace the image charges with spheres of radius R carrying a uniform volume charge density, and partially overlapping each other.

H155 With the help of an appropriate thought experiment, the solution can be found without any detailed calculation. Use the fact that any change in magnetic interaction energy depends only on the initial and final states, and that it does not depend on the process by which the final situation was reached.

H156 *a*) The magnetic field strength can be found by using the Biot–Savart law and simply summing the field components parallel to the axis of symmetry.

b) We can assume that the magnetic field very far from a plane current loop is identical to that produced by a point-like magnetic dipole. The strength of this dipole (its magnetic moment) is the product of the current in the wire and the area of the loop. So, as far as its distant magnetic field is concerned, the original circular current loop could be replaced by any other loop of the same area (πR^2), provided it carries the same current. There could be some advantage in this if the replacement loop can be handled more conveniently than the original.

H157 The magnitudes of the magnetic fields produced at the centre of the tetrahedron by currents flowing in its edges are directly proportional to the corresponding currents. Note that each of these contributions to the magnetic field vector is parallel to some (different) edge of the tetrahedron.

H158 Using the symmetry of the arrangement, prove that the magnetic field lines are concentric circles centred on the wire. When that has been done, the magnitude of the magnetic field, $|\boldsymbol{B}|$, can be found with the help of Ampère's law.

H159 Apply the superposition principle, as was done, for example, in Solution 2 on page 384. Consider first a *single* straight input wire, and imagine the current it provides being led away (to infinity) radially and uniformly in all directions, i.e. in a spherically symmetric pattern. Determine the magnetic field produced by this current distribution. Then repeat the procedure with the output current, and finally superimpose the two magnetic fields.

H160   Investigate the similarity between the magnetic field lines associated with the parallel current-carrying wires, and the equipotential surfaces associated with two parallel insulating rods that are uniformly charged with equal but opposite electric charge densities.

H161   The current distribution and the electric field, respectively, can be characterised by current-streamlines and equipotential lines, which for all practical purposes lie in one plane. These form arrays of mutually orthogonal curves (*see* solution on page 390). Rotating the chessboard through 90° reverses the two arrays of curves, but its equivalent resistance remains unaltered.

H162   The axially directed force acting on the turns of solenoid 2 can be related to the magnetic flux from solenoid 1 that passes through the curved surface of the second cylinder.

Energy considerations can also lead to a successful conclusion. Investigate the change in energy of the system if the solenoids are moved apart by a small distance Δx. It is sufficient to deal initially with the special case $I_1 = I_2 = I$, but do not forget the possibility of energy coming in to or going out from the system.

H163   The tension is the same at each point of the wire, and can be related to its radius of curvature. So, in both cases, the shape assumed by the wire is a curve with a constant radius of curvature.

H164   On the mid-plane of the circular current loops, the direction of the magnetic field is everywhere perpendicular to it, and its strength depends only on the distance from point O. Write down an expression for the force acting on the electron, and investigate the effect of any torque produced by this force on the angular momentum of the electron about some particular fixed point.

H165   Write the vector equation for the particle's motion, covering a small segment of its trajectory, and then sum the result for the complete motion.

H166   Writing the equation of motion of the pendulum, as it swings in the magnetic field, we might notice that it is very similar to the equation of motion of a simple pendulum in a rotating frame of reference (consider a *Foucault pendulum* at the North Pole).

H167   It is useful to write the equations of motion of the two electrons in vector form. The sum of the two equations of motion describes the motion of the electrons' centre of mass, and their difference shows the relative motion of the electrons around that centre of mass.

H168   Try to convert the $\triangle$ circuit in the black box into a $\curlyvee$ configuration.

H169   Find a simple equivalent circuit, including voltage sources, that has symmetry properties, and then locate points that are at the same potential. Do not forget that N can be either even or odd.

H170   The differential equations describing the changes with time in the charges on the two capacitors, i.e. the equations governing the coupled oscillations of the circuit, can be written and solved using 'brute-force' methods. But there is also a much simpler (and cunning) solution method! Imagine that a tunable *current generator* is connected to the two terminals of one of the circuit elements. What happens to the voltage difference between the connecting points when the frequency of the generator is equal to one of the resonance frequencies of the '2L2C' circuit? The correct solution may also be found by notionally replacing part of the circuit by a tunable *voltage generator*.

H171   Consider first a finite chain with n coils and n capacitors. As the chain does not contain any ohmic resistance, the phase shift between the current flowing through it and the voltage is either $+90°$ or $-90°$. In one of these cases, the chain can be replaced by a capacitor with a particular capacitance C_n, and in the other by a coil that has a suitable inductance. Suppose that the first situation holds, and determine the value of C_n for the first few n. If it turns out that C_n is negative, then the chain behaves as a coil.

Find a connection between C_n and C_{n+1}, and then determine the 'fixed (convergence) point(s)' of this recursive formula. This (or one of them) gives the equivalent capacitance of the 'infinitely long' chain.

H172   The resistances of the conductors are negligible, and so the surfaces of both wires are essentially equipotentials, with a potential difference of V between them. The current flowing through the wires is $I = V/R$. The calculation of the magnetic force (Lorentz force) that acts between the wires is quite simple, but, in order to find the electric force, the amount of electrical charge on the surfaces of the conductors needs to be determined.

H173   Note that, in the frame of reference moving together with the rotating rings, the positive ions of the metal crystal lattice are at rest but the conductive (free) electrons are moving. So there are currents in the rings while they are accelerating.

H174   In the vicinity of any edge of the plate, the current density vector must be parallel to that edge. This unusual boundary condition can be arranged if the finite plate is notionally made infinite, and suitable 'image electrodes' are added – in a similar way to that in which image charges are used in electrostatics.

H175   The charge on a capacitor that is discharging through an ohmic resistor decreases exponentially with time. Its 'half-life' depends upon both the capacitance of the capacitor and the ohmic resistance.

H176   The solution method applied to an earlier, very similar, problem (*see* page 428) cannot be used here because of the irregular shape of the chocolate figure. Neither Santa Claus's capacitance nor the 'effective resistance' of the air can be calculated.

Imagine the figure to be enclosed within a closed convex surface, and compare the rate at which charge flows out through a small part of that surface with the electric field flux produced there by the remaining charge.

H177   Show that the magnitude of the rod's acceleration and the rate of change of the enclosed magnetic flux are proportional to each other. Deduce that the sum of the rod's speed and the enclosed flux – each multiplied by suitable proportionality factors – remains constant during the motion.

H178   The braking force acting on the magnet is caused by induced eddy currents. This magnetic force is proportional to the speed of the magnet, and balances its weight.

H179   If the current in one of two circuits situated close to each other (say, that in circuit 1) varies with time, then the induced voltage V^* in circuit 2 is proportional to the rate of change of the current in circuit 1:

$$V_2^* = -M_{12}\frac{\mathrm{d}I_1}{\mathrm{d}t},$$

where M_{12} is the mutual inductance of circuit 1 with respect to circuit 2, and the negative sign indicates the polarity of the induced voltage, in accordance with Lenz's law. The mutual inductance gives the quantitative connection between the current in circuit 1 and the magnetic flux, produced by circuit 1, that passes through circuit 2. With the help of the law of conservation of energy, it can be shown that the mutual inductance is symmetrical, i.e. it does not change if the roles of the two circuits are reversed:

$$M_{12} = M_{21}.$$

H180   An ideal diode lets current flow in one direction without any resistance (a short-circuit), whereas in the opposite direction no current can flow (an open-circuit). Find a connection between the magnetic flux through the circular loop produced by the small bar magnet and the electric charge flowing through the diode. The flux can be found in an elementary way if the magnet is notionally replaced by a small circular current loop (*see* solution on page 433).

H181 Use the symmetry aspect of mutual inductance, and the fact that the system consisting of the two largest loops can be considered to be the same as that formed by the two smallest loops, but magnified by a factor of 2.

H182 The value of the magnetic flux through a superconducting ring cannot change. The following fact might prove useful: the mutual inductance M between two circuits that have self-inductances L_1 and L_2 satisfies the inequality $M \leq \sqrt{L_1 L_2}$.

H183 Note that the arrangements in figures b) and c) are equivalent to setting two or three squares, with appropriately chosen current directions, on neighbouring faces of the cube. Their currents and magnetic fields can be superimposed.

H184 Use the symmetry feature of mutual inductance, and construct the loop rule (Kirchhoff's second law) for each of the circuits.

H185 It is clear that the magnetic field of toroidal coil 1 is strong inside the toroid – but coil 2 encircles it only once (and, admittedly, in a very 'devious' way). Although the magnetic field that toroid 1 produces outside its windings is very weak, it does pass through all N_2 turns of coil 2. In the calculation of the mutual inductance of the two coils, we get the correct value (symmetrical under the interchange of N_1 and N_2) only if we take into consideration the effects of both the strong inner magnetic field and the weak outer field.

H186 Investigate the changes that take place in the pearl's angular momentum relative to P.

H187 The induced electric field, produced by the changing magnetic field in the (thin) toroidal coil, is similar to the magnetic field associated with a circular current loop. Using the similarity between Faraday's law of induction and Ampère's circuit law, determine analogous pairs of physical quantities in these two phenomena. The required solution can be found by using this analogy, together with the Biot–Savart law.

H188 The capacitor starts to rotate, and its final angular velocity will be the same in both cases.

H189 Note that electric fields have as their sources electric charges, and they must reside somewhere. A homogeneous electric field can, to a good approximation, be produced using a large parallel-plate capacitor. Analyse what is 'seen' by an observer moving inside the capacitor with a fixed velocity that is parallel to the plates.

H190 Surely, an 'experienced' reader (such as yourself?) knows that the 'official solution' to the problem is wrong. The key to the correct solution is the Biot–Savart law.

H191   Only the conduction currents in the air could produce contributions to the magnetic field around Santa, but in reality they do not do this.

H192   The speed of the electron is comparable to the speed of light, and for this reason a relativistic calculation is needed. Consequently, for example, the trajectory of the electron is not a parabola! *Warning:* Although there is no force perpendicular to the electric field acting on the electron, its velocity component in its initial direction does not remain constant, but *decreases*.

H193   First try calculating the speeds of the particles according to Newton's second law ($F = ma$). If the results are too large, roll out the relativistic formulae.

H194   A charged particle moving, without slowing down, in a plane perpendicular to a homogeneous magnetic field performs uniform circular motion. In a decay, the conservation laws of electric charge and linear momentum are obeyed. The answer does not depend upon whether or not the motions of the particles are relativistic.

H195   It is convenient to chose a system of units in which the speed of light c has unit value. In this system the total energy of a particle with mass m and linear momentum p is $E = \sqrt{p^2 + m^2}$. Using the energy and linear-momentum conservation laws, prove that, whenever the decay products are moving perpendicularly away from each other, the geometric mean of their energies is always the same. From this – and use of the general inequality between arithmetic and geometric means – the final answer can be found.

H196   The energy of an ultra-relativistic particle is proportional to the magnitude of its linear momentum. Applying the laws of conservation of energy and linear momentum, it can be shown that, if the particles' linear-momentum vectors, before and after the collision, are drawn from a common point, their ends lie on a familiar curve.

H197   Calculate the required radius of the electron's circular trajectory.

H198   According to Heisenberg's uncertainty relationship, a particle forced into a space of finite size must have some linear momentum, and so it exerts a pressure on the containing walls. An estimation of this pressure can be based on a classical physics model that treats the neutron as a small ball bouncing back and forth between walls; dimensional analysis can also be helpful.

H199   Bohr's quantum condition can be generalised for systems consisting of two particles as follows: the *total* angular momentum of the system is required to be an integral multiple of $\hbar = h/(2\pi)$.

H200   Fuel consumption can be converted in the normal way. To find a physical interpretation for its SI unit value, some imagination is needed.

Solutions

S1 *a*) Firstly, we note that the parallel trajectories of the problem define a plane in three-dimensional space and that all motions are confined to that plane. Thus, although in general the notation employed refers to three-dimensional vectors, here it describes only two-dimensional ones and the notion of a crossing point P is well defined.

Let the velocities of the two bodies be v_1 and v_2 in the original reference frame $\mathcal{K}$. Since these are parallel vectors, $v_1 \times v_2 = 0$.

If another reference frame $\mathcal{K}'$ moves with constant velocity v_0 relative to $\mathcal{K}$, the velocities in this new frame are given by

$$v_1' = v_1 - v_0 \qquad \text{and} \qquad v_2' = v_2 - v_0.$$

The two trajectories will cross each other if vectors v_1' and v_2' are *not* parallel, that is

$$v_1' \times v_2' = (v_1 - v_0) \times (v_2 - v_0) = (v_2 - v_1) \times v_0 \neq 0.$$

This condition can be satisfied if the magnitudes of the two parallel velocities are *different*, and the relative velocity v_0 of the two frames is *not parallel* to them (*see* Fig. 1).

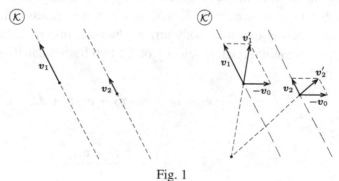

Fig. 1

b) Taking the origin of the reference frames to be the position of body (1) at time $t = 0$, the position vectors of the bodies in $\mathcal{K}$ at time t are

$$\boldsymbol{r}_1(t) = \boldsymbol{v}_1 t \qquad \text{and} \qquad \boldsymbol{r}_2(t) = \boldsymbol{v}_2 t + \boldsymbol{d},$$

where $\boldsymbol{d}$ is the relative position vector of the two bodies at $t = 0$ (*see* Fig. 2).

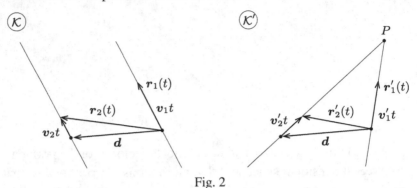

Fig. 2

In frame $\mathcal{K}'$ the position vectors are

$$\boldsymbol{r}'_1(t) = (\boldsymbol{v}_1 - \boldsymbol{v}_0)\, t \qquad \text{and} \qquad \boldsymbol{r}'_2(t) = (\boldsymbol{v}_2 - \boldsymbol{v}_0)\, t + \boldsymbol{d}.$$

If the two bodies meet at the crossing point P at time t_0, then

$$\boldsymbol{r}'_1(t_0) = \boldsymbol{r}'_2(t_0),$$

that is

$$(\boldsymbol{v}_1 - \boldsymbol{v}_2)\, t_0 = \boldsymbol{d}.$$

Since $\boldsymbol{v}_1$ and $\boldsymbol{v}_2$ are parallel to each other, this condition can only be satisfied if $\boldsymbol{d}$ is parallel to both, i.e. the starting position of body (2) lies on the (straight-line) path of body (1). This implies that in frame $\mathcal{K}$ the trajectories of the two bodies are not only parallel, but *coincident* with each other. The 'crossing' should be more accurately described as an 'overtaking'.

We can come to the same conclusion using a slightly different approach. Suppose that in reference frame $\mathcal{K}'$ one of the bodies reaches the crossing point at time t_1, whereas the other body arrives there at time t_2. Symbolically, $\boldsymbol{r}'_1(t_1) = \boldsymbol{r}'_2(t_2)$. The required relative velocity of the two frames can thus be found from the condition

$$(\boldsymbol{v}_1 - \boldsymbol{v}_0)\, t_1 = (\boldsymbol{v}_2 - \boldsymbol{v}_0)\, t_2 + \boldsymbol{d}.$$

This yields

$$\boldsymbol{v}_0 = \frac{\boldsymbol{d} + \boldsymbol{v}_2 t_2 - \boldsymbol{v}_1 t_1}{t_2 - t_1}.$$

It can be seen from this that the smaller the time difference between the bodies' arrivals at the crossing point, the larger the relative velocity v_0 required, but that the limiting condition of $t_1 = t_2 (= t_0)$ *cannot* be met for any pair of truly distinct trajectories in $\mathcal{K}$.

> *Note.* The general result of this analysis can be expressed more simply as follows. If the bodies meet (they are at the same point at the same time) when viewed in frame $\mathcal{K}'$, they also meet when viewed in frame $\mathcal{K}$. Two bodies moving parallel to each other can only meet if their trajectories coincide.

S2 Since Ann is moving directly towards Bob, his position, B in the figure, must lie on the tangent to the carousel at Ann's position A. Thus A, B and C, the centre of the carousel, must form a right-angled triangle. Using the given geometrical data, it follows that the distance between Ann and Bob at the given moment is $6\sqrt{3}$ m. It also follows that the tangential speed of the carousel is 1 m s^{-1}, and that its angular velocity is therefore $\omega = \frac{1}{6}$ rad s^{-1}.

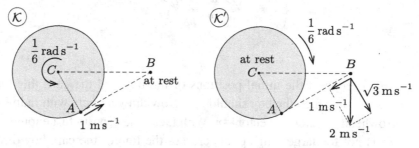

If Ann were sitting at the centre of the carousel, she would see the whole world around her rotating with the same angular speed ω, but in the opposite direction. That means she would observe Bob standing 12 m away from the centre of the carousel, but moving with a speed of $\frac{1}{6} \times 12 = 2$ m s^{-1} in a direction perpendicular to the line joining him to the centre of the carousel. Although Ann is not sitting at the centre of the carousel, but at its edge, the same conclusion applies – namely that, according to Ann, Bob's speed is 2 m s^{-1}. On the left in the figure, the relative motion can be seen from the point of view of Bob (frame of reference $\mathcal{K}$); on the right is the same situation from Ann's point of view (frame $\mathcal{K}'$).

The frame of reference $\mathcal{K}$ is inertial, but frame $\mathcal{K}'$ (i.e. the rotating carousel) is not, and so the Galilean transformation does not apply. According to Bob, Ann is moving directly towards him with a speed of 1 m s^{-1}; however, from the rotating frame $\mathcal{K}'$, Ann observes that Bob is not moving directly towards her and that his speed is different, namely 2 m s^{-1}.

Bob's velocity in frame $\mathcal{K}'$ can be resolved into two perpendicular components, as shown on the right-hand side of the figure. The magnitude of the component pointing towards Ann has the (perhaps expected) value of 1 m s^{-1}, whereas that of the component perpendicular to this is $\sqrt{3}$ m s^{-1}. This latter value can be understood as Ann observing Bob moving at right angles to her line of vision with a speed of $\omega \times 6\sqrt{3}$ m $= \sqrt{3}$ m s^{-1}.

S3 Let us solve the problem in a frame of reference fixed to the cart; here the cart is at rest but the road and meadow are moving. The cart and the boy will arrive simultaneously at some point on the road if, in this frame, the boy runs directly towards the cart. The boy's velocity can be found in the moving reference frame if we add velocity $-v$ to the boy's velocity measured in the frame fixed to the ground. The geometrical arrangement is shown in the figure.

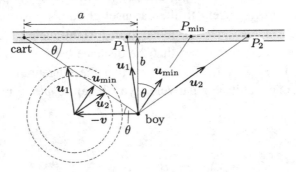

First, connect the initial positions of the boy and the cart; this line shows the direction in which the boy should run. Now draw a circle with radius u around the arrowhead of velocity vector $-v$. With luck – he can run fast enough, and the ratio b/a is not too large – this circle crosses the line of the cart–boy direction at two points. These two crossing points set the extreme directions of the boy's possible (i.e. successful) runs.

If he runs 'so as to meet' the cart with velocity vector u_1, then they meet at point P_1, or if he runs 'so as to catch up with' the cart with velocity vector u_2, then they meet at P_2. If the boy runs with speed u in any direction between these two extremes, he reaches the road before the cart arrives; he could have run more slowly. Note that in the figure each of these vectors is shown twice, once (as a velocity) when their directions are determined using the circle centred on $-v$, and once (as a position vector) to show the actual path taken by the boy.

The figure can be used to find how fast the boy has to be able to run to just catch the cart. If we were to draw a circle with a radius corresponding to this minimal speed u_{min} around the arrowhead of velocity vector $-v$, then it would have the cart–boy direction line as a tangent (i.e. the two crossing points referred to above coincide). Accordingly, the boy should run in a direction perpendicular to this line with a speed of $u_{min} = v \sin\theta$, where the angle θ can be calculated using the identical angle properties of the figure: $\tan\theta = b/a$. If the boy runs with the minimal speed u_{min} in the proper direction, then he meets the cart at point P_{min}.

S4 In a reference frame $\mathcal{K}$ fixed to the bank, we denote Joe's starting point by A, his landing position by P, and the position of the nugget by B. As recommended in the hint, we also use a reference frame $\mathcal{K}'$ that moves with the river water; in this frame, the corresponding points are A', P' and B'.

In frame $\mathcal{K}'$ the bank appears to be moving with speed v in the opposite direction to that of the original current, Joe is paddling with speed u in still water, and his hiking pace along the bank is $v + u$. By the time Joe lands, his starting point has moved to A', and the nugget to B', as shown in the figure.

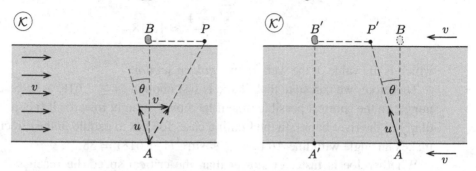

In frame of reference $\mathcal{K}'$, Joe's trajectory reminds us of light rays refracted at the interface between two different media, so we can apply Fermat's principle to find the optimal path. If Joe wants to reach the nugget in the shortest time, then he has to choose the track that a light ray would follow when passing out of a medium with refractive index $n = (v + u)/u$ and into a vacuum. In this case, with the exit ray parallel to the interface, the angle of incidence α is simply the critical angle for total internal reflection; this is given by $\sin\alpha = 1/n$ (Snell's law with the angle of refraction equal to $\pi/2$). But, from simple geometry, $\alpha = \theta$.

This means that, in frame $\mathcal{K}$, Joe must paddle in a direction that makes an angle $\pi/2 + \theta$ with the drift velocity of the water, where

$$\theta = \sin^{-1}\left(\frac{u}{v+u}\right). \tag{1}$$

This argument based on Fermat's principle is only correct if, in frame $\mathcal{K}'$, the gold nugget, travelling with speed v, passes point P' before Joe reaches it, i.e. for the value of θ found earlier,

$$v \geq u\sin\theta. \tag{2}$$

In the alternative situation ($v < u\sin\theta$), it is a waste of his time for Joe to land downstream of the nugget, and it is best for him to paddle so that in frame $\mathcal{K}$ his net velocity is perpendicular to the river bank, taking the shortest route across (and not hiking at all). The required angle θ is then

$$\theta = \sin^{-1}\left(\frac{v}{u}\right). \tag{3}$$

By equating the values of θ given by equations (1) and (2), we can obtain the critical value for u/v that separates the two strategies:

$$\frac{v}{u} = \frac{u}{v+u} \qquad \text{i.e.} \qquad u^2 - uv - v^2 = 0.$$

This yields

$$\frac{u}{v} = \frac{\sqrt{5}+1}{2} \approx 1.618,$$

which is the value of the well-known *golden section*.

Therefore, we can state that, if Joe is fast enough ($u \geq 1.618v$), he reaches the nugget in the shortest possible time if he moves straight towards it (i.e. perpendicularly to the river bank). In the limiting case, Joe has to paddle in the direction that makes an angle with line AB of $\theta_{max} \approx \sin^{-1}(1/1.618) = 38.2°$.

Whether Joe is faster or slower than the critical speed, he reaches his target as quickly as possible if he paddles in a direction making a smaller angle than θ_{max} with the line perpendicular to the river bank. Note that the required angle θ $(0 < \theta < \theta_{max})$ approaches zero if either Joe's speed approaches infinity (equation (3)) or it approaches zero (equation(1)).

> *Note.* In general, the same optimal angle θ corresponds to two different boat speeds, and we can find a connection between them by equating the sines of these angles: $u_1/(v + u_1) = v/u_2$, where u_1 is the slower and u_2 the faster paddling speed. From this equation, we get $v^2 = u_1(u_2 - v)$, i.e. the geometric mean of u_1 and $u_2 - v$ is always equal to v (the drift velocity of the river).

S5 In principle, several outcomes are possible, depending on the actual values of the friction and the speeds. In the case of small friction and a large initial speed, the disc flies across the table with virtually no change in its velocity. If the friction is large and the initial velocity is small, then the disc does not get across the conveyor belt, but 'sticks' to it, and the moving band carries the disc to the right-hand edge of the table. This latter possibility might suggest that it could be a good idea to describe the phenomenon not in a frame of reference $\mathcal{K}$ fixed to the table, but rather in a frame $\mathcal{K}'$ that moves with the conveyor belt. We will see that the solution becomes relatively simple if the coordinate system $\mathcal{K}'$ is used.

In the reference frame $\mathcal{K}'$ fixed to the moving band, the disc slides slantwise across the band, and its initial velocity is

$$v_1' = \sqrt{v_0^2 + V^2} = 5 \text{ m s}^{-1}.$$

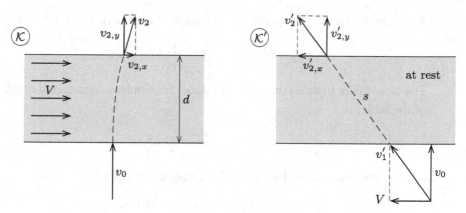

As a result of the frictional force, the disc's motion is one of uniform deceleration along a straight line, and if it does not lose all of its energy, the sliding disc exits from the band and onto the smooth table; this case is shown in the figure.

The path length of the disc across the band is

$$s = d \frac{v_1'}{v_0} = d \sqrt{1 + \frac{V^2}{v_0^2}} = 1.25 \text{ m}.$$

Its deceleration is constant and has magnitude μg, and so its final speed (from the work–kinetic energy theorem) is

$$v_2' = \sqrt{v_1'^2 - 2\mu g s} \approx 3.57 \text{ m s}^{-1}.$$

So, in practice, the disc does leave the conveyor belt. To find the exit point of the disc, we will need to know the time t_1 the uniformly decelerated disc takes to cross the band:

$$t_1 = \frac{2s}{v_1' + v_2'} \approx 0.29 \text{ s}.$$

We now return to the frame of reference $\mathcal{K}$ fixed to the table. In this frame, the band moves a distance $V t_1$ to the right, and the disc moves a distance $(V/v_0)d$ to the left, relative to the band. So, the displacement of the disc, until it exits from the belt, is

$$\Delta x_1 = V t_1 - d \frac{V}{v_0} \approx 0.125 \text{ m}$$

to the right. At the exit point, the component of the disc's velocity, perpendicular to the edge of the band, is the same in both frames of reference:

$$v_{2,y} = v_{2,y}' = \frac{v_2'}{v_1'} v_0 \approx 2.86 \text{ m s}^{-1}.$$

The velocity component parallel to the edge of the band, in the frame $\mathcal{K}$, is

$$v_{2x} = V - v'_{2x} = V - \frac{v'_2}{v'_1}V \approx 0.86 \text{ m s}^{-1}.$$

The time taken to cross the $d = 1$ m wide, frictionless region, and reach the edge of the table, is

$$t_2 = \frac{d}{v_{2,y}} = 0.35 \text{ s},$$

and the consequent displacement of the disc to the right is

$$\Delta x_2 = v_{2,x}t_2 \approx 0.301 \text{ m}.$$

Thus the total displacement of the disc to the right is $\Delta x_1 + \Delta x_2 = 0.426$ m. This is less than 1.5 m and so the exit point is on the table edge parallel to the belt (rather than on its right-hand edge) and at a distance of 42.6 cm to the right of its midpoint.

> *Note.* In the reference frame $\mathcal{K}'$ fixed to the conveyor belt, the direction and magnitude of the disc's acceleration are both constant, and so, in the frame $\mathcal{K}$ fixed to the table, the trajectory of the disc, while it is on the belt, is a parabolic arc with its axis parallel to the direction of the acceleration.

S6 *Solution 1. a)* Using the simplification given in the problem, it can be stated that the magnitude of the boy's acceleration can never be more than μg. Its direction can be chosen freely, but should that direction be constant during the turn? Or is it better to continuously change the direction of the acceleration?

The answer can be found if the motion is analysed in the v_x–v_y coordinate system, for which the axes are the east (x) and north (y) components of the velocity (*see* Fig. 1).

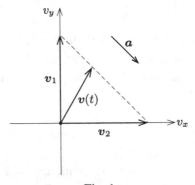

Fig. 1

Our aim is to change the initial velocity vector $v_1 = (0, v)$ to the final one $v_2 = (v, 0)$ as quickly as possible. As the acceleration is nothing more than the rate of change of velocity, i.e. the velocity of the tip of the velocity vector in the $v_x - v_y$ plane, in the optimal case, the tip of the velocity vector must move along a straight line between the end-points of vectors v_1 and v_2. Furthermore, it must do so with maximum speed (in this velocity coordinate system), corresponding to the maximal acceleration μg of the runner. So the minimal time for his turn is

$$t = \frac{|v_2 - v_1|}{\mu g} = \frac{\sqrt{2}\,v}{\mu g} \approx 7.2 \text{ s}.$$

b) The boy's acceleration vector is constant throughout, with magnitude μg, and directed to the southeast (it makes an angle of 45° with the *x*-axis), and so his trajectory during the turn (in the same way as for projectiles) is a parabola, as shown in Fig. 2. The parabola has its axis in the northwest–southeast direction, and the trajectory is completed by tangential straight-line segments at the end-points A and B of the parabolic arc.

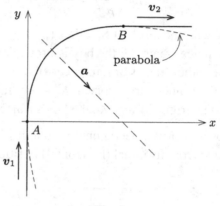

Fig. 2

Note. It can be shown that the equation of the boy's trajectory, between A and B, is

$$y = v\sqrt{\frac{2\sqrt{2}}{\mu g}}\sqrt{x} - x,$$

in the coordinate system shown in Fig. 2, which has its origin at point A.

Solution 2. *a*) We use the inertial frame of reference $\mathcal{K}'$ that moves with the boy's initial velocity v_1. While in the original frame $\mathcal{K}$ (fixed to the ice), the initial velocity vector v_1 must be changed to the velocity vector v_2, in the coordinate system $\mathcal{K}'$, the boy is initially at rest ($v_1' = 0$), and his final velocity is $v_2' = v_2 - v_1$ in a southeasterly direction (*see* Fig. 3). The magnitude of the final velocity is, from Pythagoras's theorem, $|v_2'| = \sqrt{2}\,v$.

In a Galilean transformation, the acceleration does not change, so in the frame $\mathcal{K}'$ the boy's maximum possible acceleration is also μg. It is obvious that, for the

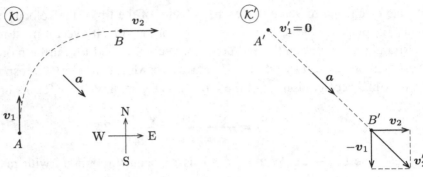

Fig. 3

fastest turn, he must accelerate towards the southeast throughout. So the minimal time for the fastest change of direction is

$$t = \frac{|v_2'|}{\mu g} = \frac{\sqrt{2}\,v}{\mu g} \approx 7.2 \text{ s},$$

which is in line with the result of Solution 1.

b) In the reference frame $\mathcal{K}'$, the boy is producing uniformly accelerated rectilinear motion. In other frames of reference – those whose velocity relative to $\mathcal{K}'$ is *not* parallel to the runner's trajectory in $\mathcal{K}'$ – the trajectory appears as a parabola. This is why, in the original frame $\mathcal{K}$, the boy's trajectory has this form.

S7 The position of the pendulum, of length ℓ, is specified by the rotational angle φ measured from its initial (horizontal) position, as shown in Fig. 1.

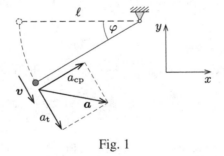

Fig. 1

The tangential acceleration is due to the tangential component of the gravitational force:

$$a_t = g \cos \varphi.$$

The speed of the bob can be found from the conservation of mechanical energy, the bob having descended vertically through a distance $\ell \sin \varphi$:

$$v = \sqrt{2g\ell \sin \varphi}.$$

So the centripetal acceleration is given by

$$a_{cp} = \frac{v^2}{\ell} = 2g \sin \varphi.$$

The x (horizontal) and y (vertical) components of the acceleration of the bob are thus

$$a_x = a_{cp} \cos \varphi + a_t \sin \varphi = 3g \sin \varphi \cos \varphi,$$
$$a_y = a_{cp} \sin \varphi - a_t \cos \varphi = 2g \sin^2 \varphi - g \cos^2 \varphi.$$

These expressions can be transformed using trigonometric identities into

$$a_x = \tfrac{3}{2}g \sin(2\varphi), \qquad a_y = \tfrac{1}{2}g - \tfrac{3}{2}g \cos(2\varphi).$$

This is the parametric form of the equation for a circle, which can also be written as

$$a_x^2 + (a_y - \tfrac{1}{2}g)^2 = (\tfrac{3}{2}g)^2.$$

The locus of the end of the acceleration vector a of the pendulum bob is a circle of radius $3g/2$ centred on $(0, g/2)$, as shown in Fig. 2. The initial release corresponds to the point $(0, -g)$ and the circle is traversed in an anticlockwise direction.

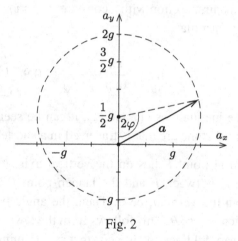

Fig. 2

Note that, during the first 'half-swing', i.e. until the pendulum is horizontal with its bob at rest for the first time, the *complete* circle is traversed. At this point the acceleration vector reverses direction and the whole circle is covered again, but in a clockwise sense.

S8 Let the length of the pendulum be ℓ. Along the arc AP the vertical acceleration of the bob is less than or equal to g (with equality only at the initial point). Correspondingly, the time required to cover this arc is clearly greater than it would be if the bob were to fall freely between the height levels of A and P:

$$t_{AP} > \sqrt{\frac{2\ell \sin 30°}{g}} = \sqrt{\frac{\ell}{g}}. \tag{1}$$

Next, let us divide the arc PB into two equal sections, PQ and QB.

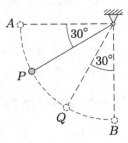

At points P and Q the speeds of the bob can be found using the law of conservation of energy:

$$v_P = \sqrt{2g\ell \sin 30°} = \sqrt{g\ell}, \qquad \text{while} \qquad v_Q = \sqrt{2g\ell \sin 60°} = \sqrt{\sqrt{3}g\ell}.$$

The bob clearly covers the $\ell\pi/6$ long arcs PQ and QB more rapidly than if it had moved along the first section with a constant speed v_P, and along the later one with speed v_Q. As a formula:

$$t_{PB} < \frac{\ell\pi}{6v_P} + \frac{\ell\pi}{6v_Q} \approx 0.92\sqrt{\frac{\ell}{g}}. \tag{2}$$

Comparing the inequalities (1) and (2), it can be seen that $t_{AP} > t_{PB}$, and so we conclude that PB is the arc that is traversed in a shorter time.

S9 The focal point F lies on the vertical axis of symmetry of the parabola. Let the distance between F and the launch point P be d, and the launch angle, measured from the *vertical*, be θ. Then, the angle between the line PF and the launch direction is also θ. This follows from the laws of reflection, the geometrical properties of parallel lines and the fact that rays coming from its focus are reflected by a parabolic mirror parallel to its axis of symmetry (*see* right-hand side of figure).

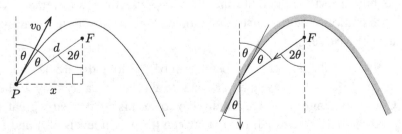

The projectile reaches the top of its flight in a time $t = (v_0 \cos \theta)/g$, and its horizontal displacement at that time is

$$x = v_0 t \sin \theta = \frac{v_0^2}{2g} \sin(2\theta).$$

The distance of the focus from the launch point can be found with the help of the right-angled triangle shown on the left-hand side of the figure:

$$d = \frac{x}{\sin(2\theta)} = \frac{v_0^2}{2g}.$$

Note that this result is independent of the launch angle. The left-hand part of the figure also shows that, if $2\theta = 90°$, i.e. if the launch angle is $45°$, then the focus and the launch point are at the same altitude.

> *Note.* Continuing the argument given above, the position of the parabola's directrix can also be deduced. As the axis of the parabolic trajectory is vertical, its directrix must be horizontal. From the definition of a parabola, all points on it are equidistant from the directrix and the focus. In particular, this applies to the launch point P; the directrix is therefore a distance $d = v_0^2/(2g)$ above P. As must be the case, this distance is independent of the launch angle.
>
> If the initial velocity v_0 were vertically upwards, then the maximum height the projectile would reach would be $v_0^2/(2g)$, the same value as the height above P of the directrix. This means that we can consider the directrix as a geometrical analogue of the total energy of projectiles. The distance between the directrix and any particular point of the parabolic trajectory is directly proportional to the kinetic energy of any projectile at that point that subsequently follows the given parabolic path.

S10 *Solution 1.* The path of the projectile can be described using the coordinate system shown in Fig. 1.

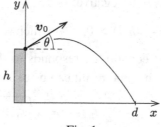

Fig. 1

The horizontal motion (along the x-axis) is uniform and

$$x(t) = t v_0 \cos \theta, \tag{1}$$

where θ is the launch angle, measured from the horizontal. The vertical uniformly accelerating motion is described by the following equation:

$$y(t) = -\frac{g}{2}t^2 + tv_0 \sin\theta + h. \qquad (2)$$

Obtaining an expression for the time from equation (1), and then inserting it into equation (2), gives the equation of the trajectory as

$$y(x) = -\frac{g}{2v_0^2 \cos^2\theta}x^2 + x\tan\theta + h.$$

Its horizontal displacement, d, from the foot of the tower when a projectile hits the ground is now found by setting $y = 0$, yielding a quadratic equation in x whose roots give the required value(s) of d.

To simplify the expressions involved, it is useful to rewrite the initial speed in the form $v_0 = \sqrt{2gH}$, where H is a constant with the dimensions of length.[15] Incorporating this change of notation and setting $y(d) = 0$ gives the following equation:

$$0 = -\frac{d^2}{4H\cos^2\theta} + d\tan\theta + h.$$

Using the trigonometric identity $1/\cos^2\theta = 1 + \tan^2\theta$ and the notation $u = \tan\theta$ reduces it to

$$(1 + u^2)d^2 - 4uHd - 4hH = 0.$$

This quadratic equation for d can, if d is given, also be written as a quadratic equation in u:

$$d^2u^2 - 4dHu + (d^2 - 4hH) = 0.$$

Any given distance d can only be reached if this equation has at least one *real* root for u, corresponding to at least one achievable launch angle θ in the range $-\pi/2 < \theta < \pi/2$. There will be such a root provided the discriminant of the quadratic equation is non-negative:

$$(2dH)^2 - d^2(d^2 - 4hH) \geq 0, \qquad \text{which reduces to} \qquad d \leq 2\sqrt{H^2 + hH}.$$

The limiting case of equality corresponds to the longest projectile distance, and using $H = v_0^2/(2g)$, the final result can be expressed as

$$d_{max} = \frac{v_0}{g}\sqrt{v_0^2 + 2gh}.$$

Solution 2(a). Let us denote the initial velocity vector by v_0, the touchdown velocity vector by v_1, the total displacement vector of the motion by r, and the gravitational acceleration vector by g (*see* Fig. 2).

---

[15] This constant H has an understandable physical interpretation: a body released from rest at this altitude would reach the ground with speed v_0.

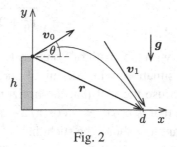

Fig. 2

The acceleration of a projectile is constant, and so

$$\frac{\boldsymbol{v}_1 - \boldsymbol{v}_0}{t} = \boldsymbol{g}, \tag{3}$$

where t is the time between the launch and the touchdown. The velocity is changing uniformly with time, and so the average velocity is equal to the arithmetic mean of the initial and final velocities, that is

$$\frac{\boldsymbol{v}_1 + \boldsymbol{v}_0}{2} t = \boldsymbol{r}. \tag{4}$$

The scalar product of equations (3) and (4) gives

$$\frac{v_1^2 - v_0^2}{2} = \boldsymbol{g} \cdot \boldsymbol{r} = gh,$$

which means that the magnitude of the touchdown velocity is

$$v_1 = \sqrt{v_0^2 + 2gh}.$$

This result is not surprising, because it follows directly from the conservation of energy.

Much less obvious is the result of taking the vector product of equations (3) and (4). Using the facts that $\boldsymbol{a} \times \boldsymbol{b} = -\boldsymbol{b} \times \boldsymbol{a}$ and $\boldsymbol{a} \times \boldsymbol{a} = \boldsymbol{0}$ for any vectors $\boldsymbol{a}$ and $\boldsymbol{b}$, we obtain

$$\boldsymbol{v}_1 \times \boldsymbol{v}_0 = \boldsymbol{g} \times \boldsymbol{r}.$$

Let us take the absolute values of both sides of this equation:

$$|\boldsymbol{v}_1 \times \boldsymbol{v}_0| = |\boldsymbol{v}_0| \cdot |\boldsymbol{v}_1| \sin \gamma = v_0 \sqrt{v_0^2 + 2gh} \sin \gamma = |\boldsymbol{g} \times \boldsymbol{r}| = gd, \tag{5}$$

where γ is the angle between the directions of the initial and final velocities. The final equality here is justified by noting that, since $\boldsymbol{g}$ is vertical, only the horizontal component of $\boldsymbol{r}$ contributes to the cross-product.

From equation (5) it can be seen that d is largest if $\gamma = \pi/2$, i.e. that vectors $\boldsymbol{v}_0$ and $\boldsymbol{v}_1$ are perpendicular to each other. The longest projectile distance is then

$$d_{\max} = \frac{v_0}{g}\sqrt{v_0^2 + 2gh},$$

an outcome in line with the result of Solution 1.

In this optimal situation, with the initial and final velocities perpendicular to each other, we can use the fact that the horizontal component of the velocity is constant to deduce that $v_0 \cos\theta = v_1 \cos(\pi/2 - \theta) = v_1 \sin\theta$. That is, for maximal range, the launch angle θ should be such that

$$\tan\theta = \frac{v_0}{v_1} = \frac{v_0}{\sqrt{v_0^2 + 2gh}}.$$

Solution 2(b). The final steps of Solution 2(a) can be carried through using graphical, rather than algebraic, methods – in essence, by considering vector cross-products in terms of the areas, rather than the vectors, that they define.

Consider the velocity vector parallelogram *KBCA* defined by the initial and final velocity vectors (*see* Fig. 3), and its relationship with equations (3) and (4).

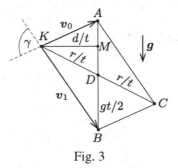

Fig. 3

Let us calculate the area T of the parallelogram in two different ways: firstly using the expression $T = |v_0| \cdot |v_1| \sin\gamma$, and then as the sum of the areas of two (identical) triangles,

$$T = AB \cdot KM = gt\,\frac{d}{t} = gd.$$

To obtain this latter expression we have used the fact that line AB is vertical, and that, from (3), its length is gt. Similarly, because of condition (4), the length of KD is r/t, and its horizontal projection is $KM = d/t$.

Equating the two different calculations of the area T we get that

$$d = \frac{1}{g}|v_0| \cdot |v_1| \sin\gamma \leq \frac{v_0}{g}\sqrt{v_0^2 + 2gh} = d_{\max},$$

which is the same result as equation (5) – and leads to the same conclusion, that, for maximal range, γ must be $\pi/2$.

The result for the launch angle θ required for maximal range also follows, because, then, the right-angled triangles KMA and BKA are similar, and so

$$\tan \theta = \frac{AM}{KM} = \frac{AK}{BK} = \frac{v_0}{v_1} = \frac{v_0}{\sqrt{v_0^2 + 2gh}}.$$

Solution 3. Denote the launch point of the projectile by P, the touchdown point by Q, and the foot of the tower (which is also the origin of the coordinate system shown in Fig. 4) by O. The length of PO is just the height of the tower, and so is equal to h.

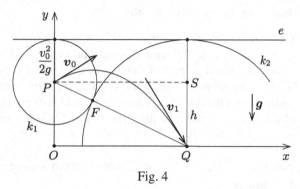

Fig. 4

According to the solution starting on page 102, if a projectile is thrown in any direction with initial speed v_0, the focal point F of its parabolic trajectory must lie on a circle k_1 that has centre P and radius $v_0^2/(2g)$. Consequently,

$$PF = \frac{v_0^2}{2g}.$$

The directrix e of the parabola is horizontal, and its distance from P is the same as that of the focus; consequently, the directrix must touch that same circle.[16] The touchdown point Q lies on the parabola, and so its distance from the directrix is the same as that from the focus F, that is

$$FQ = h + \frac{v_0^2}{2g}.$$

It follows that a circle k_2, with centre Q and radius FQ, also touches the directrix.

In general, circles k_1 and k_2 have two intersections (or none). These are the focal points of two different parabolas that have the same touchdown point. This occurs when the distance PQ is less than the sum of the distances PF and FQ. However, if the launch angle is just right, the intersection points coincide with each other, and the two circles just touch each other (at the focus of the parabola). In this case the

---

[16] Recall that, if v_0 is fixed, the position of the directrix is independent of the launch angle.

distance ℓ between the launch and touchdown points (and so also the horizontal displacement) will be as large as possible, and given by

$$\ell = PQ = PF + FQ = h + \frac{v_0^2}{g}.$$

Using Pythagoras's theorem on triangle PQS, as indicated in Fig. 4, the maximal horizontal distance covered by the projectile can be calculated:

$$d_{\text{max}} = \sqrt{\ell^2 - h^2} = \frac{v_0}{g}\sqrt{v_0^2 + 2gh},$$

in agreement with our previous results.

> *Note.* All of our results remain valid if $h < 0$ (provided $v_0^2 > 2g|h|$). In this case a possible phrasing of the same physics problem could be something like this:
>
> 'A man shovels a pile of sand from the bottom of a ditch of given depth h. In which direction should he throw the sand, if he wants it to be as far away as possible from the ditch? And how far is that? Assume that the initial speed of the sand is independent of the direction in which it is thrown, and that the sand does not hit the ditch walls.'

S11   *Solution 1.* We use the coordinate system shown in Fig. 1 (in which y is measured *downwards*).

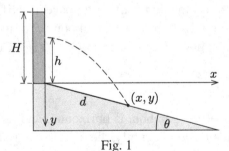

Fig. 1

The trajectory of a water jet flowing, with (horizontal) velocity v, from a hole made at height h can be obtained from the formulae for projectile motion:

$$y(x) = -h + \frac{g}{2v^2} x^2.$$

Further, according to Torricelli's law, the initial speed of the water jet is $v = \sqrt{2g(H - h)}$.

The equation of the incline is $y = x \tan \theta$, and if we substitute for y and v into the equation for the (parabolic) trajectory, we obtain the following quadratic equation for the horizontal coordinate of the point of impact:

$$\frac{x^2}{4(H - h)} - x \tan \theta - h = 0. \tag{1}$$

Since $0 \le h < H$ and the constant term $(-h)$ is negative, this equation has exactly one positive real root (x_+), but its maximum value as a function of h requires considerable calculation. Fortunately, there is an easier way to proceed.

We consider equation (1) in the context of a given position for the impact (i.e. a fixed x) and find the value(s) of h corresponding to this particular x. With x fixed, equation (1) can be rearranged as a quadratic equation in h:

$$h^2 + (x\tan\theta - H)h + \left(\frac{x^2}{4} - Hx\tan\theta\right) = 0. \tag{2}$$

This has a real (i.e. physically achievable) root if its discriminant (which happens to be factorisable) is non-negative, i.e.

$$(x\tan\theta - H)^2 - x^2 + 4Hx\tan\theta = [x(\tan\theta + 1) + H][x(\tan\theta - 1) + H] \ge 0.$$

The expression in the first square brackets is clearly positive, and so the expression in the second square brackets must not be negative. This means that x must satisfy the following inequality[17]

$$x \le \frac{H}{1 - \tan\theta}. \tag{3}$$

When the equality holds, the discriminant is zero, and the value of x is the largest possible that corresponds to a physically achievable value for h.

It follows immediately that the longest possible impact distance on the incline is

$$d = \frac{x}{\cos\theta} = \frac{H}{\cos\theta - \sin\theta}. \tag{4}$$

The corresponding height of the hole can be obtained by substituting result (3), with equality holding, into equation (2) and solving for h, obtaining the (repeated) root

$$h = \frac{1 - 2\tan\theta}{2 - 2\tan\theta} H. \tag{5}$$

As h cannot be negative, our results are valid only if $\tan\theta < 1/2$ (i.e. $\theta < 26.6°$). If $\theta > 26.6°$, then inequality (3) is not applicable. Rather, a water jet flowing from the bottom of the vessel provides the largest impact distance, with a horizontal projection of $x = 4H\tan\theta$. In this case, instead of (5) we have $h = 0$, and (4) is replaced by

$$d = \frac{4H\tan\theta}{\cos\theta}.$$

---

[17] Provided $\tan\theta < 1$, which it is, as we will see later.

Solution 2. A water jet flowing out through a hole in the wall of the vessel follows a parabolic trajectory with a horizontal directrix and a focus F that lies on the y-axis of the coordinate system shown in Fig. 2.

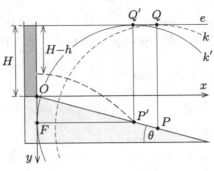

Fig. 2

In the solution starting on page 102, it is proved that, for a projectile with initial velocity v, the distance between the focus of the parabola and the launch point is $v^2/(2g)$, independent of the launch angle. Thus, for a jet originating at a height h, which has $v = \sqrt{2g(H-h)}$, the focus F lies a distance $H-h$ directly *below* the hole. But it is also the case that every point of any parabola (including the launch point) is as far from its focus as it is from its directrix e. It follows that the directrix is at a height of $H-h$ *above* the launch point. In other words, the directrix lies in the surface of the water in the vessel, whatever the value of h.

Consider an arbitrary point P on the incline, and denote the point on the directrix directly above it by Q, as shown in Fig. 2. If there is at least one possible water jet trajectory passing through P, then its focus is as far from P as P is from the directrix e. Now consider a circle, denoted by k, that has centre P and radius PQ. Suppose, firstly, that the position of the chosen point P is such that circle k has no common point with the y-axis of the coordinate system. This implies that there is no suitable launch point on the cylinder, that no parabolic trajectory could pass through P, and that no water can reach this spot.

Now, if we let P approach the vessel along the incline, at some particular point P', the associated circle k', with centre P' and radius $P'Q'$, touches the y-axis. The point at which it touches gives the focus F of the parabolic path followed by water that falls the maximum distance down the slope.

From Fig. 2 we can see that

$$P'F = OF + H \qquad \text{and} \qquad OF = P'F \cdot \tan\theta,$$

from which it follows that $P'F = H/(1 - \tan\theta)$. Further, the water jet falling at the maximal distance reaches the incline at point P', which is a distance

$$P'O = \frac{P'F}{\cos\theta} = \frac{H}{\cos\theta - \sin\theta}$$

from the bottom of the vessel.

The launch point is half-way between the directrix and the focus, and so we have that $H - h = \frac{1}{2}P'Q' = \frac{1}{2}P'F$. Using this and the result for $P'F$, we conclude that the hole should be drilled at a height

$$h = \frac{1 - 2\tan\theta}{2 - 2\tan\theta} H$$

from the bottom of the vessel.

If the incline is steep enough, it can happen that the distance OF is larger than H (corresponding to a launch point below ground). By investigating the (physically) limiting case of $OF = H$, we can obtain the critical angle θ^* for the slope of the incline:

$$\tan\theta^* = \frac{OF}{P'F} = \frac{OF}{P'Q'} = \frac{H}{2H} = \frac{1}{2}, \qquad \text{that is,} \qquad \theta^* \approx 26.6°.$$

If $\theta > \theta^*$, the best that can be done is to drill the hole at the bottom of the vessel (at O, where $h = 0$). Then, circle k' no longer touches the y-axis, but crosses it at point O. The maximal distance $P'O$ can be found with the help of conditions $OF = H$ and $P'O = P'Q'$, which produce an equation of the form

$$x^2 + (x\tan\theta - H)^2 = (x\tan\theta + H)^2$$

for the x coordinate of the impact point. This gives $x = 4H\tan\theta$ and implies that

$$P'O = \frac{4H\tan\theta}{\cos\theta}.$$

All of these conclusions are in line with those from the first solution.

S12 It is helpful to consider first a side-on photograph of an arrow in (straight-line) flight. If the exposure is not too long, only the head and fletching of the arrow are blurred. The various points of most of the shaft cannot be distinguished from each other because their images become sequentially superimposed during the exposure; consequently, most of the shaft of the arrow seems sharp.

Equally, the different parts of the spokes of the bike cannot be distinguished from each other. This is why a point on a spoke that has its velocity directed along the spoke (as a result of the combined translation and rotation of the wheel) seems more or less sharp in the photo. For the sake of clarity, Fig. 1 shows only two spokes, each in five consecutive positions with very short time intervals between them. Notice that, as well as point O (the momentary rotational axis of the wheel), point P seems comparatively sharp in this 'stroboscopic' image. Point P is not the image of any particular point of the spoke, but identifies where adjacent spoke points have the same position in the photo.

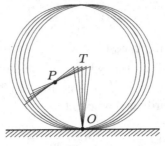

Fig. 1

What property characterises points like *P*? The crucial criterion is that the direction of the velocity vector of a point on the spoke exactly coincides with the direction of that particular spoke. Since the instantaneous centre of rotation is *O*, this direction of motion is perpendicular to *OP*. However, this is required to be the same as direction *TP*, where *T* is the centre of the wheel, as shown in Fig. 2.

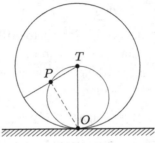

Fig. 2

Since *OPT* is a right angle, it follows from one of the theorems attributed to Thales of Miletus ('the angle in a semicircle is a right angle' and its converse) that all such points *P* lie on a circle with diameter *OT*. Where the bicycle spokes cross this (theoretical) circle define the sharp parts of the photographic image.

> *Note.* As the spokes of real bicycles are not exactly radial, but always cross each other, in practice one-half of the unblurred points are outside, and the other half are inside, the 'Thales circle' found in this solution.

S13   In order to keep our equations as simple as possible, we use units of time and distance in which the radius of the wheel and the bicycle's speed have unit value. With this choice, the numerical value of the wheel's angular velocity will also be unity.

Further, we take a coordinate system (in real space) such that the horizontal positive *x*-axis is at ground level and parallel to the bike's velocity, whereas the positive *y*-axis points vertically upwards on the finishing line. For the photo-finish picture, our coordinate system will use capital letters *X* and *Y*, with *Y* measured 'vertically' and *X*, representing time, plotted along a 'horizontal' axis.

When the hub of the bike's front wheel is at the point $(x, y) = (-1, 1)$ (and the foremost point of the wheel has just reached the finishing line at $x = 0$), the equation of a (radial) spoke that makes an angle θ with the horizontal is

$$y - 1 = (x + 1) \tan \theta.$$

After time t the spoke has rotated through an angle of $-t$, and the wheel's displacement along the positive x-axis has increased by t; the equation of the spoke at this time has become

$$y - 1 = (x - (-1 + t)) \tan(\theta - t).$$

On the photo-finish picture, all points lying on the y-axis, i.e. at $x = 0$, are recorded at a 'height' of

$$Y = y(t)\big|_{x=0} = 1 + (1 - t) \tan(\theta - t)$$

and placed horizontally at the point $X = -t$ (in accord with the unit speed).

The resulting 'curved image' of the spoke in the photo-finish image therefore has the equation

$$Y(X) = 1 + (1 + X) \tan(\theta + X).$$

The result of plotting this function for different values of θ, say at 20° intervals, is shown in the figure.

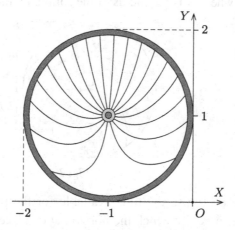

The spokes appear on the picture grossly distorted, a situation which, nevertheless, agrees very well with real electronic photo-finish pictures of bicycle races (such as can be found on the Internet, for example).

The rim and the tyre of the wheel appear in the image as undistorted circles, because a rotation transforms a circle into a circle, and the translation of the wheel is properly taken into account by the 'electronic shift' of the image points.

S14   The problem is equivalent to the following question. For how long may we put one of our fingers, held stationary with respect to the ground, between the spokes of the rolling wheel without touching them?

The motion of the spokes of the rolling cartwheel is quite complicated, and an analytic solution to the problem is difficult. Instead of attempting this, let us tackle the question using a graphical method. To do this, we stick a paper disc onto the base of a convenient circular cylinder (e.g. a jar or a tin), and, while the cylinder is rolled along a table, the tip of a felt pen (stationary with respect to the table) is held against the paper disc, as shown in Fig. 1.

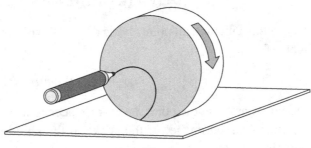

Fig. 1

What we will get are the curves shown in Fig. 2, the different curves corresponding to different heights of the pen tip above the surface of the table. In the actual figure shown, the height of the pen has been changed between 0 and $2R$ in eight steps of $R/4$, where R is the radius of the rolling cylinder.

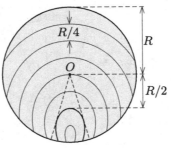

Fig. 2

In order to maximise the time available for the passage of the bolt, we have to choose the curve for which the horizontal displacement of the rolling cylinder is maximal, but subject to the constraint that the pen tip remains between two successive 'spokes' (on the paper disc, the longest curve that lies entirely between two radii with an angle of $360°/12 = 30°$ between them). We may choose only from those curves that do not cross any spokes, or, more precisely, if they do, then only its length between the crossing points may be considered. The required optimal curve can be identified as the one that is just touched by the two successive 'spokes'. As seen in Fig. 2, this is a curve very close to the one for which the pen height was $R/2$. From this heuristic approach, we can conclude that the optimal result is obtained if the pen tip is at a height of approximately $0.5R$ above the level of the table.

Thus, the crossbow bolt should be fired between two spokes at a distance of $0.5R$ from the axis of the wheel. At this height, the length of a horizontal chord of the wheel is $2\sqrt{1 - (0.5)^2}\,R \approx 1.7R$; the bolt may remain between the spokes for as long as it takes the axis to move this distance. The time for this is

$$t = \frac{1.7 \times 0.5 \text{ m}}{15 \text{ m s}^{-1}} \approx 0.057 \text{ s}.$$

So the speed of the bolt should be at least

$$v = \frac{0.2 \text{ m}}{0.057 \text{ s}} \approx 3.5 \text{ m s}^{-1}.$$

Notes. 1. We can find a more accurate result using calculus, together with the numerical solution of a trigonometrical equation. It shows that the optimal position for firing the bolt is at a distance of $0.524R$ from the axis of the wheel. So, the graphical method gave a good approximation.

2. In practice, gravity will have some effect on the result, in that a certain amount of vertical displacement will take place during the flight. This can be minimised by aiming slightly above horizontal to ensure that the bolt is at the top of its flight at the midpoint of its passage through the wheel. For the calculated transit time of $t = 57$ ms, the minimal vertical displacement of the bolt is $\Delta h = \frac{1}{2}g(t/2)^2 \approx 4$ mm; this is negligible compared to the length of the bolt and the radius of the wheel.

S15 *a*) The change in gravitational potential energy is the same whichever path is followed between A and B, and so – if there is no friction – the final kinetic energy, and hence the final speed, of the bob must be the same in the two cases.

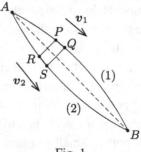

Fig. 1

In order to compare the transit times, consider two small arcs, denoted by PQ and RS, which are reflections of each other in the line AB, and are therefore of equal *length*, as shown in Fig. 1. As segment RS is 'lower' than PQ, the small bob's speed

v_2 along this segment of path (2) is larger than the speed v_1 along the corresponding segment PQ of path (1); that is, the bob traverses the arc RS in a *shorter time* than it would take to cover arc PQ. The same argument, and conclusion, applies to every pair of corresponding segments, and hence also for the whole motion. In other words, the bob reaches point B more quickly by following trajectory (2).

b) When there is significant friction, the final speeds can be compared using the work–kinetic energy theorem. The gravitational potential energy lost during the descent is the same for both trajectories, but the work done against friction may be different.

Again, consider two small segments of arc, one on each curve, and denoted by XY and VW (*see* Fig. 2). This time, they are defined to be reflections of each other in the midpoint C of the straight line AB; with this choice, their lengths and slopes are identical.

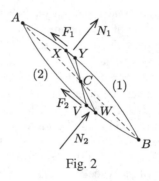

Fig. 2

The centripetal acceleration of the sliding bob is provided by the (directional) sum of the normal force exerted by the surface and the component of the gravitational force that is perpendicular to the velocity. Because of the identical slopes of the two small segments in question, these perpendicular components of the gravitational force are the same in both cases.

On both trajectories, the centripetal acceleration is directed towards the centre of the corresponding arc: on path (1) it is directed obliquely 'downwards', and on path (2) obliquely 'upwards'. As a result, on segment XY the normal force N_1 has to be less than the perpendicular component of the gravitational force, whereas on segment VW the normal force N_2 has to be greater than it. This argument is independent of which pair of small segments – centrally symmetrical about C – are chosen. So, for each point pair on the two trajectories, it is true that $N_1 < N_2$; and the same inequality is valid for the magnitudes of the resistive frictional forces, $F_1 < F_2$. The work done against friction is larger along path (2), and the corresponding final bob speed will be smaller.

The above arguments are clearly irrelevant if the friction is so large that either the bob does not have enough initial potential energy to complete the whole of path (2), or it cannot even start to move on path (1).

S16 There are several forces acting on the bike, but in this problem we need to take into consideration only their components parallel to its velocity. For gravity, this component is $\pm mg \sin \theta$, where θ is the angle of the incline above the horizontal; the sign depends upon the direction of travel. The drag force is proportional to the square of the speed, because it is proportional to the kinetic energy of the air (along a unit length of the road) disturbed by the bike.[18]

The forces exerted on the pedals by the rider's feet produce a torque, which – through the transmission system – is converted into a backwardly directed force exerted on the road by the rear-wheel tyre. In reaction to this force, the road exerts a frictional force on the wheel.

Denote the cyclist's top speeds, uphill, downhill and on a flat road, by v_1, v_2 and v_3, respectively. In each case the bike's speed is constant and therefore its acceleration is zero. The equation of motion for the bike moving uphill is

$$F_1 - mg \sin \theta - kv_1^2 = 0, \tag{1}$$

where F_1 is proportional to the 'effort' produced by the rider's legs, and is the frictional force that pushes the bike forwards; k is a constant dependent on, among other things, the density of the air. Similar equations can be written for when the bike moves downhill or along the flat:

$$F_2 + mg \sin \theta - kv_2^2 = 0, \tag{2}$$

$$F_3 - kv_3^2 = 0. \tag{3}$$

a) If the biker's 'maximal effort' is interpreted to mean that the *frictional reaction force* generated as the result of his or her efforts is the same in each of the three cases, then

$$F_1 = F_2 = F_3. \tag{4}$$

The addition of equations (1) and (2) yields

$$F_1 + F_2 - k(v_1^2 + v_2^2) = 0.$$

Comparing this result with (3) and (4) gives immediately that

$$2kv_3^2 - k(v_1^2 + v_2^2) = 0,$$

---

[18] The average speed of the air that is put into motion is roughly equal to the cyclist's speed.

from which we can deduce that

$$v_3 = \sqrt{\frac{v_1^2 + v_2^2}{2}} = \sqrt{\frac{12^2 + 36^2}{2}} \text{ km h}^{-1} \approx 27 \text{ km h}^{-1}.$$

Note. We get a similar result even if we assume that, in addition to the forces considered above, rolling friction and friction in the bike's bearings and sprockets are significant. In this case, the equation of motion (1) is modified to

$$F_1 - mg \sin \theta - kv_1^2 - k^* - \mu mg \cos \theta = 0,$$

where k^* and μ are coefficients characterising bearing and rolling friction, respectively. The other two equations are similarly modified.

From these modified equations of motion it can be deduced that

$$v_3^2 = \frac{v_1^2 + v_2^2}{2} - \frac{mg}{k} \mu (1 - \cos \theta). \tag{5}$$

Bearing in mind the fact that roads suitable for cycling are not really steep (that is, $\cos \theta \approx 1$), and, moreover, the coefficient of rolling friction is usually very small, we see that the last term on the right-hand side of equation (5) is negligible.[19] If this final term can be neglected, our previous result is recovered.

b) If the phrase 'maximal effort' is interpreted to mean that the cyclist produces his or her maximum *power*, then instead of (4) we should write equalities for the various products of force and speed:

$$F_1 v_1 = F_2 v_2 = F_3 v_3.$$

From the first equality, using (1) and (2), we obtain

$$mg \sin \theta = k \frac{v_2^3 - v_1^3}{v_1 + v_2}. \tag{6}$$

From the equality $F_1 v_1 = F_3 v_3$, together with (1) and (3), it follows that

$$mg \sin \theta = k \frac{v_3^3 - v_1^3}{v_1}. \tag{7}$$

Comparing (6) and (7) we get an expression for the required speed:

$$v_3 = \sqrt[3]{\frac{v_1 v_2 (v_1^2 + v_2^2)}{v_1 + v_2}} = \sqrt[3]{\frac{12 \times 36 \times (12^2 + 36^2)}{48}} \text{ km h}^{-1} \approx 23.5 \text{ km h}^{-1}.$$

Notes. 1. It is worth remarking that the rider (as part of the moving system) can produce only *internal* forces, forces which – without the (*external*) frictional force – could not keep the bike moving. However, the frictional force does *no*

---

[19] Unless mg/k is very large; the terminal speed of free fall of a human body is about 200 km h^{-1} and this would suggest that the reduction in the calculated value of v_3 might be as much as 2 % for $\mu = 0.05$ and $\theta = 10°$.

work, because the lowest point of the wheel, which rolls without slipping, does not move relative to the ground. The work is done at the application points of the internal forces, namely at the pedals.

2. For bikes with gearshifts the 'pedal force' required can be either increased or decreased, as the rider wishes; in such cases, interpretation *a*) of 'maximal effort' has no realistic meaning.

3. The top speed for cycling on a flat road lies between v_1 and v_2, and is, in some way, their mean value. For interpretation *a*), this mean (Q) is the root mean square (r.m.s.), while the mean for case *b*) with the identical powers is $\sqrt[3]{G^2 Q^2 / A}$, where A is their arithmetic mean and G their geometric one.

S17 The equation of motion for a biker (bike plus rider!) of total mass m coasting down a slope of angle θ with speed v is

$$ma = mg(\sin\theta - \mu\cos\theta) - kv^2,$$

from which (using the condition $a = 0$) the terminal speed on the very long slope can be found:

$$v_{max} = C\sqrt{\frac{m}{k}}. \tag{1}$$

Here C is a constant, and has the same value for both Ann and Bob. As the ratio m/k is $\frac{110}{60} \times \frac{1}{1.5} = 1.22$ times larger for Bob than for Ann, his terminal speed is also larger.

After a biker has left the slope and moved onto the horizontal road with initial speed $v_0 = v_{max} \sim \sqrt{m/k}$, he or she is decelerated by friction in the bearings and air drag. It is more convenient to investigate the speed decrease as a function of *horizontal position*, x, rather than of time. In accord with the work–kinetic energy theorem, it is true for both bikers that

$$\frac{d(mv^2/2)}{dx} = -\mu mg - kv^2(x).$$

The solution to this equation gives the total path length over which the initial speed v_0 decreases to zero. It is convenient to introduce the dimensionless variable $f(x) = v^2(x)/v_0^2$, which decreases from 1 to 0 during the horizontal motion.

Making this change of variable yields

$$\frac{mv_0^2}{2}\frac{df}{dx} = -\mu mg - kv_0^2 f(x),$$

which can be rearranged, using equation (1) and the fact that $v_0 = v_{max}$, to read

$$\frac{df(x)}{dx} = -\frac{k}{m}[2f(x) + K]. \tag{2}$$

Here K is a constant that is independent of both k and m.

Bob's value for k/m is smaller than Ann's, and so from equation (2) it follows that, along any particular segment of the road Δx, and for the same value of $f(x)$, the magnitude of the loss ($|\Delta f|$) in the ratio of squared speeds is *smaller* for Bob than it is for Ann. This means that Bob coasts a *greater* distance before stopping than Ann does and so wins the 'free-wheel bike race'.

S18   If the feather is moving in a stationary orbit, then that orbit must be a circle centred on the rotation axis. This is because, relative to a rotating frame of reference centred on the end of the crop, the feather is *at rest*. Because of the infinitesimally small mass of the feather, we can ignore gravity, and the forces determining its motion are much simplified: the tension in the thread and the air drag force balance each other.

Let the length of the crop be R, and that of the thread be L. The air drag force is directly opposed to the velocity. It follows that the thread tension, and so the thread itself, must also lie along this direction. In other words, the straight line of the thread is a tangent to the feather's orbit.

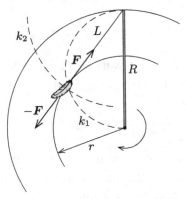

The radius r of the circular trajectory can be found by considering the intersection of the Thales circle k_1 (with diameter R)[20] and the circle k_2 which has a radius L and is centred on the moving end of the crop (*see* figure):

$$r = \sqrt{R^2 - L^2}.$$

If the thread is longer than the crop, then the feather has *no* stable (stationary) trajectory.

S19   The equation of motion of a pearl of mass m moving under gravitational and viscous forces is

$$m\frac{\mathrm{d}\boldsymbol{v}(t)}{\mathrm{d}t} = m\boldsymbol{g} - k\boldsymbol{v}(t).$$

---

[20] See the solution that appears on page 112.

The coefficient k in the expression for the viscous drag can be found from the equation for the terminal velocity:

$$0 = m\boldsymbol{g} - k\boldsymbol{v}_1, \qquad \text{giving} \qquad k = \frac{mg}{v_1}.$$

Here, and later, we use the fact that $\boldsymbol{g}$ and $\boldsymbol{v}_1$ are parallel to convert from vector expressions to scalar ones, and vice versa. After substitution for k, the equation of motion becomes

$$\frac{d\boldsymbol{v}(t)}{dt} = \boldsymbol{g} - \frac{g}{v_1}\boldsymbol{v}(t). \tag{1}$$

Rather than working directly with the pearl's velocity, it is more convenient to consider the *difference* between it and the terminal velocity, i.e.

$$\boldsymbol{u}(t) = \boldsymbol{v}(t) - \boldsymbol{v}_1. \tag{2}$$

Writing $\boldsymbol{v}(t) = \boldsymbol{v}_1 + \boldsymbol{u}(t)$ in equation (1), we obtain

$$\frac{d\boldsymbol{u}(t)}{dt} = \boldsymbol{g} - \boldsymbol{g} - \frac{g}{v_1}\boldsymbol{u}(t) = -\text{constant} \cdot \boldsymbol{u}(t). \tag{3}$$

This means that the magnitude of vector $\boldsymbol{u}(t)$ gradually (but not uniformly) decreases to zero from an initial value of $v_2 - v_1$, while its direction remains unchanged. We can therefore write

$$\boldsymbol{u}(t) = (\boldsymbol{v}_2 - \boldsymbol{v}_1)\lambda(t), \tag{4}$$

where $\lambda(0) = 1$, and $\lambda(t)$ decreases to zero after a sufficiently long time.[21]

a) Returning to the original question, we combine (2) and (4) to obtain the pearl's velocity in the form

$$\boldsymbol{v}(t) = [1 - \lambda(t)]\boldsymbol{v}_1 + \lambda(t)\boldsymbol{v}_2.$$

Geometrically, this means that the pearl's velocity vector moves along the straight-line segment that joins the horizontal initial velocity vector $\boldsymbol{v}_2$ to the terminal velocity vector $\boldsymbol{v}_1$; in Fig. 1 this is the line AB.

It can be seen from the figure that the minimal speed $v_{\min}$ corresponds to the perpendicular distance from the vector origin O to the line segment AB. Since the right-angled triangles OPB and AOB are similar, we have

$$\frac{OP}{OB} = \frac{AO}{AB}.$$

This, together with help from Pythagoras's theorem, gives as the final answer

---

[21] The elementary solution to differential equation (3) shows that $\lambda(t)$ is a decreasing exponential function with decay constant g/v_1 – similar to that governing the time dependence of radioactive decay.

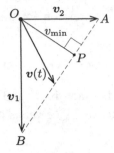

Fig. 1

$$v_{min} = OP = \frac{v_1 v_2}{\sqrt{v_1^2 + v_2^2}}.$$

b) If the initial velocity of the pearl is not horizontal, but makes an angle θ with and below the horizontal, then A will be some point on the circle that has centre O and radius v_2.

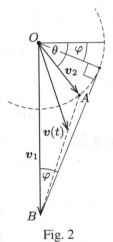

Fig. 2

For $v_2 < v_1$, as shown in Fig. 2, the speed of the pearl will monotonically increase provided

$$\theta > \varphi = \sin^{-1}\left(\frac{v_2}{v_1}\right).$$

Note. If $v_2 > v_1$, it is obvious that the speed of the pearl must decrease during some part of the descent. It can be proved that if $\theta < \sin^{-1}(v_1/v_2)$, the speed decreases initially, but later increases; if θ is greater than this critical angle, $v(t)$ always decreases.

S20 Suppose that the small ball is not lifted from the table-top just after the release. If this supposition is correct, then the acceleration of the smaller ball is horizontal at that moment. Denote this acceleration by a, as in the figure.

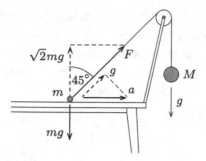

As *M* is much greater than *m*, the larger ball is practically in free fall, and its downward acceleration is *g*. It follows that the length of thread between the smaller ball and the pulley also decreases with acceleration *g*. But this acceleration of the thread must be equal to the component of the small ball's acceleration that lies in that same direction, i.e.

$$a \cos 45° = g, \tag{1}$$

implying that $a = \sqrt{2}\,g$.

As this horizontal acceleration is produced by the horizontal component of the tension *F* in the thread, this component must have a magnitude of $\sqrt{2}\,mg$. Further, since the thread is aligned at 45° above the horizontal, the upward vertical component of *F* must have the same value. This means that the ball of mass *m* experiences both a vertically upward force of $\sqrt{2}mg$ and a downward gravitational force of *mg*. Since the surface of the table can exert only upwardly directed normal forces, the net vertical component of the smaller ball's acceleration cannot be zero. This contradicts our initial supposition, which must therefore be wrong; thus, the smaller ball *is* lifted from the table-top immediately after the release.

> *Note.* Equation (1) indicates that $a > g$, and this might seem surprising; $a = g \cos 45°$ might have seemed a more plausible result. However, it is sometimes difficult to visualise accelerations at a particular instant, as there are no 'visible' distances involved. To express the situation in 'more concrete' terms, we may consider it relative to a set of *x–y* axes, where the point at which the thread first touches the pulley is $(0, h)$ and the small mass is at $(-x, 0)$, with $0 \le x \le h$. Then, the speed at which the mass–pulley segment of thread shortens is clearly gt ($t \ge 0$) and we can write
>
> $$\frac{\mathrm{d}}{\mathrm{d}t}[(-x)^2 + h^2]^{1/2} = -gt.$$
>
> From this, after some careful calculus, we find that
>
> $$a(t) = \frac{\mathrm{d}^2 x}{\mathrm{d}t^2} = \frac{(x^2 + h^2)^{1/2}}{x}\,g - \frac{h^2}{x^3}\,g^2 t^2,$$
>
> yielding $a = \sqrt{2}\,g$ when $t = 0$ and $x = h$ (corresponding to the 45° angle of the problem), thus confirming result (1).

S21   Consider the limiting case, when the angular frequency ω of the vibration is such that the pearl of mass m is on the brink of flying away from the rod. In this special situation, the pearl would not slide down to the pivot even if there were no disc, so the only forces acting on the pearl are the normal force N due to the rod and the gravitational force mg (*see* figure).

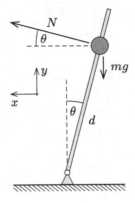

The horizontal displacement of the pearl, as a function of time, is

$$x(t) = -d \sin \theta \approx -\theta(t)d = -\theta_0 d \sin \omega t,$$

where we have used the fact that the angular displacement $\theta(t)$ is small. Accordingly, in the horizontal direction, the pearl's motion is harmonic to a good approximation, and so its acceleration is

$$a_x(t) = \theta_0 \omega^2 d \sin \omega t.$$

This acceleration is provided by the horizontal component of the normal force N exerted by the rod:

$$ma_x(t) = N \cos \theta(t) \approx N.$$

From this, we can find the vertical component of the normal force as a function of time:

$$N_y = N \sin \theta(t) \approx N\theta(t) = m\theta_0^2 \omega^2 d \sin^2 \omega t.$$

Under this time-varying upward force N_y, the pearl will separate from the disc and slide up the rod when the time-averaged value is larger than the gravitational force. The average value of the $\sin^2$ function over a whole number of (half-)cycles is $\frac{1}{2}$, and so[22]

$$\langle N_y \rangle = m\theta_0^2 \omega^2 d \langle \sin^2 \omega t \rangle = \frac{1}{2} m\theta_0^2 \omega^2 d.$$

---

[22] This is a similar procedure to that used to determine the Joule heat produced in a.c. circuits.

It follows that the condition for the pearl to fly off the end of the rod is

$$\frac{1}{2}\theta_0^2\omega^2 d > g,$$

and, from this, that the required frequency f is

$$f = \frac{\omega}{2\pi} > \frac{1}{2\pi}\sqrt{\frac{2g}{\theta_0^2 d}}.$$

The distance dependence of the critical frequency ($\sim 1/\sqrt{d}$) shows that the required frequency *decreases* as the distance from the pivot increases, and so if the pearl does separate from the disc at some frequency, then, since its time-averaged acceleration is always upwards, it will continue to do so and eventually leave the rod. Finally, we note that the correct result is $\sqrt{2}$ times larger than the naive answer.

S22 The forces acting on the block of mass m are the gravitational force mg, the normal reaction force N of the board and the frictional force F (kinetic or static). The direction of the frictional force may or may not change during one cycle of the board's vibration. The equations of motion, in the directions perpendicular to and parallel to the incline are

$$N - mg\cos\alpha = 0 \quad \text{and} \quad F + mg\sin\alpha = ma.$$

The positive direction of the acceleration a has been chosen to be down the slope (*see* Fig. 1).

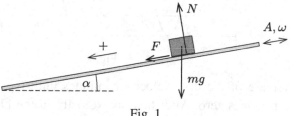

Fig. 1

The connection between the magnitudes of the static frictional force and the normal force is the inequality $F \le \mu N$, while that for kinetic friction is the simple equation $F = \mu N$. The acceleration that the frictional force produces in the block is largest when the block is sliding, and the (relative) velocity of the block *with respect to the board* is in the negative direction. In this case

$$a_{\max} = g(\sin\alpha + \mu\cos\alpha),$$

and, inserting the data, we get $a_{\max} \approx 5.6$ m s^{-2}. The maximal acceleration of the board, resulting from its harmonic oscillation, is $A\omega^2 = 250$ m s^{-2}, which is more than 40 times larger than $a_{\max}$. We conclude that the block starts to slide *almost immediately* after the start of the vibrations. We will see later that the block does not adhere to the board at any point in the ensuing motion. Consequently, the only frictional force acting on the block is a *constant-magnitude* kinetic one – though its direction is constantly reversing.

It follows from the previous paragraph that the acceleration of the block has only two different values during its motion:

$$a_\pm = g(\sin\alpha \pm \mu\cos\alpha), \tag{1}$$

and since, from the given data, μ (0.4) $>$ $\tan\alpha$ (0.18), a_+ is positive and a_- is negative. The a_+ phase lasts until the (signed) speed of the board is larger than the speed of the block, and the situation is just the opposite for the a_- phase.

In the graph shown in Fig. 2, the velocities of the board and the block are plotted as a function of time. The latter is piecewise linear with two different gradients, namely a_+ and a_-. As $|a_+| > |a_-|$, the average velocity of the block over one period (the 'drift velocity'), increases all the time, and the block drifts downwards on the board.

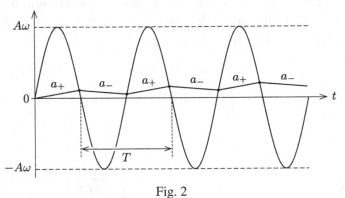

Fig. 2

The increase of the drift velocity continues until the average acceleration of the block becomes zero. After that, the velocity of the block fluctuates around a constant value v_{drift} (*see* Fig. 3). Because of the relatively large frequency of the board's vibrations, this steady (stationary) state of the motion is achieved quite quickly, and so, in our estimation of the total time for the motion, that for the initial transient stage can be neglected.

The condition for a steady drift is that the average acceleration over a complete cycle is zero:

$$\langle a \rangle \equiv \frac{a_+ t_+ + a_- t_-}{T} = 0, \tag{2}$$

where the definitions of t_+ and t_- will be clear from Fig. 3. Of course, we also have

$$T = t_+ + t_-. \tag{3}$$

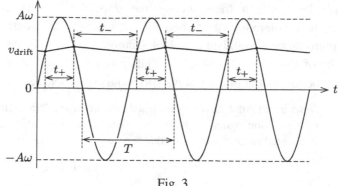

Fig. 3

From equations (1)–(3), we can find the time interval t_+:

$$t_+ = \frac{a_-}{a_- - a_+} T = \left(1 - \frac{\tan \alpha}{\mu}\right) \frac{T}{2}. \tag{4}$$

The drift velocity can be calculated from the condition that the acceleration of the block changes direction when the block and the board have the same velocity. Neglecting the fluctuations in the drift velocity, which are small compared to its value, we see from Fig. 3 that

$$v_{\text{drift}} \approx A\omega \sin\left(\frac{\pi}{2} - \omega\frac{t_+}{2}\right) = A\omega \cos\left(\omega\frac{t_+}{2}\right).$$

Finally, after result (4) has been inserted into this expression, it yields

$$v_{\text{drift}} = A\omega \cos\left[\left(1 - \frac{\tan \alpha}{\mu}\right)\frac{\pi}{2}\right] = A\omega \sin\left(\frac{\pi \tan \alpha}{2\mu}\right).$$

Using the given numerical data, we find that $v_{\text{drift}} \approx 0.32$ m s^{-1}, and so the estimated time of the block's motion is

$$t = \frac{L}{v_{\text{drift}}} \approx 18.8 \text{ s.}$$

The remaining task is to prove that the block does not in fact adhere to the board. Adherence requires two conditions to be satisfied: one of them is that, at the relevant moment, the velocities of the block and the board must be the same; the other is that, at that same time, the magnitude of the board's acceleration must be less than $|a_+|$ or $|a_-|$, according to whether the board is moving downwards or upwards, respectively. The first condition is fulfilled each time the board reverses direction, but this momentary 'sticking' is not considered a real adherence in our model calculation. At those moments the kinetic frictional force changes direction, but this reversing process is so fast that the displacement of the block during the brief time it takes is negligible.

It can be seen from the velocity–time graph that the two conditions can occur only if the acceleration of the board is very small, i.e. when its speed is near its maximum (almost $A\omega$). But the block cannot attain such a speed, because the drift velocity of ≈ 0.32 m s^{-1}, which is smaller than the required speed of ≈ 0.5 m s^{-1}, sets in sooner. So the block always slides during its journey down the incline.

> *Note.* In our solution it was assumed that the time taken for the first transient stage of the motion (whereas the block's speed attains the drift velocity v_{drift}) is short. A detailed calculation shows that the range of this time interval is
>
> $$\tau \approx \frac{A\omega}{\mu g \cos \alpha} \approx 0.13 \text{ s},$$
>
> and so the error introduced into our estimation by ignoring it is really quite small, some 1–2 %.

S23 Consider the work done as the length of the cord is changed by $\Delta\ell < 0$, i.e. the loose end of the cord is lifted through a distance $|\Delta\ell|$. This work is equal to the change in the total mechanical energy of the pendulum bob. If the energy is expressed in terms of the current length ℓ of the pendulum and the linear amplitude A of its swing, it can be shown, as below, that the angular amplitude ($\Phi = A/\ell$) of the motion increases due to the work done, but that A decreases.

In what follows we make use of the very slow change in ℓ (measured from the ceiling to the bob), i.e. for a single swing its fractional change $|\Delta\ell|/\ell \ll 1$. With this condition, the other quantities characterising the motion also change very slowly, and (to first order) can be considered constant during a single swing. Further, the tension F in the cord can be replaced by its time-averaged value $\overline{F}$.

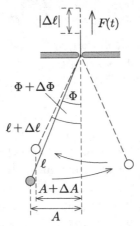

The instantaneous angular displacement $\varphi(t)$ and velocity $v(t)$ of the pendulum can be described as functions of time in terms of simple harmonic motion:

$$\varphi(t) = \Phi \sin(\omega t) \quad \text{and} \quad v(t) = \Phi\ell\omega \cos(\omega t), \tag{1}$$

where $\omega = \sqrt{g/\ell}$ is the angular frequency corresponding to the current length of the pendulum.

It is true that only the component $mg\cos\varphi$ of the gravitational force mg contributes to the tension in the cord, but this tension also provides the centripetal acceleration mv^2/ℓ. So the tension in the cord (for small-angle swings) is

$$F(t) = mg\cos\varphi + \frac{mv^2}{\ell} \approx mg\left(1 - \frac{\varphi^2}{2}\right) + \frac{mv^2}{\ell}.$$

Using equations (1) and the expression for ω we get:

$$F(t) = mg + mg\Phi^2\left(-\frac{\sin^2(\omega t)}{2} + \cos^2(\omega t)\right),$$

and because $\overline{\sin^2\varphi} = \overline{\cos^2\varphi} = 1/2$, its time-averaged value is

$$\overline{F} = mg\left(1 + \frac{\Phi^2}{4}\right). \tag{2}$$

The total mechanical energy of the pendulum is (taking the zero of the gravitational potential energy to be at the ceiling)

$$E = -mg\ell + \frac{1}{2}mv_{\text{max}}^2 = -mg\ell + \frac{1}{2}m\Phi^2\ell^2\omega^2,$$

which can be expressed, using $\omega = \sqrt{g/\ell}$, as

$$E = mg\ell\left(-1 + \frac{\Phi^2}{2}\right). \tag{3}$$

When the length of the pendulum is changed by a small amount $\Delta\ell$, and the resulting change in the angular amplitude is $\Delta\Phi$, the small change in the mechanical energy can be calculated using expression (3) as follows:

$$\Delta E = mg\Delta\ell\left(-1 + \frac{\Phi^2}{2}\right) + mg\ell\Phi\Delta\Phi. \tag{4}$$

Let us now apply the work–kinetic energy theorem for a single swing of the pendulum (remembering, when needed, that $\Delta\ell$ is negative):

$$\overline{F}|\Delta\ell| = \Delta E,$$

which can be written using equations (2) and (4) as

$$-mg\Delta\ell\left(1 + \frac{\Phi^2}{4}\right) = mg\Delta\ell\left(-1 + \frac{\Phi^2}{2}\right) + mg\ell\Phi\Delta\Phi,$$

After some rearrangement this yields

$$\frac{\Delta\Phi}{\Phi} = -\frac{3}{4}\frac{\Delta\ell}{\ell} > 0, \qquad \text{since } \Delta\ell < 0. \tag{5}$$

This inequality shows that the maximal angular extent of the swings increases slowly with the decrease in the pendulum length. However, by contrast, the horizontal amplitude $A = \ell\Phi$ decreases, because

$$\frac{\Delta A}{A} = \frac{\ell\Delta\Phi + \Phi\Delta\ell}{\ell\Phi} = \frac{\Delta\Phi}{\Phi} + \frac{\Delta\ell}{\ell} = +\frac{1}{4}\frac{\Delta\ell}{\ell} < 0.$$

Note. Let us multiply both sides of the equality in equation (5) by $\ell^3\Phi^4$. After some rearrangement, we get

$$4\ell^3\Phi^3\Delta\Phi + 3\ell^2\Phi^4\Delta\ell = 0.$$

This can be written in the alternative form

$$\Delta(\ell^3\Phi^4) = 0,$$

showing that, although the length of the pendulum is decreasing slowly, the quantity $\ell^3\Phi^4$ *does not change*.

From this 'conservation law', it follows that (to a good approximation) the ratio of the kinetic energy of the pendulum $E^* = E + mg\ell$ (the total energy without its gravitational component) to its angular frequency ω remains constant during a gradual change of the pendulum length:

$$\frac{E^*}{\omega} = \frac{\frac{1}{2}mg\ell\Phi^2}{\sqrt{g/\ell}} = \frac{m\sqrt{g}}{2}\sqrt{\ell^3\Phi^4}.$$

This kind of quantity (one that hardly changes during a slow change of a parameter) is called an *adiabatic invariant*. Such invariants are of great importance in quantum theory. The quantised energy levels of 'vibrating systems' (quantum harmonic oscillators), parallelling the swings of a classical pendulum, are given by formulae of the form

$$E_n = (n + \tfrac{1}{2})\hbar\omega \qquad (n = 0, 1, 2, \ldots).$$

The adiabatic invariant character of the quotient E/ω guarantees that, in the course of a slow change of an external parameter (ℓ in our case), corresponding to external work being done on the system, the quantum number n remains constant. The 'index' n of the quantum state can be changed only by a disordered energy transfer, that is by heat (*see* Solution 2 on page 463).

S24   Let us denote the position vectors of the carabiners, measured from an arbitrary point O, by r_1, r_2, r_3 and r_4, and that of the mountaineer's equilibrium position by r_0. The condition for equilibrium is

$$m\mathbf{g} + \sum_{i=1}^{4} k_i(r_i - r_0) = 0,$$

which gives

$$r_0 = \frac{m\mathbf{g} + \sum k_i r_i}{\sum k_i}.$$

If the mountaineer is displaced from this point to the position $r = r_0 + x$, then the net force acting on his body is

$$F(x) = mg + \sum_{i=1}^{4} k_i(r_i - r_0 - x),$$

which can be simplified using the first equation above to read

$$F(x) = -x \sum_{i=1}^{4} k_i.$$

From this, it can be seen that the mountaineer attached to these idealised springs (with zero unstretched lengths) moves as though a single spring with spring constant $\sum_{i=1}^{4} k_i$ is pulling him back to the equilibrium position. It is somewhat surprising that this result does not depend on the mass of his body, or on the positions of the fixed carabiners, and not even on the direction of the initial displacement.

The period of the motion is

$$T = 2\pi \sqrt{\frac{m}{\sum\limits_{i=1}^{4} k_i}} = 1.6 \text{ s}.$$

S25 First of all, we investigate whether the rubber eraser will start moving at all. It will do so provided that $mg \sin \alpha > \mu mg \cos \alpha$, i.e. $\mu < \tan \alpha = \tan 45° = 1$. This is clearly the case, since $\mu = 0.6$. So, the eraser will start moving.

A calculation of the work done against friction can be carried out, to any given degree of accuracy, only by using a computer. The trouble is that the determination of how the normal force acting on the eraser varies with position is difficult. However, it is certain that the frictional force is always larger than its initial value of $\mu mg \cos \alpha = \mu mg \cos 45° = \mu mg/\sqrt{2}$. This is because, after the initial release, the angle with the horizontal made by the slope on which the eraser moves decreases, and, in addition, the track has to provide a centripetal force for the moving eraser.

The path to reach the lowest point of the track would be one-eighth of a circle, with a length of $R\pi/4$. The work done against friction can now be *underestimated* by taking the normal force as if it always had its initial value:

$$|W_f| > |W_{und}| = \frac{\mu mg}{\sqrt{2}} \frac{R\pi}{4} \approx 0.333 mgR.$$

The gravitational potential energy difference between the initial position and the bottom of the track is

$$\Delta E_p = mgR(1 - \cos \alpha) = mgR(1 - \cos 45°) \approx 0.293 mgR.$$

It can be seen that $|W_f| > \Delta E_p$, i.e. the work to be done against friction is clearly larger than what can be provided by the gravitational potential energy. So, the rubber eraser cannot reach the very lowest part of the track.

> *Note.* One might be wondering how large the coefficient of friction might be while still letting the eraser slide to the bottom of the track. A complete description of the motion is given by the following differential equation:
>
> $$mR\ddot{\varphi} = -mg\sin\varphi + \mu m(g\cos\varphi + R\dot{\varphi}^2),$$
>
> where φ is the angle the radius vector to the eraser makes with the vertical, and has an initial value of $\pi/4$.
>
> Treating the dimensionless quantity $y(\varphi) = (R/g)\dot{\varphi}^2$, which is proportional to the kinetic energy of the eraser, as a function of φ, the equation of motion takes the form
>
> $$\tfrac{1}{2}y'(\varphi) - \mu y(\varphi) = -\sin\varphi + \mu\cos\varphi.$$
>
> With the boundary condition $y(0) = 0$ (i.e. the eraser just stops at the bottom of the track), the solution to the above equation is
>
> $$y(\varphi) = \frac{6\mu}{4\mu^2 + 1}\sin\varphi + \frac{2 - 4\mu^2}{1 + 4\mu^2}(\cos\varphi - e^{2\mu\varphi}).$$
>
> The speed of the rubber eraser is also zero when $\varphi = \pi/4$ (at the start of the track), and this condition is also met if the (critical) value of the coefficient of friction satisfies the following equation:
>
> $$\frac{1 + 3\mu - 2\mu^2}{1 - 2\mu^2} = \sqrt{2}\,e^{\mu\pi/2}.$$
>
> Numerical solution of this equation yields $\mu \approx 0.37$. This value *does not depend* on either the radius R of the track or the gravitational acceleration g.

S26 The very slow motion of the box means that there is no need to concern ourselves with kinetic energy. Because, at the end of the motion, the box arrives back at its initial position, the total change in gravitational potential energy is zero. So, the energy used, W, is entirely dissipated in working against the frictional force F:

$$10\,\text{J} = W = 2FL = 2(\mu mg\cos\theta)\left(\frac{h}{\sin\theta}\right) = \frac{2\mu mgh}{\tan\theta}, \qquad (1)$$

where the length of the incline is denoted by L and its inclination by θ.

We are given that the coefficients of static and kinetic friction are equal, and also that the box does not spontaneously slide on the incline. The latter implies that

$$\mu mg\cos\theta \geq mg\sin\theta, \qquad \text{that is} \qquad \mu \geq \tan\theta.$$

Putting this result for the coefficient of friction into equation (1) yields the inequality

$$10 \text{ J} \geq 2mgh,$$

from which, using the approximation $mg \approx 10$ N, we get $h \leq 0.5$ m. So the height of the incline is at most half a metre.

Note. It can be shown that, if the box is pulled up the incline, then the work done (*see* figure) is

$$W_{\text{up}} = mgL \sin \theta + \mu mgL \cos \theta = mgh + \mu mgx.$$

Correspondingly, the work done when the box moves down the incline is

$$W_{\text{down}} = -mgL \sin \theta + \mu mgL \cos \theta = -mgh + \mu mgx.$$

So, the total work W done against friction on the incline is related to the length x of its base (shown in the figure) by $W = 2\mu mgx$; from this it follows that $\mu x = 0.5$ m.

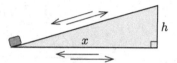

This means that, with a given value for the total work done, the length of the incline's base depends only on the coefficient of friction (it does not depend upon the angle of the incline): $x = (0.5/\mu)$ m. Now, the height h of the incline is given by $x \tan \theta = 0.5 \tan \theta / \mu \leq 0.5$, since the 'no-slide' condition showed that $\mu \geq \tan \theta$. Thus the maximum possible value of h is 0.5 m.

S27 Denote the masses of the magnets by m_1 and m_2, and the distance between their centres of mass by x. During the motion, x decreases from an initial value of $d + d_0$ to d_0. Denote the interaction energy of the magnets at this general position by $-W(x)$, with its zero level chosen to be in the initial configuration, i.e. $W(d + d_0) = 0$.

When the centres of mass of the magnets are a distance x apart, the total work the magnets have done upon each other, through their interaction forces, is just $W(x)$. In principle, this function can be calculated using the characteristic properties of the magnets, and applying the laws of magnetostatics – fortunately, the explicit form of this function is not needed in the following.

When the magnet with mass m_1 is released, its speed can be calculated using the law of conservation of energy:

$$\frac{1}{2}m_1 v_1^2 = W(x),$$

and so

$$v_1(x) \equiv -\frac{dx}{dt} = \sqrt{\frac{2W(x)}{m_1}},$$

from which the time lapse until the collision is

$$T_1 = -\int_{d+d_0}^{d_0} \frac{1}{v_1(x)} \, dx = \sqrt{m_1} \int_{d_0}^{d+d_0} \frac{1}{\sqrt{2W(x)}} \, dx. \tag{1}$$

Similarly, we get the time interval between the release of the second magnet and the subsequent collision:

$$T_2 = \sqrt{m_2} \int_{d_0}^{d+d_0} \frac{1}{\sqrt{2W(x)}} \, dx. \tag{2}$$

When both magnets are released simultaneously, their common centre of mass (CM) remains at rest during the motion. The common CM divides the distance between the magnets into two parts in the ratio of their masses (*see* figure) and the same ratio holds for the speeds of the bodies:

$$\frac{v_1}{v_2} = \frac{m_2}{m_1}. \tag{3}$$

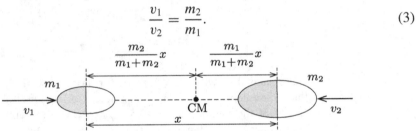

According to the law of conservation of energy, we must have

$$\frac{1}{2}m_1 v_1^2 + \frac{1}{2}m_2 v_2^2 = W(x),$$

from which – using (3) – the speed of one of the bodies (say, the one on the left) is

$$-\frac{m_2}{m_1 + m_2} \frac{dx}{dt} \equiv v_1(x) = \sqrt{\frac{2W(x)}{m_1(1 + m_1/m_2)}}.$$

From this, the time interval until the collision can be calculated in the same way as previously:

$$T_3 = \sqrt{\frac{m_1 m_2}{m_1 + m_2}} \int_{d_0}^{d+d_0} \frac{1}{\sqrt{2W(x)}} \, dx. \tag{4}$$

In equations (1), (2) and (4), we have three ratios that are all equal to the same (unevaluated) integral of the same function, so we can write

$$\frac{m_1}{T_1^2} = \frac{m_2}{T_2^2} = \left(\frac{m_1 m_2}{m_1 + m_2}\right) \frac{1}{T_3^2}.$$

From this multiple equality, we can now eliminate the two masses, obtaining the time interval in question in the form

$$\frac{1}{T_3^2} = \frac{1}{T_1^2} + \frac{1}{T_2^2},$$

that is

$$T_3 = \frac{T_1 T_2}{\sqrt{T_1^2 + T_2^2}} = 0.48 \text{ s}.$$

> *Note.* The result is the same for any force law between the bodies, provided they are not rotated. If the force has an inverse-square dependence, as in Newton's universal gravitational law acting on point-like or spherical bodies, or a linear one as in Hooke's law, then the time intervals can be calculated explicitly by evaluating the integrals involved. In each case the result vindicates our more general argument.

S28   If the gravitational attraction of the ball were much larger than that due to the Earth, then the two liquid surfaces would coincide with the same equipotential surface of the ball's gravitational field. These surfaces are spheres centred on the ball; so the length of the liquid column in the left arm would clearly be greater than the corresponding length in the right arm.

When the gravitational effects of both the Earth and the ball are significant, the equipotential surfaces are neither horizontal planes (as measured in the Earth's frame) nor spheres centred on the ball. Rather, the free liquid surfaces coincide with an equipotential surface that is intermediate between these two – in general terms, an 'incline' with an increasingly downward inclination from left to right (*see* figure). We can therefore conclude that the level of the liquid in the left-hand arm will rise, whereas that in the right will sink.

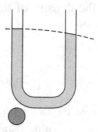

It is possible to strengthen the qualitative reasoning given above with a semi-quantitative analysis, as follows.

The original situation (in which the liquid levels in the two arms are the same) cannot be the new equilibrium state because we can find another arrangement, only marginally different from the original, in which the energy of the system is lower.

Let us transport, hypothetically, a small amount of the liquid (say, a layer of height ε) from the top of the right arm to the top of the left one. The mass Δm of this amount of liquid is proportional to ε, and so the increase in its gravitational potential energy in the essentially uniform field of the Earth (using the well-known formula mgh, with 'h' increasing by $\varepsilon/2$) is proportional to ε^2.

However, the action also changes the (negative) gravitational energy of the same amount of liquid in the field of the heavy ball. As the ball, of mass M, is very close to the tube, this field is not uniform and the change in potential energy is given by

$$-GM\Delta m \left(\frac{1}{r_2} - \frac{1}{r_1} \right),$$

where r_1 and r_2 are the distances of the transferred liquid from the centre of the ball before and after the transportation, respectively. As $r_2 < r_1$, this change represents a decrease in the potential energy, and because $\Delta m \propto \varepsilon$, this decrease is proportional to ε.

Since, in this thought experiment, ε can be made arbitrarily small, the quadratic gain can be made smaller than the linear loss, whatever the relative sizes of the ball and Earth. Therefore, the energy of the whole system decreases in such a process. This shows the non-equilibrium nature of the original situation, and why the liquid level moves upwards in the left arm until it reaches a position where any further mass realignment would not decrease the total energy.

S29   Suppose that our aim is to produce, at the origin of our coordinate system, the greatest possible gravitational field along the x-axis. Consider a small piece of plasticine of volume ΔV whose position vector is of length r and makes an angle φ with the x-axis. It produces a gravitational field

$$\Delta g = G\frac{\varrho \Delta V}{r^2}$$

at the origin (where ϱ is the density of the plasticine). Its x component is $\Delta g_x = \Delta g \cos \varphi$, and so the contribution per unit volume of the plasticine to the required gravitational field component is

$$\frac{\Delta g_x}{\Delta V} = G\varrho \, \frac{\cos \varphi}{r^2}.$$

This specific contribution is the same for all parts of the plasticine for which the fraction $\cos \varphi / r^2$ has the same value. Such 'level surfaces' can be described in a polar coordinate system (with the x-axis as the polar axis) by the equation

$$r(\varphi) = a\sqrt{\cos \varphi}, \tag{1}$$

where the constant a characterises and differentiates the various 'level surfaces' (*see* figure). The larger the value of a, the further the surface is from the origin on average, and the smaller the specific contribution, $\Delta g_x/\Delta V$, to the required field made by points on that surface.

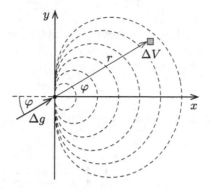

Imagine that a given (solid) volume of plasticine is initially moulded so that its surface has the shape of a 'level surface', and denote this surface by S. Changing the surface in any way must involve moving some part of the plasticine from the interior to the exterior of surface S; inevitably its contribution to the required field decreases. This means that the gravitational acceleration at the origin is as large as possible if the plasticine is moulded to the shape given by equation (1).

Notes. 1. By using the volume V of the plasticine, the coefficient a in equation (1) can be evaluated. Transforming the polar coordinates (r, φ) into Cartesian coordinates, we get

$$x = r \cos \varphi = a(\cos \varphi)^{3/2}, \qquad y = r \sin \varphi = a(\cos \varphi)^{1/2} \sin \varphi.$$

In terms of these, the volume of the plasticine (as a solid of revolution) is

$$V = \int_0^a \pi y^2(x) \, dx = -\frac{3\pi}{2} a^3 \int_{\pi/2}^0 (\cos \varphi)^{3/2} \sin^3 \varphi \, d\varphi = \frac{4\pi}{15} a^3.$$

From this, a can be expressed in terms of V, and the polar coordinate equation describing the optimal shape for the plasticine is

$$r(\varphi) = \sqrt[3]{\frac{15V}{4\pi}} \sqrt{\cos \varphi}.$$

2. A more formal analysis gives the same result for equation (1), though the mathematics called upon (the calculus of variations with constraints) is more advanced than that we normally use.

We first note that the distribution must be symmetric about an axis passing through the origin, because otherwise an improvement could be made by moving some material to a different azimuth. For a similar reason, all material must be on the same side of a plane containing the origin.

As previously, we take the x-axis as this axis of symmetry, and the plane as $\varphi = \pi/2$. Considering only the non-cancelling axial component of the gravitational force, i.e. with the $\cos \varphi$ factor included, we have

$$d^2 g = G\varrho \, \frac{2\pi\rho}{\rho^2 + x^2} \frac{x}{(\rho^2 + x^2)^{1/2}} \, dx \, d\rho.$$

Integrating this with respect to ρ from 0 to $R(x)$ (the radius of the surface at the current value of x) gives

$$dg \propto \left[\frac{x}{x} - \frac{x}{(R^2 + x^2)^{1/2}} \right].$$

So we need to choose $R(x)$, which gives the surface shape we seek, so as to maximise

$$\int_0^X \left[1 - \frac{x}{(R^2 + x^2)^{1/2}} \right] dx,$$

subject to $\int_0^X R^2(x) \, dx = $ constant. A secondary boundary condition, which need not concern us here, is that $R(x > X) = 0$.

Using the calculus of variations with a constraint (handled via a Lagrange multiplier λ), but with no derivatives involved, yields

$$\frac{xR}{(R^2 + x^2)^{3/2}} = 2\lambda R.$$

Clearly $R = 0$ gives a minimum and can be discarded, leaving $x \propto (R^2 + x^2)^{3/2}$.

But if the (cylindrically symmetric) surface is expressed in *spherical polar coordinates* (r, φ), we have that $R^2 + x^2 = r^2$ and $x = r \cos \varphi$, and so it follows that

$$r = a(\cos \varphi)^{1/2},$$

so reproducing equation (1). The value of the constant a is determined, as before, by the total volume of plasticine available.

S30   Around a very long cylinder, the gravitational field has cylindrical symmetry, the direction of the field (sufficiently far from the ends of the cylinder) is radial, i.e. perpendicular to the axis of the cylinder, and its magnitude at any point depends only on the distance of the point from the cylinder's axis.

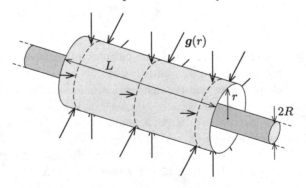

We draw analogies between a mass m (the 'gravitational charge') and an electric charge q, between the gravitational constant G and $1/(4\pi\varepsilon_0)$, and between the gravitational acceleration $g = F/m$ and the electric field strength $E = F/q$. In both cases, F is the force acting upon the 'charge' (real or analogous).

Using these, and the analogy between Newton's universal gravitational law and Coulomb's law, we can say that the number of gravitational field lines – the product of the area of a closed surface and the component of the gravitational acceleration g that is perpendicular to it – is $4\pi Gm$, where m is the total mass contained within the surface. Accordingly, for a coaxial cylinder of length L and radius r $(r > R)$ that surrounds the planet, we have

$$g \cdot 2\pi r L = 4\pi G \cdot \pi R^2 L \varrho, \qquad \text{that is} \qquad g(r) = \frac{2\pi G\varrho R^2}{r}. \qquad (1)$$

Here, ϱ denotes the average density of the planet, which is assumed to have a cylindrically symmetric mass distribution.

a) Using the form of the gravitational field strength, the first cosmic velocity can be found. The condition for the orbital motion of a point-like body with mass m around a circular trajectory of radius r is

$$mg(r) = m\frac{v_{c,1}^2}{r},$$

from which using (1), and the data supplied in the question, the orbital velocity is determined as

$$v_{c,1} = R\sqrt{2\pi G\varrho} = 9.7 \text{ km s}^{-1}.$$

This result is independent of the radius of the circular orbit (so long as it is much less than the length of the planet). In particular, it applies to the case $r = R$, and so this value is also the first cosmic velocity for this 'sausage' planet.

It is $\sqrt{3/2}$ times larger than the first cosmic velocity for Earth, which is given by

$$v_{c,1}^{\text{Earth}} = \sqrt{\frac{GM_{\text{Earth}}}{R}} = R\sqrt{\frac{4}{3}\pi G\varrho} = 7.9 \text{ km s}^{-1}.$$

b) The period of a satellite orbiting in a circle of radius r is $2\pi r/v_{c,1}$, so if the rotational period of the planet around its axis is $T_0 = 1$ day, then the radius of the trajectory of a 'geostationary' communications satellite is

$$r_0 = \frac{T_0 v_{c,1}}{2\pi} = R\sqrt{\frac{T_0^2 G\varrho}{2\pi}} = 1.33 \times 10^8 \text{ m}.$$

In the case of the Earth, the corresponding distance is

$$r_0^{\text{Earth}} = R\sqrt[3]{\frac{T_0^2 G\varrho}{3\pi}} = 4.2 \times 10^7 \text{ m}.$$

The relationship between the two distances can be written in the form

$$r_0 = \sqrt{\frac{3(r_0^{\text{Earth}})^3}{2R}}.$$

That they are not proportional to each other is accounted for by the fact that one case is effectively two-dimensional, whereas the other is three-dimensional. So the 'geostationary' communications satellite orbits at an altitude of $r_0 - R \approx$ 127 000 km above the surface of the long, cylindrical 'sausage' planet.

c) The second cosmic velocity (i.e. the escape velocity) is very large, and its precise value depends upon the actual length of the planet. For a planet of infinite length, the escape velocity is also infinite. This is because escape from a gravitational field with an r^{-1} strength dependence is *impossible* using a finite amount of energy.

In order to prove this, consider a series of distances increasing in geometric progression: $r_n = r_0 \alpha^n$ ($\alpha > 1$ and, let us say, $r_0 = R$). The energy required to reach the altitude r_n from the altitude r_{n-1} is

$$E(r_{n-1} \rightarrow r_n) = \int_{r_{n-1}}^{r_n} \frac{k}{r}\, dr = k \ln\left(\frac{r_0 \alpha^n}{r_0 \alpha^{n-1}}\right) = k \ln \alpha,$$

and is independent of n. As n increases, no matter how many times the force decreases, the path length increases in the same ratio. It follows that the energy required to move from the height r_0 to the height r_N must be $E(r_0 \rightarrow r_N) = N \times E(r_0 \rightarrow r_1)$. It is apparent that any finite energy input will only be enough to reach a finite altitude.

When the planet has a finite length H, the field strength is proportional to $1/r$, if the positions considered are very far from the ends of the cylinder and $r \ll H$. But if r is comparable to H, then the nature of the gravitational field changes, and in the case of $r \gg H$ it has the usual r^{-2} dependence. From such a planet, a finite initial speed may be sufficient for an escape to be made.

> *Note.* With the help of the integral calculus, it can be shown that the second cosmic velocity (the escape velocity) is approximately $\sqrt{2\ln(H/R)}$ times larger than the first cosmic velocity. (For the Earth this ratio is $\sqrt{2}$.) This logarithmic factor is not too large even if $H \gg R$ (e.g. for $H = 10R$ the ratio $v_{c,2}/v_{c,1} \approx 2.1$ and for $H = 1000R$ it is still only 3.7).

S31   Denote the radius of the asteroid (which is equal to the tunnel's length) by R and its density by ϱ, the end-points of the tunnel by A and B, and the position of the mine captain by P (*see* Fig. 1). The tunnel and the radii corresponding to its end-points form an equilateral triangle, and so the distance of the tunnel's midpoint Q from the centre of the asteroid is $d = \sqrt{3}R/2$.

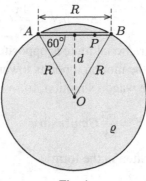

Fig. 1

Consider the situation when the wagon of mass m, which started from point A without any initial speed, has already covered a distance x in the tunnel, and its radial distance from the centre O of the asteroid is r (*see* Fig. 2). The gravitational force acting on it is

$$F = Gm\frac{4\pi}{3}r^3\varrho \cdot \frac{1}{r^2},$$

where we have used the well-known fact that, inside a homogeneous spherical shell, the gravitational field is zero, and so the effect of all titanium lying further from O than r can be ignored.

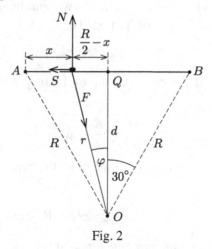

Fig. 2

The normal force exerted by the rails (perpendicular to the direction of motion) is

$$N = F\cos\varphi = F\frac{d}{r} = \frac{4\pi}{3}Gm\varrho d = \text{constant}.$$

As the wagon experiences no acceleration perpendicular to the rails, the kinetic frictional force (S) is also constant during the motion, with

$$S = \mu N = \frac{4\pi}{3} Gm\varrho\mu d.$$

The direction of the frictional force is opposite to that of the wagon's velocity, and so in the first part of the motion it points towards A, and in the second towards B.

The equation of the wagon's motion, from the start to the turning point, is

$$ma = F\sin\varphi - S = \frac{4\pi}{3}Gm\varrho(r\sin\varphi - \mu d) = \frac{4\pi}{3}Gm\varrho\left(\frac{R}{2} - x - \mu d\right),$$

which can also be written in the form

$$a = -\omega^2(x - x_0),$$

using the following notation:

$$x_0 = \frac{R}{2} - \mu d \qquad \text{and} \qquad \omega^2 = \frac{4\pi}{3}G\varrho.$$

This is an equation for simple harmonic motion around the midpoint x_0, and its period is

$$T = \frac{2\pi}{\omega} = \sqrt{\frac{3\pi}{G\varrho}}.$$

As the period T depends only on the known density of the material of the asteroid (titanium), it can be found numerically without having values for the radius of the asteroid or the coefficient of friction. The result is

$$T = 5597 \text{ s} \approx 1.55 \text{ h}.$$

The distance of the centre of the oscillatory motion from the midpoint of the tunnel is

$$\frac{R}{2} - x_0 = \mu d,$$

and so that of the reversing point P of the wagon, measured from its starting point, is

$$\overline{AP} = 2x_0 = R - 2\mu d, \tag{1}$$

whereas from the midpoint of the tunnel it is

$$\overline{PQ} = 2x_0 - \frac{R}{2} = \frac{R}{2} - 2\mu d.$$

To cover this phase of the motion takes a time $T/2$.

After the wagon turns back, the direction of the frictional force changes, so the centre of the oscillatory motion (the equilibrium position) will now be on the B

side of the midpoint Q of the tunnel, at a distance μd from it. As the wagon finally stops at the midpoint of the tunnel, it follows that

$$2\mu d = \overline{PQ} = \frac{R}{2} - 2\mu d,$$

that is

$$\mu = \frac{R}{8d} = \frac{2}{8\sqrt{3}} \approx 0.14,$$

the value that appeared in the expert's report. He was also able to calculate from (1) that the mine captain was standing at a distance of

$$\overline{AP} = 2x_0 = R - 2\mu d = R - \frac{R}{4} = \frac{3}{4}R$$

from the starting point A of the wagon, i.e. midway between Q and B.

The total time of the mine wagon's motion is equal to the sum of two half-periods, that is, $T = 1.55$ h.

S32 *Solution 1.* First of all, we have to find the gravitational fields in both the trial bore and the cavity.[23]

First, consider the trial bore, which is of negligible volume compared to that of the whole planet. Let the density of titanium (the density of the planet) be ϱ. The magnitude of the gravitational field at radius r is the same as it would be if only the sphere of radius r below it were present (and its mass were concentrated at its centre):

$$g(r) = G\frac{m(r)}{r^2} = G\frac{(4/3)\pi\varrho r^3}{r^2} = \frac{4\pi G\varrho}{3}r = Br,$$

where $B = (4\pi G\varrho)/3$ is a constant. Thus the gravitational field is directly proportional to the distance from the centre of the planet, and always points towards it.

Next, the gravitational field in the spherical cavity, which extends from the surface of the planet to its centre, has to be determined. This can be done using a 'cunning' application of the principle of superposition. Imagine the cavity to be filled with a mixture of 'normal' titanium and 'negative-density' titanium. The gravitational fields of the complete planet and the sphere of 'negative titanium' have to be added. In Fig. 1, r is the vector position, relative to O, of an arbitrary point P in the cavity, c is the position of O relative to the cavity's centre, and $r + c$ that of P relative to the latter.

---

[23] A calculation of the gravitational fields can be found in the predecessor of this book: see 'Problem 110' in P. Gnädig, G. Honyek & K. F. Riley, *200 Puzzling Physics Problems* (Cambridge University Press, 2001).

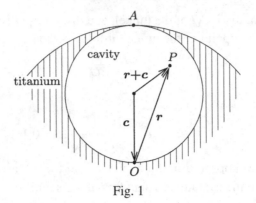

Fig. 1

From our earlier calculation, the gravitational field at a point inside a homogeneous spherical body is proportional to the radial distance of the point from the sphere's centre, the coefficient of proportionality being the constant previously denoted by $-B$. This can be applied to both the complete homogeneous planet and the cavity, to obtain the field acting when they are superposed. The 'lack of matter' in the cavity is represented by a negative density. The resultant gravitational field (*see* Fig. 2) is

$$\mathbf{g} = -B\mathbf{r} + B(\mathbf{r} + \mathbf{c}) = B\mathbf{c}.$$

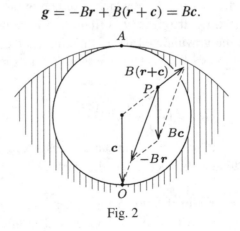

Fig. 2

This gravitational field is a constant (in both magnitude and direction), regardless of the position of point P inside the cavity. This means that there is a homogeneous gravitational field in the cavity, with a magnitude $B|\mathbf{c}| = BR/2$, which is the same as the gravitational acceleration in the middle of the trial bore (which coincides with the centre of the cavity). Now we can start the calculation of the pressure at point O.

a) The condition for equilibrium of an air layer of thickness Δr in the trial shaft at a radius of r is

$$\Delta p = -\varrho_{\text{air}}(r)g(r)\Delta r, \tag{1}$$

where

$$\varrho_{air}(r) = \frac{p(r)M}{R_0 T_0}.$$

In the above expression for the density, R_0 is the universal gas constant and M is the molar mass of air. Using the linear dependence of the gravitational field, it follows that

$$\Delta p = -\frac{p(r)M}{R_0 T_0} B r \Delta r,$$

that is

$$-\frac{\Delta p}{p} = \frac{MB}{R_0 T_0} r \Delta r.$$

If we now let the small changes become infinitesimal, we can integrate both sides of the equation:

$$-\int_{p_O}^{p_A} \frac{dp}{p} = \frac{MB}{R_0 T_0} \int_0^R r \, dr.$$

After integrating we have

$$\ln \frac{p_O}{p_A} = \frac{MB}{R_0 T_0} \frac{R^2}{2},$$

and finally from this

$$p_O = p_A \exp\left(\frac{MBR^2}{2R_0 T_0}\right) = p_A \exp\left(\frac{2\pi G\varrho MR^2}{3R_0 T_0}\right).$$

b) In the case of the spherical cavity, we can also use equation (1), with the same expression for the air density, but we have to remember that the gravitational field is homogeneous inside the cavity ($g = BR/2$):

$$\Delta p = -\varrho_{air}(r)g\Delta r = -\frac{p(r)M}{R_0 T_0}\frac{BR}{2}\Delta r.$$

We have to carry out the same manipulations, i.e. separate the variables,

$$-\frac{\Delta p}{p} = \frac{MBR}{2R_0 T_0}\Delta r,$$

and then integrate both sides of the equation:

$$\ln \frac{p_O'}{p_A'} = \frac{MBR^2}{2R_0 T_0}.$$

It will be seen that the ratio of the pressures at points O and A is the same in both cases. This is not just a coincidence, but a consequence of the fact that the value of the homogeneous gravitational field in the spherical cavity is equal to the average of that in the trial shaft, where the gravitational acceleration is directly proportional to the radius.

Solution 2. We can use the Boltzmann distribution in the comparison of the pressures at points O and A:

$$\frac{p_A}{p_O} = \exp\left(-\frac{\Delta E_{\text{pot}}}{R_0 T_0}\right),$$

where ΔE_{pot} is the gravitational potential energy difference of one mole of air between points A and O.

The gravitational field in the trial shaft is $\mathbf{g} = -B\mathbf{r}$, and so the gravitational potential is

$$U_{\text{grav}} = \frac{Br^2}{2} + \text{constant}.$$

The potential energy difference of one mole of gas between points A and O is therefore

$$\Delta E_{\text{pot}} = M\frac{BR^2}{2},$$

which can be inserted into the expression for the pressure ratio:

$$\frac{p_A}{p_O} = \exp\left(-\frac{MBR^2/2}{R_0 T_0}\right),$$

which is the same ratio as that obtained in Solution 1.

Note that, for the spherical cavity, the gravitational potential difference between points A and O is the same as it was for the thin trial shaft. It is as a result of this that

$$\frac{p_A}{p_O} = \frac{p'_A}{p'_O}.$$

To understand the equality of the potential differences in the two cases, from the point of view of superposition, we note that it can be considered as the sum of the contribution of the whole solid planet (as calculated above for the trial shaft) and the contribution of the half-size sphere with 'negative density' (which converts it to the case of the mined cavity). But the surface of a homogeneous sphere is an equipotential (even for a sphere with negative density!), and so it makes the same contribution to the potentials at both points, and therefore adds nothing to their difference.

S33   Harmonic motion occurs whenever the restoring force acting on a body is proportional to its displacement from equilibrium. This statement holds not only for one-dimensional motions but also for motions in a plane, provided the force does not depend upon the displacement's direction (isotropy). Such a situation is relevant to a conical pendulum undergoing small oscillations.

In a plane, superpositions of oscillations with identical frequencies generally produce elliptical trajectories, with the centre of attraction at the middle of the ellipse, in exactly the same way as the 'sun' is at the centre of each planet's orbit in Wonderland. If, there, the 'gravitational law' is that the attractive force experienced by a planet is proportional to its distance from the 'sun', then the trajectories of the planets would fit the orbits in Mr Tompkins dream.

In Wonderland, gravitational attraction is due to a central force (just as it is in our real world), and so the law of conservation of angular momentum is obeyed, even there, and its inevitable consequence is the 'sweeping of equal areas' theorem. This means that Kepler's second law is the same in Wonderland as it is in our world.

The period of harmonic oscillatory motion does not depend on the associated amplitude, and this is why the orbital periods of the planets in Wonderland are *independent* of the sizes of their orbits; hence, they are all equal. (This statement is only true if we assume that, in Wonderland, the equivalence of gravitational and inertial mass holds, and that, further, the gravitational force is directly proportional to the product of the two mutually attracting masses.)

> *Note.* It can be shown that there are only two central forces for which the trajectories of all finite motions (i.e. ones that are restricted to finite regions of space) are closed. These central fields are governed either by Newton's universal inverse square law or by a law parallelling the restoring force involved in spatial harmonic motion.

S34 Consider first the maximal impact speed with which a comet can slam into the Earth. The larger the velocity of the comet relative to the Sun, the larger is its total (gravitational potential plus kinetic) energy. For a body, bound to the Sun (and moving along an ellipse), this total energy must be negative, and cannot reach zero. So, for a comet approaching the orbit of the Earth, we must have

$$-G\frac{mM_S}{r} + \frac{1}{2}mv_0^2 < 0,$$

where m and M_S are the masses of the comet and the Sun, respectively, r is the radius of the Earth's orbit, and v_0 is the speed of the comet relative to the Sun. If the comet arrives from very far away (the major axis of its ellipse is very long), then the total energy can reach almost zero, and so

$$v_0^{\max} \approx \sqrt{\frac{2GM_S}{r}} = 42.1 \text{ km s}^{-1}.$$

The orbital speed of the Earth is just $1/\sqrt{2}$ times the value given above, i.e. 29.8 km s^{-1}. *Relative to Earth* the comet arrives in the vicinity of our planet with the maximal speed if its velocity is opposite to the orbital velocity of the Earth (and approaching it). In this situation, in the Earth's frame of reference, the impact speed of the comet can be as high as

$$v_{rel} = 42.1 \text{ km s}^{-1} + 29.8 \text{ km s}^{-1} = 71.9 \text{ km s}^{-1},$$

but no higher.

But, in addition, we have to take into consideration the fact that the gravitational attraction of the Earth, of mass M_E, further increases the impact velocity calculated above. The work done by the gravitational field (in attracting the comet of mass m from infinity to the surface of the Earth, which has radius R) increases the kinetic energy (and speed) of the comet:

$$0 - \left(-G\frac{mM_E}{R}\right) = \frac{1}{2}mv_{impact}^2 - \frac{1}{2}mv_{rel}^2,$$

from which the maximal impact speed is

$$v_{impact}^{max} = \sqrt{v_{rel}^2 + 2G\frac{M_E}{R}} \approx \sqrt{(71.9)^2 + (11.2)^2} \text{ km s}^{-1} = 72.8 \text{ km s}^{-1}.$$

> *Note.* An even larger value can be found by taking into consideration the fact that the speed of the comet is greater if the impact occurs when the Earth is near the Sun. Here the velocity of the comet relative to the Sun is 42.4 km s^{-1}, and the orbital velocity of the Earth is 30.3 km s^{-1}, so we get an enhanced result of 73.6 km s^{-1} for the maximal impact speed.

We now come to consider the smallest possible impact speed. The relative velocity of the comet and the Earth (before the considerable effect of the gravitational attraction of the Earth is taken into account) could be arbitrarily small. To bring this about, it is necessary 'simply' to align the speeds and trajectories of the comet and Earth as closely as possible.

If such a, hypothetical, comet approached the Earth slowly, then the gravitational attraction of the Earth would accelerate it, and the comet would explode on the Earth.[24] The impact speed can be calculated as before, but we must now insert zero for v_{rel}:

$$v_{impact}^{min} = \sqrt{2G\frac{M_E}{R}} \approx 11.2 \text{ km s}^{-1},$$

which is the well-known escape velocity for the Earth. So, any comet orbiting around the Sun that crashes into the Earth will do so with at least this speed.

---

[24] There are no such dangerous comets following the orbit of the Earth. If there were, we would surely have observed them a long time ago.

Note. The impact speeds determined above are calculated relative to the centre of the Earth, not to its surface. Taking account of the fact that, because of the rotation of the Earth, points on the Equator are moving with a speed of approximately 0.5 km s^{-1}, the maximal impact speed relative to the surface (of an atmosphere-free Earth) is 0.5 km s^{-1} larger than the value calculated earlier, and the minimal impact speed is the same amount less.

S35 As they are following parabolic trajectories, the comets must have initial total energies of zero. Because of the energy lost during the collision (as heat, etc.), the pieces of debris, all with the same initial speed, but many different directions, are going to be captured into a variety of elliptical orbits.

The length of the major axis of an elliptical orbit depends only on the orbiting body's specific energy (that is to say, on its energy per unit mass). As all the pieces of debris start with the same specific energy, the lengths of the major axes of their orbits have a common value, $2a$, say. Fig. 1 shows, as a heavy solid line, a possible orbit for one piece of debris; it passes through P and has S as one of its foci.

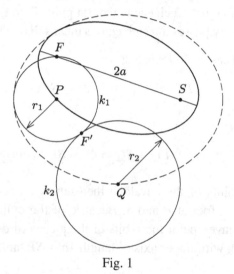

Fig. 1

Since the situation must have cylindrical symmetry about the line SP, it is sufficient to work in an arbitrary plane that contains it. As already noted, one focus is common to all the ellipses; this is the Sun S. The distance (but not the direction) of the other focus, as measured from P, is also fixed; it is

$$r_1 = 2a - SP,$$

because, for any point on an ellipse, the sum of the distances from the two foci must be equal to the length of the major axis. So, the other foci (not S) of the ellipses all lie on a circle with centre P and radius r_1. This circle is marked as k_1

in Fig. 1, and contains the particular point F that is appropriate to the (previously mentioned) sample trajectory.

Consider now an arbitrary point Q on the plane. If an elliptical trajectory is to go through Q, then its focus must be at a distance

$$r_2 = 2a - SQ$$

from Q. Draw a circle (k_2 in Fig. 1) with radius r_2 and centred on point Q. Three cases are now possible:

1. If circle k_2 crosses circle k_1, then point Q is on two different 'eligible' ellipses, and the two intersection points give the second foci of the two corresponding trajectories.
2. If circle k_1 and circle k_2 have no common point, then none of the possible orbits of the debris pieces can go through Q; point Q lies outside the required envelope.
3. In the limiting case that separates the first two possibilities, i.e. when circles k_1 and k_2 just touch each other (at point F' in Fig. 1), only a single elliptical trajectory passes through Q; this means that Q lies on the required envelope.

In the third case it follows that

$$PQ = r_1 + r_2 = (2a - SP) + (2a - SQ),$$

that is

$$PQ + SQ = 4a - SP = \text{constant},$$

and so the points of the envelope themselves lie on an ellipse (the dashed line in Fig. 1), with foci at P and S. Because of the cylindrical symmetry, the three-dimensional envelope of the orbits of the pieces of debris is a spheroid (ellipsoid of revolution), with major axis of length $4a - SP$, and foci at P and S (*see* Fig. 2).

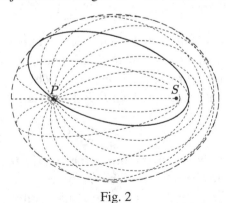

Fig. 2

S36 Denote by v_0 the speed of the space probe in the fixed reference frame of the Sun after it has 'left' the gravitational field of the Earth. (In practice, this means that the distance between the probe and Earth is much larger than the Earth's radius, but even this distance is negligible compared to the Sun–Earth distance.) Our task is to find the minimal value of v_0 that is sufficient for the probe to leave the Solar System.

> *Note.* When account is taken of the orbital speed $v_E \approx 30$ km s^{-1} of the Earth and of its rotational 'willingness to help', the space probe can be launched with a speed of
>
> $$v_L = \sqrt{(v_0 - 30 \text{ km s}^{-1})^2 + (11.2 \text{ km s}^{-1})^2} - 0.5 \text{ km s}^{-1},$$
>
> which is smaller than v_0 (*see*, for instance, the solution given on page 147). As v_0 is a monotonic function of v_L, instead of finding the minimal v_L, it is enough to find the minimally sufficient value of v_0.

Denote the orbital radius of the Earth by R ($R = 1$ AU), and that of the planet used for the gravitational slingshot by $R_p = xR$. With an appropriate launch date, the space capsule is going to reach some particular point on the planet's orbit at the same time as the planet does (*see* Fig. 1). The tangential and radial components of the probe's velocity at that point (v_1 and v_2, respectively) can be calculated using the conservation of angular momentum and energy (but without, for the moment, taking into consideration the effect of the planet):

$$mRv_0 = mxRv_1, \qquad \text{that is} \qquad v_1 = \frac{v_0}{x}, \tag{1}$$

and

$$\frac{1}{2}mv_0^2 - G\frac{Mm}{R} = \frac{1}{2}m(v_1^2 + v_2^2) - G\frac{Mm}{xR}. \tag{2}$$

Here, the mass of the probe is denoted by m, and that of the Sun by M.

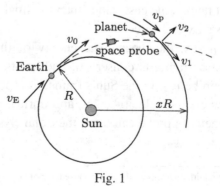

Fig. 1

It still remains to calculate the orbital speeds of the Earth and the planet. From the centripetal–gravitational force balance for a circular orbit, the speed of the Earth is given by

$$G\frac{M}{R^2} = \frac{v_E^2}{R}, \qquad \text{that is} \qquad v_E^2 = \frac{GM}{R}. \tag{3}$$

Similarly, the speed of the planet is found to be

$$v_p = \sqrt{\frac{GM}{xR}} = \frac{1}{\sqrt{x}} v_E. \tag{4}$$

From (1), (2) and (3) it follows that

$$v_2 = \sqrt{v_0^2 \left(1 - \frac{1}{x^2}\right) - 2v_E^2 \left(1 - \frac{1}{x}\right)}, \tag{5}$$

which is real if the expression in the square root is non-negative, i.e. the following inequality is obeyed:

$$\frac{v_0}{v_E} \geq g(x) \equiv \sqrt{\frac{2x}{1 + x}}. \tag{6}$$

If it is not, the space probe cannot reach the orbit of a planet x astronomical units from the Sun.

Let us imagine ourselves to be at rest in a coordinate frame moving with the planet. From here, the space probe will appear to have a velocity component directed away from the Sun of v_2, and one perpendicular to this (in the tangential direction) of $v_1 - v_p$. Thus, the relative velocity of the probe and the planet is

$$v_{rel} = \sqrt{(v_1 - v_p)^2 + v_2^2}. \tag{7}$$

The velocity of the space probe after its 'near miss' with the planet can – if the parameters of the collision[25] are adjusted carefully – be made to be in any arbitrary direction. In particular, it can be arranged that, after the 'fly-by', the velocities of the probe and planet are parallel (*see* Fig. 2). In terms of leaving the Solar System, this is the most favourable case, since then the final speed of the probe, relative to the Sun, has its maximal value.

During the fly-by, the mechanical energy of the probe, as measured in the planet's reference frame, is conserved, and so its speed as it leaves the planet behind will again be v_{rel}. In the Sun's frame, the speed of the probe is $v_{rel} + v_p$. If this exceeds the escape velocity $\sqrt{2}\, v_p$ appropriate to the particular orbital radius of the planet, then the probe can leave the Solar System. The condition for this is clearly

---

[25] For example, the minimum height of the probe above the planet's surface.

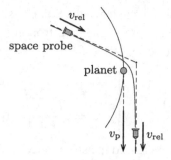

Fig. 2

$$v_{\rm rel} = \left(\sqrt{2} - 1\right)v_{\rm p}. \qquad (8)$$

Using equations (1), (4), (5), (7) and (8), we can obtain a quadratic equation for v_0, and its positive root is, for $x > 1$:

$$\frac{v_0}{v_{\rm E}} = \frac{1}{\sqrt{x^3}} + \sqrt{\frac{1}{x^3} + 2 - \frac{\sqrt{8}}{x}} \equiv f(x). \qquad (9)$$

We now have two conditions on the ratio $v_0/v_{\rm E}$: equation (6) to ensure the probe reaches the planet; and equation (9) to ensure that, having got there, it has acquired enough energy to complete the escape from the Solar System.

Now, although the two functions involved in these constraints both have asymptotic values of $\sqrt{2}$, they have different values at $x = 1$, viz. $f(1) = \sqrt{2}$, while $g(1) = 1$. Clearly, for x only a little above 1, $f(x) > g(x)$ and equation (9) is the limiting constraint. However, there is a minimum finite value X (> 1) for which $g(x) > f(x)$ for all $x > X$ (though not by much!). It can be found by setting $f(x) = g(x)$; some careful algebra then yields $X = 2 + \sqrt{8}$. For values of x greater than $2 + \sqrt{8}$, equation (6) becomes the limiting constraint. Fig. 3 illustrates these algebraic results.

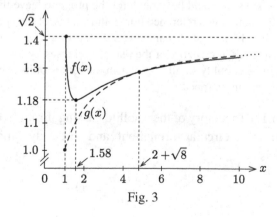

Fig. 3

Thus the space probe can leave the Solar System with the help of a gravitational slingshot from a planet positioned x astronomical units from the Sun if its initial launch speed is at least

$$\left(\frac{v_0}{v_E}\right)_{min} = \begin{cases} f(x) & \text{for } 1 \le x \le 2+\sqrt{8}, \\ g(x) = \sqrt{\dfrac{2x}{x+1}} & \text{for } x > 2+\sqrt{8}. \end{cases} \tag{10}$$

The orbital radius (expressed in astronomical units) of a planet producing the most efficient gravitational assist is given by the value of x that minimises expression (10).

If $x \approx 1$ or $x \gg 1$, then, according to (10), $v_0 \approx \sqrt{2}\, v_E$, i.e. the required launch speed is equal to that for a programme that does not include a gravitational slingshot. This is easily understood in qualitative terms, as follows. The Earth (or any other celestial body orbiting near the Earth) could not give the probe a significant helpful impulse, because their relative tangential velocity would be very small. A very distant celestial body (e.g. Pluto) could not be an efficient 'slingshot' centre because, if the probe had got that far 'under its own steam', it would already have nearly left the Solar System, and would hardly need any further assistance.

By solving $f'(x) = 0$, that is $x = \left(9 + \sqrt{81 - 24\sqrt{8}}\right)/8$, or by plotting expression (10), it can be shown that the minimal launch speed corresponds to a value of x of 1.58 (*see* Fig. 3). It is interesting to note that this value is very close to the orbital radius of Mars (1.52 AU), so the (almost) optimal 'slingshot manoeuvre' is realisable in practice.

> *Notes.* 1. Calculations using the orbital radius of Mars show that the minimal initial speed of the space probe, relative to the Sun, is $v_0 = 1.18 v_E \approx 35.2$ km s^{-1}. Without the slingshot effect, this speed would have to be $\sqrt{2}\, v_E \approx 42.1$ km s^{-1}. This difference is even more striking if the launch speeds v_l relative to the Earth's surface are compared. It can be shown that the minimal launch speed would have to be 16.2 km s^{-1} without the slingshot manoeuvre, but with its help only 11.9 km s^{-1} would be enough for the probe to leave the Solar System.
>
> 2. In the Sun's reference frame, the energy of the space probe is not conserved. During the slingshot manoeuvre, the space probe gains energy from the planet. As a result, the energy of the planet decreases by the same amount, thus causing an imperceptibly small *decrease* in the radius of its orbit, and an equally minimal *increase* in its speed.

S37 *a*) The trajectory of the satellite passing through the (thin) atmosphere can be assumed to be circular throughout, and so the dynamical condition governing the motion is

$$G\frac{mM}{r^2} = m\frac{v^2}{r}, \tag{1}$$

where m and M are the masses of the satellite and Earth, respectively, r is the distance of the satellite from the centre of the Earth, and v is the speed of the satellite. From this we get the expression for v as

$$v = \sqrt{\frac{GM}{r}},$$

which shows that, if the altitude (the orbital radius) of the satellite decreases, then its speed increases, i.e. the satellite *speeds up* as the result of any air drag. This surprising fact is usually known as the *astronautical paradox*.

The speed increase of the satellite may also be understood dynamically with the help of the (not-to-scale) figure. The drag force F_d is directly opposed to the velocity, and the gravitational force is directed towards the centre of the Earth. However, because of the (slow) decrease in the satellite's height, the latter is *not* perpendicular to the satellite's velocity. The non-zero tangential component of the gravitational force is in the same direction as the velocity. If (as will be shown in part *b*) to be the case) this component is larger than the drag force, then the satellite will speed up, rather than slow down.

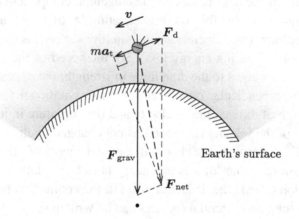

b) The (formal) power of the drag force is negative, $F_d \cdot v = -F_d v$, and this is to be equated with the rate of total energy loss of the satellite:

$$\frac{dE_{total}}{dt} = -F_d v. \tag{2}$$

The total energy is the sum of the kinetic energy and the gravitational potential energy:

$$E_{total} = E_{kin} + E_{pot} = \frac{1}{2}mv^2 + \left(-G\frac{mM}{r}\right).$$

As a result of equation (1), the connection between the kinetic and potential energy is always the following:

$$E_{\text{pot}} = -G\frac{mM}{r} = -mv^2 = -2E_{\text{kin}}, \tag{3}$$

and so we can write $E_{\text{total}} = -E_{\text{kin}}$. It follows that any *decrease* in the total energy of the satellite results in the same *increase* in its kinetic energy.

With the help of this equation, statement (2) can be transformed to

$$-\frac{dE_{\text{kin}}}{dt} = -F_{\text{d}}v,$$

and using the usual expression for the kinetic energy:

$$F_{\text{d}}v = \frac{d(\frac{1}{2}mv^2)}{dt} = mv\frac{dv}{dt}.$$

The expression dv/dt on the right-hand side is just the tangential acceleration a_{t} of the satellite (along the trajectory), so

$$F_{\text{d}} = ma_{\text{t}} \qquad \text{or using vector notation} \qquad -\boldsymbol{F}_{\text{d}} = m\boldsymbol{a}_{\text{t}}. \tag{4}$$

In accordance with Newton's second law, the expression ma_{t} must be equal to the tangential component of the net force $\boldsymbol{F}_{\text{net}}$ exerted on the satellite. Here, this is the sum of the drag force and the tangential component of the gravitational force (*see* figure). It then follows from equation (4) that the tangential component of the gravitational force (increasing the satellite's speed) is exactly twice as large as the drag force (causing energy loss). So, the speed of the satellite is increased by a force that is equal to the drag force in strength, but *oppositely* directed.

The perpendicular component of the gravitational force causes the change in direction of the satellite's velocity and the curvature in its trajectory, but the magnitude of the velocity (the speed) is not changed by this.

c) Consider again the change in the total energy of the satellite, but now expressing it in terms not of kinetic energy but of gravitational potential energy. Using equation (3) and the deduction from it, the connection between the small changes in the total and potential energies can be written as

$$\Delta E_{\text{total}} = \tfrac{1}{2}\Delta E_{\text{pot}}. \tag{5}$$

The change of the potential energy can be expressed in terms of the small change of the orbital radius Δr as follows:

$$\Delta E_{\text{pot}} = \Delta\left(-G\frac{mM}{r}\right) = G\frac{mM}{r^2}\Delta r. \tag{6}$$

We know that, during a single revolution (lasting $2\pi r/v$), the distance of the satellite from the centre of the Earth is decreased by $\varepsilon = 100$ m, so the rate of change of the orbital radius of the satellite is

$$\frac{\Delta r}{\Delta t} = -\frac{\varepsilon}{2\pi r/v}. \tag{7}$$

Using equations (2), (5), (6) and (7), the drag force acting on the satellite can be expressed as

$$F_d = \frac{1}{4\pi} G \frac{mM}{r^3} \varepsilon.$$

Now using equation (1), and the fact that the drag force has the form $F_d = c\varrho v^2$, the atmospheric density at the altitude of the satellite is found to be

$$\varrho = \frac{1}{4\pi c} \frac{m}{r^2} \varepsilon.$$

Inserting the given data for ε, c and m, and the value of the orbital radius $r = 6370 \text{ km} + 200 \text{ km} = 6570 \text{ km}$ gives

$$\varrho = 4.0 \times 10^{-10} \text{ kg m}^{-3}.$$

Notes. 1. This very low density of the atmosphere at these altitudes (in the so-called thermosphere, stretching from 85 km to 500 km) is not at all unusual. The density can be characterised by the so-called *mean free path*, which describes the mean distance covered by the particles of the air between successive collisions. The value of this at room temperature and at normal atmospheric pressure near the surface of the Earth is 70 nm, whereas at an altitude of 200 km the mean free path is greater than 200 m.

2. The International Space Station (ISS) also orbits in the thermosphere, at altitudes of 330–420 km. Because of the air drag, the ISS is losing energy continuously; that is why from time to time it is necessary to 'boost' the ISS into a higher orbit (with the help of space craft launched from Earth).

3. In the first decades of space flights, the density of air in the thermosphere was measured by the orbital altitude decrease of satellites, in just the way described in this problem. In this region, the local air density changes with time.

S38 Denote the mass of the Earth by M, its radius by R and its rotational angular velocity by Ω; denote the corresponding quantities for the Moon by m, r and ω, respectively. If the Moon–Earth distance is L, then the centre of mass of the system is a distance $mL/(m+M)$ from the centre of the Earth, and a distance $ML/(m+M)$ from the centre of the Moon. Both of the celestial bodies orbit the common centre of mass, with an angular velocity ω equal to that of the Moon's rotation around its own axis.

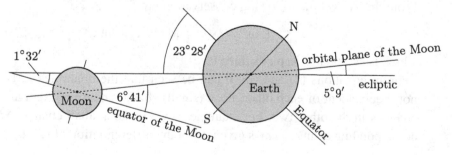

For the sake of simplicity, in the following estimations, we neglect the angles of inclination of the plane of the Earth's Equator, the orbital plane of the Moon, and the plane of the Moon's equator, and proceed as though all of these motions take place in the same plane. The actual values of these angles of inclinations relative to each other can be seen in the (not-to-scale) figure.

a) If we neglect angular-momentum effects due to the Sun, the Earth–Moon system can be considered as a closed system in which the law of conservation of angular momentum is obeyed. The total angular momentum of the system consists of the spin angular momenta and the orbital angular momenta of the celestial bodies:

$$J_{\text{total}} = J_{\text{spin}}^{\text{Earth}} + J_{\text{orbital}}^{\text{Earth}} + J_{\text{spin}}^{\text{Moon}} + J_{\text{orbital}}^{\text{Moon}} = \text{constant}.$$

Denoting the moments of inertia (or rotational inertia) of the Earth by I^{Earth} and of the Moon by I^{Moon} (both of them relative to axes passing through their own centres of mass), the total angular momentum of the system is

$$J_{\text{total}} = I^{\text{Earth}}\Omega + M\left(\frac{m}{m+M}L\right)^2 \omega + I^{\text{Moon}}\omega + m\left(\frac{M}{m+M}L\right)^2 \omega.$$

At this stage it is worthwhile to compare the orders of magnitudes of these terms. Inserting the given data, we find that the orders of magnitude of the first and last terms are similar, but that the values of the second and the third terms are *several orders of magnitude* smaller than either. So, henceforth, the orbital angular momentum of the Earth and the spin angular momentum of the Moon will be justifiably neglected, with the total angular momentum of the system being approximated as the sum of the spin of the Earth and the orbital angular momentum of the Moon:

$$J_{\text{total}} \approx I^{\text{Earth}}\Omega + mL^2\omega = \text{constant}. \tag{1}$$

We are told that the trajectory of the Moon around the centre of mass of the system can be considered as a circle (to a good approximation), and so, for the Moon, Newton's equation of motion has this form:

$$G\frac{mM}{L^2} = m\frac{M}{m+M}L\omega^2.$$

From this we get the following conservation law:

$$L^3\omega^2 = G(m+M) = \text{constant}, \tag{2}$$

which is essentially Kepler's third law.

From equations (1) and (2), it can be seen that the quantities L, Ω and ω are not independent of each other, i.e. a (small) change in any of them also produces changes in the other two. For example, if we make a small change ΔL in L, the corresponding change in ω is given, from the differentiation of (2), by

$$3L^2\omega^2\Delta L + 2L^3\omega\Delta\omega = 0 \qquad \text{yielding} \qquad \Delta\omega = -\frac{3\omega}{2L}\Delta L. \qquad (3)$$

Similarly, equation (1) yields

$$\Delta\Omega = -\frac{mL\omega}{2I^{\text{Earth}}}\Delta L. \qquad (4)$$

It can be seen that an increase of the Earth–Moon distance causes slowing down of the orbital motion of the Moon and a decrease of the Earth's rotation speed. These two equations are enough to compare the changes in the kinetic energies of the Earth and the Moon. Using astronomical data, it can be shown that not only the angular momentum of the Earth but also its kinetic energy come mainly from the Earth's rotation around its axis, and that the Earth's translational kinetic energy relative to the common centre of mass of the Earth–Moon system is negligible. For the Moon the situation is just the reverse: the Moon's kinetic energy comes mostly from its orbital motion, and its rotational kinetic energy is not large. Accordingly,

$$E_{\text{kinetic}}^{\text{Earth}} \approx \tfrac{1}{2}I^{\text{Earth}}\Omega^2, \qquad E_{\text{kinetic}}^{\text{Moon}} \approx \tfrac{1}{2}mL^2\omega^2.$$

If the rotational angular velocity of the Earth is changed by a small amount $\Delta\Omega$, the Moon's orbital angular velocity by $\Delta\omega$, and the Earth–Moon distance by ΔL, then the changes of the kinetic energies are

$$\Delta E_{\text{kinetic}}^{\text{Earth}} = I^{\text{Earth}}\Omega\Delta\Omega,$$
$$\Delta E_{\text{kinetic}}^{\text{Moon}} = mL\omega^2\Delta L + mL^2\omega\Delta\omega.$$

These equations can be derived in the same way as (3) and (4) were. Now, using (3) and (4), the energy changes can be rewritten in the form

$$\Delta E_{\text{kinetic}}^{\text{Earth}} = -\tfrac{1}{2}mL\omega\Omega\Delta L, \qquad \Delta E_{\text{kinetic}}^{\text{Moon}} = -\tfrac{1}{2}mL\omega^2\Delta L.$$

To answer question *a*), we note that, since $\dot{E} = dE/dt$, the ratio of the rates of change of the kinetic energies is equal to the ratio of the changes of kinetic energies. This gives a very simple and perhaps surprising result:

$$\frac{\dot{E}_{\text{kinetic}}^{\text{Earth}}}{\dot{E}_{\text{kinetic}}^{\text{Moon}}} = \frac{\Delta E_{\text{kinetic}}^{\text{Earth}}}{\Delta E_{\text{kinetic}}^{\text{Moon}}} = \frac{\Omega}{\omega} = \frac{27.3 \text{ day}}{1 \text{ day}} = 27.3.$$

It can be seen that the tidal forces are decreasing the kinetic energy of the Earth much more rapidly, in absolute terms, than that of the Moon. Another curiosity is that our answer does not depend upon the Earth's moment of inertia, and so any preliminary assumption about the mass distribution within the Earth is not needed.

> *Note.* The total mechanical energy of the Earth–Moon system, i.e. the sum of the kinetic and gravitational potential energies, does not remain constant with time; because of dissipative internal friction in the rocks and oceans of the two bodies, it slowly decreases.

b) The change in the rotational period of the Earth can be related to the change in the angular velocity:

$$\Delta T^{\text{Earth}} = \frac{2\pi}{\Omega + \Delta\Omega} - \frac{2\pi}{\Omega} \approx -\frac{2\pi}{\Omega^2}\Delta\Omega.$$

Using equation (4), the change in the length of an Earth day is related to ΔL by

$$\Delta T^{\text{Earth}} = \frac{\pi m L \omega}{I^{\text{Earth}}\Omega^2}\Delta L.$$

Estimating the moment of inertia of the Earth using the formula $\frac{2}{5}MR^2$ for a sphere with homogeneous mass distribution, we get a time increase of $\Delta T^{\text{Earth}} \approx 18$ μs per year. In reality, the Earth's crust has a smaller density than the Earth's core, and so its moment of inertia is smaller than the 'homogeneous sphere' estimate. Using, instead, the experimental value of the Earth's moment of inertia of 8×10^{37} kg m^2 (*see* Appendix), the day length increase is a little bit larger, namely 21 μs per year. According to measurements made using the most accurate atomic clocks, the yearly length increase in the Earth's day is 15 μs, which is reassuringly close to the value estimated (considering the approximations applied).

c) At the end of the synchronisation, Ω and ω have a common value, say ω_0, and so the conservation of angular momentum in relations (1) can be written in the form

$$I^{\text{Earth}}\Omega + mL^2\omega = I^{\text{Earth}}\omega_0 + mL_0^2\omega_0, \tag{5}$$

where L_0 is the stabilised value of the Earth–Moon separation. From Kepler's third law, equation (2), we also have

$$L^3\omega^2 = L_0^3\omega_0^2. \tag{6}$$

Using (5) and (6) to eliminate L_0, we get the following equation for ω_0:

$$I^{\text{Earth}}\Omega + mL^2\omega = I^{\text{Earth}}\omega_0 + mL^2\frac{\omega^{4/3}}{\omega_0^{1/3}}. \tag{7}$$

This is a quartic (fourth-degree) equation for the unknown angular velocity ω_0 in terms of the current parameters of the Earth–Moon system. An accurate value for ω_0 can be determined numerically, but a good approximation to the accurate result can be found by noting that the first term on the right-hand side can be neglected compared to the second term. Whether or not this neglect is justified will have to be the subject of a final check.

This approximation gives the following expression for the angular momentum of the synchronised Earth–Moon system:

$$\omega_0 \approx \omega^4\left(\frac{I^{\text{Earth}}\Omega}{mL^2} + \omega\right)^{-3}.$$

Inserting known data (including the measured value for the Earth's moment of inertia), we conclude that the synchronised angular velocity is $\omega_0 \approx 1.53 \times 10^{-6}$ s^{-1}. Using this approximate value makes the second term on the right-hand side of equation (7) about 280 times larger than the first term; so neglect of the latter seems well-founded.

At the end of the synchronisation, the length of the *new* Earth day will be

$$T_0^{\text{Earth}} = \frac{2\pi}{\omega_0} \approx 47 \text{ (old) days,}$$

which is more than a month and a half. The ratio of the future Earth–Moon distance to the present one can be found using (6):

$$\frac{L_0}{L} = \left(\frac{\omega}{\omega_0}\right)^{2/3} \approx 1.45.$$

Notes. 1. The dynamical explanation of the Earth's slowing rotation and of the Earth–Moon synchronisation is as follows. The Moon exerts a larger gravitational force on the points of the Earth that are closer to it than on the ones which are more remote. As a result, the Earth 'expands' (just to a very minor extent), and it is transformed into an ellipsoid of revolution (*see* figure). But because of the relatively fast rotation of the Earth around its own axis and because of the internal friction in the oceans and rocks on Earth, the major axis of the ellipsoid does not point directly at the Moon. At any instant, it has already passed through that direction – the major axis is 'in a hurry' compared to the line that joins the Earth to the Moon. The Moon's gravitational field exerts a torque on these 'bulges', and that torque acts in a sense that slows, rather than speeds up, the Earth's rotation.

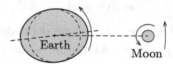

2. There are a few examples of celestial body pairs in the Solar System in which perfect synchrony has already been established. One of the most widely known examples is the dwarf planet Pluto and its largest moon called Charon.

S39 *Solution 1.* The ISS orbits the Earth along a circular trajectory of radius R, with angular velocity ω_0, and an orbital period T of about 92 min. The radius $R \approx 6700$ km is only a little greater than the radius of the Earth, for which the first cosmic velocity[26] is $R\omega_0 \approx 7.7$ km s^{-1}.

We describe the astronaut's motion using the distance $r(t)$ measured from the centre of the Earth, which has mass M, and the rotational angle $\varphi(t)$. In this polar coordinate system (with all the motions confined to the plane $\theta = \pi/2$), the radial equation of motion is

---

[26] See part *a*) of the problem on page 9.

$$\ddot{r} = -\frac{GM}{r^2} + r\omega^2, \tag{1}$$

while the law of conservation of angular momentum can be written in the form

$$r^2\omega = R^2\omega_0, \tag{2}$$

where $\omega(t) = \dot{\varphi}(t)$. Here, use has been made of the fact that, because of the direction of his jump, the astronaut's angular momentum was not changed by it.

Equation (1) is also valid for the ISS, but, for it, $r(t) \equiv R$ and $\omega(t) \equiv \omega_0$, from which it follows that

$$\omega_0 = \sqrt{\frac{GM}{R^3}}.$$

The initial conditions for the astronaut's motion are

$$r(0) = R, \qquad \dot{r}(0) = v_0, \qquad \varphi(0) = 0, \qquad \dot{\varphi}(0) = \omega(0) = \omega_0. \tag{3}$$

We also note that $v_0 \ll R\omega_0$, and this is why[27] the motions of the astronaut and the ISS, which he is shoving off from (and showing off to), deviate only a little from each other. The 'smallness' of the deviation can be characterised by the dimensionless number

$$\epsilon = \frac{v_0}{R\omega_0} \approx 1.3 \times 10^{-5}.$$

Henceforth, the order of magnitude of dimensionless quantities will be expressed in terms of ϵ and its powers.

The coordinates characterising the astronaut's motion can be written in the form

$$r = R(1 + \alpha) \qquad \text{and} \qquad \omega = \omega_0(1 + \beta), \tag{4}$$

where $\alpha(t)$ and $\beta(t)$ are small, time-dependent, dimensionless quantities. Expressing the initial conditions in terms of them:

$$\alpha(0) = 0, \qquad \dot{\alpha}(0) = \frac{v_0}{R} = \epsilon\omega_0, \qquad \beta(0) = 0. \tag{5}$$

Inserting the expressions from (4) into equations (1) and (2), and retaining only zeroth- and first-order terms in α or β, yields

$$\ddot{\alpha} = \omega_0^2(3\alpha + 2\beta), \tag{6}$$

$$2\alpha + \beta = 0. \tag{7}$$

Eliminating β from (6) and (7) shows that α satisfies the familiar equation for simple harmonic motion:

$$\ddot{\alpha} = -\omega_0^2\alpha.$$

---

[27] Leastways, the adventurous astronaut desperately hopes it is.

Its solution satisfying the initial conditions (5) is

$$\alpha(t) = \epsilon \sin \omega_0 t, \qquad \text{that is} \qquad r(t) = R + R\epsilon \sin \omega_0 t. \tag{8}$$

It follows that, after time $T/4 = \pi/(2\omega_0) \approx 23$ min, the astronaut has moved radially away from the trajectory of the ISS by $R\epsilon \approx 87$ m.

However, this is *not* his maximal distance from the ISS, because, from (4) and (7), we also have

$$\dot{\varphi}(t) = \omega(t) = \omega_0(1 - 2\alpha) = \omega_0(1 - 2\epsilon \sin \omega_0 t),$$

which can be integrated, with the initial condition $\varphi(0) = 0$, to yield

$$\varphi(t) = \omega_0 t + 2\epsilon(\cos \omega_0 t - 1). \tag{9}$$

Now at a time $T/2 = \pi/\omega_0 \approx 46$ min after his jump, equation (8) shows that the astronaut has returned to his initial orbital radius. But, in contrast, (9) shows that at that same time he lags behind the ISS in orbital angle by 4ϵ corresponding to a separation from it of $4R\epsilon \approx 350$ m.

Considering equations (8) and (9) together shows that, geometrically relative to the ISS, the astronaut moves in an ellipse with minor axis $2R\epsilon$ and major axis $4R\epsilon$.

As the motions of the astronaut and the ISS are both periodic with the same frequency, after one period, of $T = 2\pi/\omega_0 = 92$ min, they will again be at the same position. This means that the astronaut can survive his stunt, and rejoin the ISS still alive, if he has got enough oxygen to last 92 min.

Solution 2. The problem can also be solved using Kepler's laws of planetary motion. After his jump, the astronaut of mass m has a total energy of

$$E = -\frac{GMm}{R} + \frac{m}{2}(R^2\omega_0^2 + v_0^2) = -\frac{GMm}{2R}(1 - \epsilon^2),$$

where $\epsilon = v_0/(R\omega_0)$ is the small, dimensionless constant used in Solution 1 (*see* above), and his angular momentum is equal to

$$J = mR^2\omega_0,$$

the same value as before his jump. His trajectory is – according to Kepler's first law – an ellipse that has one of its foci at the centre of the Earth F (*see* Fig. 1, which is grossly not to scale).

We now apply the conservation of energy and angular momentum to the astronaut when he is at each of the ends of the major axis of the ellipse. At these points (which, in the usual notation, are at distances of $a + c$ and $a - c$ from the focus), denote his velocities by v_1 and v_2, respectively:

$$-\frac{GMm}{a+c} + \frac{m}{2}v_1^2 = -\frac{GMm}{2R}(1 - \epsilon^2),$$
$$m(a+c)v_1 = mR^2\omega_0,$$

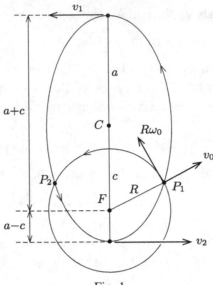

Fig. 1

which, on eliminating v_1 and using $\omega_0 = \sqrt{GM/R^3}$, becomes

$$-\frac{R}{a+c} + \frac{1}{2}\frac{R^2}{(a+c)^2} + \frac{1-\epsilon^2}{2} = 0.$$

Similarly, from the other end of the major axis, we have

$$-\frac{R}{a-c} + \frac{1}{2}\frac{R^2}{(a-c)^2} + \frac{1-\epsilon^2}{2} = 0.$$

These are quadratic equations for $x = R/(a \pm c)$, i.e.

$$-x + \frac{x^2}{2} + \frac{1-\epsilon^2}{2} = 0,$$

and they have two roots, $1 \pm \epsilon$, so

$$x_1 = 1 - \epsilon = \frac{R}{a+c} \qquad \text{and} \qquad x_2 = 1 + \epsilon = \frac{R}{a-c}.$$

It follows directly from these that the semi-major axis of the ellipse is

$$a = \frac{R}{1-\epsilon^2} \approx R,$$

the distance between the focus and the centre of the ellipse is

$$c = \frac{R\epsilon}{1-\epsilon^2} \approx R\epsilon,$$

and the semi-minor axis of the ellipse is

$$b = \sqrt{a^2 - c^2} = \frac{R}{\sqrt{1 - \epsilon^2}} \approx R.$$

It also follows that, since $c/a = \epsilon$, the graphical interpretation of the dimensionless parameter used in our analysis is that it is just the *eccentricity* of the ellipse, which describes how 'different from a circle' it is.

Since $\epsilon \ll 1$, the major axis of the astronaut's orbit is approximately equal to that of the ISS (the diameter of its circular trajectory). This is why – in accordance with Kepler's third law – their orbital periods are very nearly the same, namely about 92 min. After this interval they will again be at the same position, i.e. they will meet. If the astronaut's oxygen supply lasts for at least this length of time, then he can rejoin the ISS safely.

It can be seen from the parameters of the ellipse that the astronaut's maximal radial deviation from the circular orbit of the ISS is $a + c - R \approx R\epsilon \approx 87$ m. However, the distance between the astronaut and the ISS can be larger than this, because the two orbiting bodies are separating tangentially as well. In Fig. 2 the circular orbit of the ISS (k_1) and the elliptical orbit (k_2) of the astronaut moving away from the ISS are marked. The ellipse can be considered a circle to a very good approximation, and its centre C is ϵR away from the centre F of the circular orbit of the ISS, i.e. from the focus of the ellipse. (In Fig. 2, the scale of the ellipse is accurate, but the distance $CF = \epsilon R$ is shown greatly magnified.)

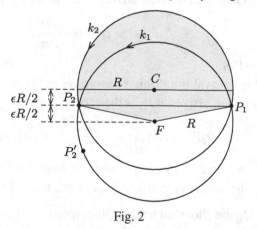

Fig. 2

The astronaut jumps away from the ISS at the point P_1. Their orbits will cross each other at point P_2 after a half 'revolution', but the two bodies do not meet, because the astronaut, moving along the ellipse arc k_2, arrives at P_2 later than the ISS, moving along the circle k_1. The time lag (and from this the distance between the two bodies) can be calculated with the help of Kepler's third law. The position vectors of both bodies sweep out equal areas during equal intervals of time, because their orbital periods and the surface areas of their orbits are the same (to a high degree of accuracy). This areal velocity is

$$\dot{A} = \frac{R^2\pi}{T} = \frac{R^2}{2}\omega_0.$$

To first order in ϵ the area swept out by the position vector of the astronaut moving along the arc k_2 (the shaded area in Fig. 2) consists of a semicircle, a 'rectangle' and a triangle:

$$A_1 = \frac{R^2\pi}{2} + \frac{\epsilon R}{2}2R + \frac{1}{2}\frac{\epsilon R}{2}2R,$$

and the time for this motion is

$$t_1 = \frac{A_1}{\dot{A}} = \frac{1}{2}T + \frac{3\epsilon}{2\pi}T.$$

Similarly, the orbiting ISS sweeps out the area

$$A_2 = \frac{R^2\pi}{2} - \frac{\epsilon R^2}{2},$$

and, to do so, it needs a time of

$$t_2 = \frac{A_2}{\dot{A}} = \frac{1}{2}T - \frac{\epsilon}{2\pi}T.$$

It can be seen that the astronaut arrives at the point P_2 a time interval of

$$\Delta t = t_1 - t_2 = \frac{2\epsilon}{\pi}T$$

later than the ISS.

During this interval the ISS has moved on to the point P_2'. Their distance apart when the astronaut arrives at P_2 is therefore

$$P_2P_2' = R\omega_0\Delta t = R\frac{2\pi}{T}\frac{2\epsilon}{\pi}T = 4\epsilon R \approx 350 \text{ m}.$$

It can be shown by a further application of Kepler's laws that this distance is the furthest that the astronaut ever gets away from the ISS before rejoining it.

S40   *a)* On the 'floor' of a space ship, rotating with angular velocity ω, a force of $mR\omega^2$ needs to be exerted on a body of mass m for it to remain at rest relative to the space ship. So, in the rotating reference frame, each body 'feels' an apparent gravitational field with 'gravitational acceleration' $g = R\omega^2$. This is equal to the given value of g if the angular velocity is

$$\omega = \sqrt{\frac{g}{R}} = 1 \text{ rad s}^{-1}.$$

The astronaut, slowly climbing up through a distance of $R/2$, must exert a force on the pole that decreases linearly with height, and so the work he does can be calculated using the average force:

$$W = \frac{1}{2}\left(mg + \frac{mg}{2}\right)\frac{R}{2} = \frac{3}{8}mgR = 3 \text{ kJ}.$$

His kinetic energy changes at the same time, since his (tangential) initial speed of $R\omega$ decreases to one-half of that value. The corresponding kinetic energy change is

$$\Delta E_{\text{kin}} = \frac{1}{2}m\left[\left(\frac{R\omega}{2}\right)^2 - (R\omega)^2\right] = -\frac{3}{8}mR^2\omega^2 = -\frac{3}{8}mgR = -3 \text{ kJ}.$$

This *does not* tally with the work done during the astronaut's climb because, although their absolute values are the same, their signs are not. He did some work, but gained no energy – in fact he lost some! Whatever became of the 'missing' 6 kJ?

The answer is that we need to recognise that it is the astronaut and the space ship that *together* form a closed system. The mass of the space ship is much larger than that of the astronaut, and so during his climb any change in the angular velocity of the space ship is barely noticeable. Indeed, in the above calculation, the astronaut's final tangential speed and the effective gravity were both calculated using the initial angular velocity ω, and any variation was tacitly ignored.

However, if we do take into account the slight change $\Delta\omega$ in the space ship's angular velocity, the change in its rotational kinetic energy can be written as

$$\Delta E_{\text{rot}} = \frac{1}{2}I[(\omega + \Delta\omega)^2 - \omega^2] \approx I\omega\,\Delta\omega,$$

where I is its moment of inertia.

The magnitude of $\Delta\omega$ can be calculated using the conservation of angular momentum:

$$I\omega + mR(R\omega) = I(\omega + \Delta\omega) + m\frac{R}{2}\frac{R\omega}{2}.$$

On the right-hand side of the above equation, a (second-order) small term, involving the product of m and $\Delta\omega$, has been neglected, since $I \gg mR^2$. It follows that

$$\Delta\omega = \frac{3mR^2\omega}{4I},$$

and that the change in the space ship's rotational kinetic energy is

$$\Delta E_{\text{rot}} = I\omega\,\Delta\omega = \frac{3}{4}mR^2\omega^2 = \frac{3}{4}mgR = +6 \text{ kJ}.$$

We are going to be all right! This is just the answer we needed for the work–kinetic energy principle to be obeyed, even in space:

$$W = \Delta E_{\text{kin}} + \Delta E_{\text{rot}}.$$

b) Measured in an inertial frame of reference moving with the centre of mass of the space ship, the velocity of the astronaut has magnitude $v_A = R\omega/2$ when he is at the top of the climbing pole (at point A in the figure).

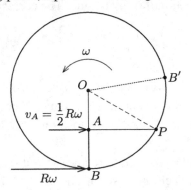

When he falls from the pole, his velocity remains constant, and so his (point-like) body undergoes uniform rectilinear motion, and he finally reaches point P on the gym floor. The time required for this motion (his 'free fall', relative to the gym) is

$$t_{AP} = \frac{\overline{AP}}{v_A} = \frac{\sqrt{3}}{\omega} \approx 1.7 \text{ s}.$$

During his 'free fall', the base B of the climbing pole travels to the point B', a 'distance', measured along the floor, of

$$s_{BB'} = R\omega t_{AP} = \sqrt{3}R \approx 17.3 \text{ m}.$$

However, the astronaut has only reached P, which is

$$s_{BP} = R\cos^{-1}\frac{R/2}{R} = R\frac{\pi}{6} \approx 10.5 \text{ m}$$

from B.

The other astronauts in the gym (observers rotating with the space ship) see that, when the astronaut falls from the top of the pole, he does not hit the floor at the bottom of it, but at a spot

$$\Delta s = s_{PB'} = s_{BB'} - s_{BP} \approx 6.8 \text{ m}$$

away, in the 'backward' direction relative to the direction of rotation. The observing astronauts, all having studied at least some college-level physics, and sitting at rest in the gym, ascribe this 'short-fall' to the effect of *Coriolis force*. The astronaut's 'free fall' relative to the space ship (i.e. using a rotating frame of reference) can be calculated only in a rather sophisticated way, while the trajectory and the time dependence of the motion are very simply found in an inertial reference frame.

S41 At first sight, we might think that, in energy terms, only a simple single change occurs during the boy's escapade. The boy is raised $h = 20$ m in the Earth's gravitational field; the escalator is in exactly the same state as it was before the run. The change in the boy's gravitational potential energy is $\Delta E_{\text{grav}} = mgh = 9.8$ kJ; the work he has done must be equal to this.

But this argument is false! As the boy runs up the steps, he pushes down on them with an average force of mg. If the steps were at rest, the normal force acting on them would do no work, but, as the steps (as part of the escalator) are moving, there is a displacement in the direction of the force. Consequently, the boy is doing positive work on the steps – and 'helping' the escalator to operate.

The downward velocity component of the steps is exactly twice as large as the vertical component of the boy's (net) upward speed. As the force involved has magnitude mg in both cases, the work done on the escalator requires twice the power needed for that done against gravity. So, the work he does on the escalator is $W_{\text{esc}} = 2mgh$, and the total work done is

$$W_{\text{boy}} = W_{\text{esc}} + \Delta E_{\text{grav}} = 3mgh = 29.4 \text{ kJ}.$$

In summary, one-third of the total work done increases the gravitational potential energy of the boy's body, and two-thirds of it helps to reduce the electricity bill for the escalator.

> *Note.* The correct answer can also be found by using a reference frame fixed to the escalator. In it, the steps are at rest, but the top of the escalator rises with some constant speed, while the mischievous boy races after it with a one-and-a-half times greater speed. In this frame of reference, the boy must climb to triple the actual height h of the escalator, and so the work he has to do is $3mgh$.

S42 *a)* If the mass ratio M/m is sufficiently large, then it can always be arranged, using fine tuning of the impact parameters (and some good luck!), that D will bounce from each of the discs of mass $\mu = M/N$ in turn. To a good approximation, its direction is monotonically changed by an angle of π/N at each bounce. However, if the common mass μ of the discs initially at rest is smaller than a certain critical value, then it is not possible, even with ideal impact parameters, to make D turn through an angle of π/N at each collision.

We first investigate the maximal angle through which a disc of mass m can be scattered by another with mass μ and initially at rest. Denote by v_0 the initial velocity of the incoming disc in a reference frame $\mathcal{K}$ fixed to the air-hockey table (*see* Fig. 1).

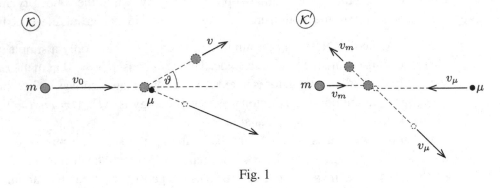

Fig. 1

It is convenient also to describe the collision in the centre-of-mass reference frame $\mathcal{K}'$ of the two discs; it moves with a velocity of $mv_0/(m + \mu)$ relative to the table. In this frame, the initial velocities of the discs are

$$v_m = \frac{\mu}{m + \mu} v_0 \quad \text{and} \quad v_\mu = \frac{m}{m + \mu} v_0. \tag{1}$$

In the coordinate system $\mathcal{K}'$ the total linear momentum of the two discs is zero, so they are always moving in directions directly opposed to each other, but with the same magnitudes of linear momentum. Because the collisions are elastic, the law of conservation of mechanical energies requires that the magnitude of the momentum of each disc (and so also its speed) remains unaltered during the collision, and only its direction can change (*see* Fig. 1).

We can transform back to the reference frame $\mathcal{K}$ fixed to the table if we add the relative velocity of the centre-of-mass system to the velocity vectors in frame $\mathcal{K}'$. After the collision, the velocity vector of D is directed to some point on the circle shown in Fig. 2. This vector makes the largest possible angle with the initial velocity direction if it is a tangent to the circle; for future reference we denote its magnitude by v.

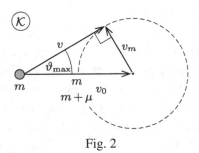

Fig. 2

It can be seen in Fig. 2 that

$$v_m = \frac{m}{m + \mu} \, v_0 \sin \vartheta_{max},$$

and from this and (1) we can deduce that the largest possible angle, ϑ_{max}, through which disc D can be scattered is given by

$$\sin \vartheta_{max} = \frac{\mu}{m} = \frac{M/N}{m}. \tag{2}$$

A successful realisation of the miraculous motion of the disc is possible if $\vartheta_{max} \geq \pi/N$. As we are investigating the limiting case of $N \to \infty$, and $\sin \vartheta_{max}$ is therefore very small, the approximation $\sin \vartheta_{max} \approx \vartheta_{max}$ can be used, and the required condition on the masses involved becomes

$$\frac{\pi}{N} \leq \frac{M/N}{m} \qquad \text{from which} \qquad \frac{M}{m} \geq \pi.$$

Note. It can be shown (after lengthy calculations) that, even if the motions involved are *relativistic*, the largest possible angular deviation ϑ_{max} of a particle of mass m scattered elastically from a particle of mass μ initially at rest is still given by formula (2).

b) Henceforth, only the case of critical mass ratio, i.e. when $M/m = \pi$, will be investigated. It can be seen from Fig. 2 that, for the maximal angular deviation, the connection between the speeds of disc D before and after the collision can be expressed as follows:

$$v^2 = \left(\frac{m}{m + \mu} \right)^2 v_0^2 - v_m^2 = \frac{m^2 - \mu^2}{(m + \mu)^2} v_0^2.$$

Because $\mu \ll m$, speed v can be approximated as follows

$$v = m \left(1 - \frac{\mu^2}{m^2} \right)^{1/2} m^{-1} \left(1 + \frac{\mu}{m} \right)^{-1} v_0$$

$$= \left(1 - \frac{\mu}{m} + \frac{\mu^2}{m^2} - \frac{1}{2} \frac{\mu^2}{m^2} + \cdots \right) v_0$$

$$\approx \left(1 - \frac{\mu}{m} \right) v_0.$$

So the speed of the disc decreases by a factor of $(1 - \mu/m)$ at each bounce, and the ratio of the final to initial speed (after N bounces) is

$$\frac{v_{final}}{v_{initial}} = \left(1 - \frac{M}{Nm} \right)^N \approx e^{-M/m} = e^{-\pi} \approx 0.043.$$

Use of the exponential approximation (as given in the hint) when evaluating the multiple product is justified by the fact that N is very large.

In summary, after the N collisions, disc D finishes up with approximately 4 % of its initial speed.

S43   In their centre-of-mass frame, the (common) speed of the balls is reduced by a factor k at each collision, and so approaches zero through a geometric sequence of values. The relative speed of the balls also decreases with each collision. Since the relative speed of the balls is the same in any reference frame, it follows that the relative speed also decreases to zero in the laboratory frame. So, after a sufficient number of collisions, their relative speed will be close to zero and the two balls will be swinging in phase.

As the net linear momentum of the system cannot change in a collision, and the repeated interchange between kinetic and gravitational energy serves only to reverse the sign of the momentum at each subsequent collision, the magnitude of the total momentum of the balls moving in unison must be equal to what it was just before the very first collision.

As we now show, the maximal speed of a ball is directly proportional to the amplitude of its swing. Equating the initial gravitational energy of the raised ball to its kinetic energy just before the first collision, we have

$$mg\ell(1 - \cos\theta_0) = \frac{1}{2}mv_0^2.$$

For the small initial angular displacement $\theta_0 \approx d/\ell$, this gives

$$\frac{1}{2}mg\ell\theta_0^2 \approx \frac{1}{2}mv_0^2 \quad \longrightarrow \quad v_0 \approx \sqrt{\ell g}\,\theta_0 = \omega d,$$

where $\omega = \sqrt{g/\ell}$ is the angular frequency of the pendulum; the other ball is, of course, at rest. The law of conservation of linear momentum then shows that, when the two balls are moving in phase, their maximal speed will be $v_0/2$, and so the (common) amplitude of their swings will be $d/2$. When this stage has been reached, no more collisions will occur, and it will be air resistance that eventually brings the balls to rest.

Notes. 1. It can be proved that, in the centre-of-mass frame, the speeds of the balls following the nth collision are

$$v_n^{(CM)} = k^n\frac{v_0}{2} \quad \text{and} \quad u_n^{(CM)} = -k^n\frac{v_0}{2}.$$

This translates into the laboratory frame as

$$v_n = \frac{v_0}{2}\left(1 + k^n\right) \quad \text{and} \quad u_n = \frac{v_0}{2}\left(1 - k^n\right).$$

After an odd-numbered collision, the left-hand ball moves with the greater speed v_n; and after an even-numbered one, the right-hand ball does so; the other ball moves with speed u_n.

2. If one of the end balls of the 'executive toy' known as Newton's cradle (*see* figure) is initially displaced through a distance d and released, then, after a long time, all of the n balls forming the cradle will swing with a common amplitude. By this stage, the initial linear momentum will have been evenly distributed among the n balls of the toy, and so the common amplitude of their swings will be d/n.

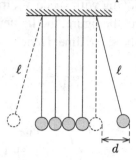

S44 *Solution 1.* Denote the velocity of the ball with mass M after the nth collision by V_n, the velocity of the ball with mass m by v_n, and the distance between the position of the nth collision and the wall by L_n. The position and velocity data for the first two collisions are shown in the figure.

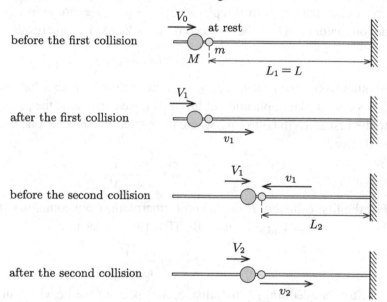

During the collisions, linear momentum and mechanical energy are conserved. Let us write these conservation laws for the $(n+1)$th collision:

$$\frac{1}{2}M(V_n^2 - V_{n+1}^2) = \frac{1}{2}m(v_{n+1}^2 - v_n^2), \tag{1}$$

$$M(V_n - V_{n+1}) = m(v_{n+1} - (-v_n)). \tag{2}$$

Dividing the corresponding sides of these two equations by each other, and making a minor rearrangement of the result, we get

$$V_n + v_n = v_{n+1} - V_{n+1}. \tag{3}$$

The heavier ball, moving with speed V_1, travels a distance $L_1 - L_2$ between the two collisions shown in the figure; the lighter one covers a path of length $L_1 + L_2$ with speed v_1 during the same time. Hence

$$\frac{L_1 - L_2}{V_1} = \frac{L_1 + L_2}{v_1},$$

from which

$$L_2 = \frac{v_1 - V_1}{v_1 + V_1} L_1.$$

In the same way, a connection can be found between the distances L_{n-1} and L_n associated with the nth collision:

$$L_n = \frac{v_{n-1} - V_{n-1}}{v_{n-1} + V_{n-1}} L_{n-1}. \tag{4}$$

From equation (3) with n replaced by $n - 1$, the expression $v_{n-1} + V_{n-1}$ in the denominator of (4) can be replaced by $v_n - V_n$ to yield the result

$$L_n(v_n - V_n) = L_{n-1}(v_{n-1} - V_{n-1}). \tag{5}$$

In other words, the product $L_k(v_k - V_k)$ has the same value, whatever the value of k.

As a particular application of this result, we can equate the products appropriate to the first and Nth collisions, where the Nth collision is the one that occurs closest to the wall:

$$L_N = L_1 \frac{v_1 - V_1}{v_N - V_N}. \tag{6}$$

For the first collision, $L_1 = L$, and furthermore, from equation (3), we have $v_1 - V_1 = v_0 + V_0 = V_0$ (since $v_0 = 0$). Thus (6) reduces to

$$L_N = L \frac{V_0}{v_N - V_N}. \tag{7}$$

After the (defining) Nth collision, the speed of the heavier ball will be zero or very close to zero.[28] Even if it never stops completely, we can be sure that the ball will move 'away from the wall' after the next collision. So, for practical purposes, we can take V_N to be zero.

---

[28] The condition $M \gg m$ means that the speed of the large ball decreases only in very small steps.

As the speed of the large ball is zero (or approximately zero) when it is closest to the wall, its total initial kinetic energy will have been given to the smaller ball:

$$\frac{1}{2}MV_0^2 = \frac{1}{2}mv_N^2,$$

from which it follows that

$$v_N = \sqrt{\frac{M}{m}}\, V_0.$$

Substituting this into (7), and taking into account the condition $V_N \approx 0$, we finally obtain the required minimum distance as

$$L_{\min} = L_N = \sqrt{\frac{m}{M}}\, L.$$

Solution 2. In the following, the notation used in Solution 1 will be retained, and we use $\mathcal{V}$, $\mathcal{V}_1$ and $\mathcal{V}_2$ to signify volumes (not velocities or speeds), though V_0 is still a velocity. From the laws of conservation of mechanical energy and linear momentum, it follows that, after the first collision, the velocity of the lighter ball is

$$v_1 = \frac{2MV_0}{M+m} \approx 2V_0,$$

and so its kinetic energy is

$$E_1 = \frac{1}{2}mv_1^2 \approx 2mV_0^2.$$

From here on, the series of collisions of the balls is conceptualised as the larger ball acting as a movable piston *adiabatically* compressing gas in a small tube, which has an initial volume of AL. The gas contains only a single 'molecule' (namely, the small ball), and the internal energy of this gas is equal to the kinetic energy of that ball (initially E_1). The number of degrees of freedom of this gas is one, because the 'molecule' can move only in one dimension (along the horizontal rod). The cross-sectional area A of the tube can be chosen arbitrarily; its size is unimportant in the following.

> *Note.* The description of this phenomenon as adiabatic compression of an ideal gas that consists of only a single particle cannot be taken too seriously, because (unlike the normal situation) the gas contains only one 'molecule', and the 'piston' (i.e. the large ball) cannot exert a continuous force on the gas, its input impulses being spasmodic. Although it is true that these impulses are rare at the beginning, later, as the speed of the small ball increases, they become more frequent, and finally almost continuous. The problem caused by the notion of a single-particle gas can be avoided if a very large number of rods is imagined, with an initially stationary ball of mass m on each; a single (perforated!) 'piston' with a very large mass is then made to collide with the small balls. This prescription

comes closer to the normal description used in the kinetic theory of gases, but it does not solve the problem of the initially spasmodic nature of the input provided by the 'piston'.

Because of the defects described above, results calculated using this 'gas model' cannot be considered as accurate predictions, and should only be used as *order-of-magnitude* estimates.

For adiabatic processes

$$p_1 \mathcal{V}_1^\gamma = p_2 \mathcal{V}_2^\gamma, \tag{8}$$

where $\gamma = (f+2)/f$, and, because here $f = 1$, we have $\gamma = 3$. The internal energy of the gas is

$$E = \frac{1}{2} f p \mathcal{V},$$

and so the ratio of the internal energies in the initial and final (most compressed) states is

$$\frac{E_1}{E_2} = \frac{p_1 \mathcal{V}_1}{p_2 \mathcal{V}_2}.$$

Using (8), this ratio can also be written in the form

$$\frac{E_1}{E_2} = \frac{\mathcal{V}_1^{1-\gamma}}{\mathcal{V}_2^{1-\gamma}} = \frac{\mathcal{V}_2^2}{\mathcal{V}_1^2} = \left(\frac{L_2}{L_1}\right)^2. \tag{9}$$

Initially, $L_1 = L$ and $E_1 = 2m V_0^2$; finally, $L_2 = L_{\text{min}}$ (the minimal distance we seek) and

$$E_2 = \frac{1}{2} M V_0^2.$$

This last equation follows because the speed of the 'piston' when it is nearest to the wall is zero (to a good approximation), and so all of its initial kinetic energy has been transferred to the single 'gas particle'.

Substituting in (9) for the three known quantities gives the minimal distance as

$$\frac{L_{\text{min}}}{L} = \sqrt{\frac{E_1}{E_2}} = \sqrt{\frac{2m V_0^2}{\frac{1}{2} M V_0^2}} = 2\sqrt{\frac{m}{M}}.$$

This result differs from the 'exact' outcome of Solution 1 by a factor of 2; for the reasons given earlier, this is not surprising.

S45   After it has covered a path of length x, denote the instantaneous velocity of the drop by v, its radius by r and its mass by m; each of v, r and m varies with x.

Along a small element of path length Δx, the drop collects the fog particles from a volume of size $\pi r^2 \Delta x$, the water content of which increases the radius of the drop by Δr. If the (water) density of the fog is ϱ_{fog}, then we have that

$$\pi r^2 \Delta x \cdot \varrho_{fog} = 4\pi r^2 \Delta r \cdot \varrho_{water},$$

from which we can deduce that

$$\Delta r = \frac{1}{4} \frac{\varrho_{fog}}{\varrho_{water}} \Delta x.$$

So, the size increase of the drop is proportional to the path length covered. If x is measured from an appropriately chosen notional origin, the radius can be taken as proportional to the path length:

$$r(x) = \frac{1}{4} \frac{\varrho_{fog}}{\varrho_{water}} x.$$

The mass of the drop expressed in terms of its radius is

$$m(r) = \frac{4\pi}{3} r^3 \varrho_{water}.$$

From these last two equations, the rate of change of the mass can be expressed as follows:

$$\frac{\Delta m}{m} = 3 \frac{\Delta r}{r} = 3 \frac{\Delta x}{x} \qquad \text{or, in differential form,} \qquad \frac{dm}{dx} = \frac{3m}{x}.$$

Newton's equation of motion for the water drop is

$$\frac{d(mv)}{dt} = mg - \lambda \pi r^2 v^2, \tag{1}$$

where λ is a constant.[29]

We next transform the left-hand side of the equation of motion, as follows:

$$\frac{d(mv)}{dt} = m\frac{dv}{dt} + v\frac{dm}{dt}$$

$$= m\frac{dv}{dx}\frac{dx}{dt} + v\frac{dm}{dx}\frac{dx}{dt}$$

$$= mv\frac{dv}{dx} + v^2\frac{dm}{dx}$$

$$= m\frac{d(v^2/2)}{dx} + \frac{3m}{x}v^2.$$

---

[29] More explicitly, λ is given by $c_d \varrho_{air}/2$, where the factor 2 is conventional and c_d is the so-called drag coefficient relating to the aerodynamic resistance of the moving body. Experimentally, for a sphere in a fluid flow that is neither very fast nor very slow, c_d varies between about 0.1 for laminar flow and 0.5 when the flow is turbulent.

Now, using the earlier expressions for $m(r)$ and $r(x)$ (and some careful algebra), equation (1) can be written in the form:

$$m\frac{d(v^2/2)}{dx} = g - \left(3 + \frac{3\lambda}{\varrho_{fog}}\right)\frac{v^2}{x},$$

or, in terms of the drag coefficient (*see* footnote 29),

$$\frac{d(v^2/2)}{dx} = g - \left(3 + \frac{3c_d}{2}\frac{\varrho_{air}}{\varrho_{fog}}\right)\frac{v^2}{x}. \tag{2}$$

As the water drop has uniform acceleration a, after path length x has been covered, its speed must be

$$v(x) = \sqrt{2ax}. \tag{3}$$

Inserting this into both sides of equation (2), we obtain

$$\frac{d(ax)}{dx} = g - \left(3 + \frac{3c_d}{2}\frac{\varrho_{air}}{\varrho_{fog}}\right)2a,$$

which can be rearranged to give an upper limit for the acceleration:

$$a = \frac{g}{7 + 3c_d\varrho_{air}/\varrho_{fog}} < \frac{1}{7}g. \tag{4}$$

Notes. 1. The speed of the drop is given by (3) only if its acceleration is uniform throughout. In fact, a solution that uses a different initial condition (the drop starts from rest) shows that, though the acceleration is not uniform initially, it quickly attains the constant value given by (4).

2. The maximal acceleration $a = g/7$ can be reached only in the limiting case of $\varrho_{fog} \gg \varrho_{air}$. Such a fog density cannot be produced, even with carburation (the mixing of hydrocarbons with air to make a suitable explosive mixture). In real fogs formed under natural circumstances, just the opposite is the case, with $\varrho_{fog} \ll \varrho_{air}$.

S46 We start by assuming that the required manoeuvre is possible, and then test whether it violates any of the known laws of physics. So we assume that the tension in the thread is such that the statuette moves with constant acceleration a, but does not tip over, neither forwards nor backwards. In other words, the statuette's motion is purely translational, with uniform acceleration and no rotation.

Newton's equations of motion for the statuette's centre of mass are (using the notation shown in the figure)

$$T - F = ma,$$
$$mg - N = 0,$$

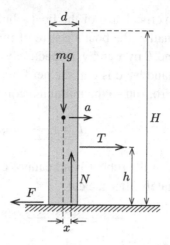

and the equation for the (absence of) rotational motion is

$$Nx + T\left(\frac{H}{2} - h\right) - F\frac{H}{2} = 0.$$

Here we have assumed that all the surface forces distributed across the base of the statuette can be replaced by a single vertical force N whose line of action is a distance x (as yet unspecified) from the symmetry axis of the statuette. Because the statuette is sliding, the equation $F = \mu N$ also holds.

For all these equations to hold, the value of the (signed) distance x must be

$$x = \mu h - \frac{a}{g}\left(\frac{H}{2} - h\right).$$

However, the line of action of N must not lie outside the base of the statuette, i.e. the condition

$$|x| \le \frac{d}{2}$$

must be met. From this and the expression for x above, and using the numerical values given, we get the following inequalities for the statuette's acceleration:

$$a \ge \left(\frac{2\mu h - d}{H - 2h}\right) g \approx 1.6 \text{ m s}^{-2}$$

and

$$a \le \left(\frac{2\mu h + d}{H - 2h}\right) g \approx 11.4 \text{ m s}^{-2}.$$

So the statuette *can be pulled* across the table without turning over; to do this, a considerable, but not arbitrarily large, acceleration (and hence force) is required.

S47   Since the vertical size of the boat's hull is small compared to its length, to a good approximation the boat's centre of mass lies on the line AB. Denote its distances from A and B by x and y, respectively, as shown in Fig. 1.

Before the left-hand band is cut, the net force and torque acting on the boat, of mass m, are both zero, and so the initial tension in the right-hand band is

$$T = \frac{x}{x+y}\, mg.$$

Because the length of the rubber band cannot change instantaneously, this is also the tension immediately after the cut.

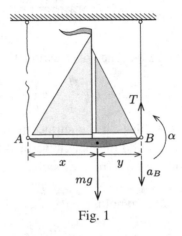

Fig. 1

Using the notation of Fig. 1, the equations of the boat's initial vertical and rotational motions are

$$mg - T = ma,$$
$$yT = I\alpha.$$

Here a is the downward acceleration of the boat's centre of mass, I is its moment of inertia about that centre of mass, and α is its angular acceleration.

The *downward* acceleration of point B is the signed sum of the acceleration of the centre of mass and the tangential acceleration ($y\alpha$) due to the hull's rotation:

$$a_B = a - y\alpha = g - \frac{x}{x+y}\,g - \frac{y}{I}\frac{x}{x+y}\,mg = \frac{y}{x+y}\left(1 - \frac{mxy}{I}\right)g.$$

From this, it can be seen that point B starts to move *upwards* if

$$I < mxy.$$

We will now prove that this inequality always holds.

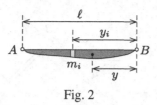

Fig. 2

Consider the boat's hull, of length $\ell = x + y$, as divided into many point-like parts with masses m_i and positions y_i, with the y_i $(i = 1, 2, \ldots)$ measured from point B as shown in Fig. 2. Using this notation, the total mass of the boat and the position of its centre of mass can be written as follows:

$$m = \sum m_i, \qquad y = \frac{\sum m_i y_i}{m}.$$

Its moment of inertia I about the centre of mass can be expressed, with the help of the parallel axis theorem, as

$$I = \sum m_i y_i^2 - my^2.$$

Using the last two of these equations and the fact that $\ell = x + y$, the inequality to be proved can be written successively as follows:

$$\sum m_i y_i^2 - my^2 < m(\ell - y)y,$$

$$\sum m_i y_i^2 < \ell \sum m_i y_i,$$

which must be valid, since each of the y_i is less than ℓ.

So, it has been proved that, irrespective of the mass distribution of the boat (specifically, of the position of its centre of mass and the magnitude of its moment of inertia), when one of the rubber bands is cut, then the end of the other band is *bound* to start moving *upwards*.

> *Note.* The conclusion – i.e. that when one of the rubber bands is cut, then the end of the other band starts moving upwards – holds even if the boat does not start off 'on an even keel'.

S48 Denote the common mass of the rods by m, and their lengths by ℓ, with the centres of the two upper rods shown in the figure labelled as A and C, and their common end-point as B. The centre of mass of the whole system is denoted by O.

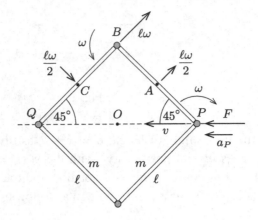

Let the external force acting at point P be $\mathbf{F}$, and, as we are concerned only with the *initial* acceleration of Q, consider the implications of the work–energy theorem applied over an arbitrarily short time interval t following the start of the motion. In this context, the linear and rotational accelerations of the rods can be considered constant during the part of the motion being investigated. At the end of the short time t, we have

$$F \cdot \frac{1}{2} a_P t^2 = E_{\text{rotational}} + E_{\text{translational}}.$$

The total kinetic energy of the system can be expressed in terms of the velocity $v = a_P t$ of point P and the *common* angular velocity $\omega = \alpha t$ of the rods. At any given moment, the angles through which the rods have rotated are all the same in magnitude (but not in sense). Consequently, their angular velocities, and their accelerations, must also have common values, ω and α, respectively.

As point A is moving, with angular velocity ω, around a circle of radius $\ell/2$ centred on P, its velocity v_A relative to the table-top is the vector sum of the velocity of point P and the tangential velocity from the rotation around P. So, the square of its velocity is given by

$$v_A^2 = \left(v - \frac{\ell\omega}{2\sqrt{2}} \right)^2 + \left(\frac{\ell\omega}{2\sqrt{2}} \right)^2 = a_P^2 t^2 + \frac{\ell^2 \alpha^2 t^2}{4} - \frac{a_P \ell \alpha t^2}{\sqrt{2}},$$

where the factors of $1/\sqrt{2}$ arise from $\cos 45°$ or $\sin 45°$. It is important to note that, in the figure, the velocity of point P relative to the table-top is indicated, but that the velocities of points A and B are relative to P, and the velocity shown for C is relative to B.

Similarly, the square of the velocity of point C can be calculated from the velocity of B relative to P (its magnitude is $\ell\omega$) and the tangential velocity, of magnitude $\frac{1}{2}\ell\omega$, of C around B:

$$v_C^2 = \left(v - \frac{\ell\omega}{\sqrt{2}} - \frac{\ell\omega}{2\sqrt{2}}\right)^2 + \left(\frac{\ell\omega}{\sqrt{2}} - \frac{\ell\omega}{2\sqrt{2}}\right)^2 = a_P^2 t^2 + \frac{5\ell^2\alpha^2 t^2}{4} - \frac{3a_P\ell\alpha t^2}{\sqrt{2}}.$$

Now, the moment of inertia of a uniform rod about its centre of mass is $I = m\ell^2/12$, and so the work–energy theorem reads:

$$F\frac{a_P}{2}t^2 = 4 \times \frac{1}{2}\frac{m\ell^2}{12}\omega^2 + 2 \times \frac{1}{2}mv_A^2 + 2 \times \frac{1}{2}mv_C^2,$$

and, using the formulae above, we get the following relationship (after dividing through by mt^2):

$$F\frac{a_P}{2m} = \frac{5}{3}\ell^2\alpha^2 + 2a_P^2 - \frac{4}{\sqrt{2}}a_P\ell\alpha. \tag{1}$$

The acceleration of the centre of mass O of the four rods is equal to the component in the direction PQ of B's acceleration. At the initial moment, when both the angular velocity and the centripetal acceleration are zero, this can be written as

$$a_O = a_P - \frac{\ell\alpha}{\sqrt{2}}. \tag{2}$$

However, in accordance with Newton's second law, this acceleration can also be expressed in terms of the external net force acting on the system as

$$a_O = \frac{F}{4m}. \tag{3}$$

From equations (1)–(3), it follows that $a_O = \frac{2}{5}a_P$.

But it is also the case that a_O can be expressed as $a_O = \frac{1}{2}(a_P + a_Q)$, which, together with $a_O = \frac{2}{5}a_P$, implies that $a_Q = -\frac{1}{5}a_P$. It is perhaps a somewhat surprising result that joint Q starts moving towards joint P, and therefore in a direction directly opposed to that of the external force.

S49 First of all, we investigate the motion the system would have if the heavy weight were not attached to it. The forces affecting the cylinder are the gravitational force acting at its centre of mass and the tensions in the cords; denote the sum of the latter by T (*see* figure).

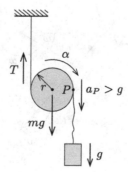

The acceleration of the centre of mass of the cylinder of mass m, radius r and moment of inertia I can be calculated from the translational equation of motion,

$$mg - T = ma,$$

and that describing the rotational motion,

$$rT = I\alpha,$$

where $\alpha = a/r$ is the angular acceleration of the cylinder.

From these equations, the acceleration of the centre of mass of the cylinder is found to be

$$a = \frac{mr^2}{mr^2 + I} g.$$

As the maximum value for I is mr^2, corresponding to the cylinder being a hollow shell, we have that $0 < I \le mr^2$, and so

$$\frac{1}{2} g \le a < g.$$

The point of the cylinder that is denoted by P in the figure starts moving with an acceleration $a + \alpha r = 2a$, i.e. with at least g. When the weight is attached to the cylinder and released, it can move with an acceleration of at most g. However, the part of the cord that touches P starts to move with an acceleration of at least g. So the cord becomes slack and the weight goes into *free fall*!

S50   Suppose firstly that the translational motion is the first to stop, i.e. the disc still has a significant angular velocity ω^* when the velocity of its centre of mass has decreased to zero. We investigate the motion a very short time before this happens, when the disc is rotating with angular velocity $\omega > \omega^*$, and the speed of its centre of mass is $v \ll R\omega^*$, where R is the radius of the disc.

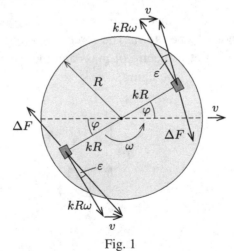

Fig. 1

Consider two identical, small pieces of the disc positioned symmetrically about its centre, as shown in Fig. 1, in which the disc is moving to the right with speed v. The two radius vectors to the pieces each make an angle φ with the direction of motion of the disc. Since the disc presses uniformly on the ice, the magnitudes of the frictional forces (denoted by ΔF) acting on these pieces are identical. The direction of a frictional force is directly opposed to that of the relevant piece's velocity relative to the ice. For this reason, it is not perpendicular to the corresponding radius vector, but deviates from it by a *small* angle ε. From geometrical considerations in Fig. 1, the approximate magnitude of angle ε is

$$\varepsilon \approx \frac{v \cos \varphi}{kR\omega}, \tag{1}$$

where kR is the length of the appropriate radius vector ($0 < k \leq 1$).

The translational motion of the disc is decelerated by the net component of the frictional forces that is in the direction of, but opposed to, the disc's velocity. In the figure, the left-hand piece contributes a retarding force of

$$\Delta F \cos(\tfrac{1}{2}\pi - \varepsilon - \varphi) = \Delta F \sin(\varphi + \varepsilon),$$

whereas the right-hand piece produces an *accelerating* force of

$$\Delta F \cos(\tfrac{1}{2}\pi + \varepsilon - \varphi) = \Delta F \sin(\varphi - \varepsilon).$$

So the net retarding force is

$$\Delta F_{\text{trans}} = \Delta F \sin(\varphi + \varepsilon) - \Delta F \sin(\varphi - \varepsilon) = 2\Delta F \cos \varphi \sin \varepsilon.$$

Since $\varepsilon \ll 1$, we have $\sin \varepsilon \approx \varepsilon$, and so using (1), we can write

$$\Delta F_{\text{trans}} \approx 2\varepsilon \Delta F \cos \varphi = \frac{2\Delta F \cos^2 \varphi}{k} \frac{v}{R\omega}. \tag{2}$$

From (2) it can be seen that, for each value of φ and k, the retarding force ΔF_{trans} is directly proportional to the fraction $v/(R\omega)$. Consequently, so are the net force and the deceleration of the disc.

So the equation governing the motion of the centre of mass of the disc can be written in this form:

$$\frac{\mathrm{d}v}{\mathrm{d}t} = -C \frac{v(t)}{\omega(t)}, \tag{3}$$

where the positive constant C depends on the coefficient of friction, the radius of the disc and the gravitational acceleration. In this equation, v and ω are both time-dependent variables .

If the constant ω^* (the angular velocity when the centre of the disc stops) were inserted in place of $\omega(t)$, equation (3) would have the same form as one governing nuclear decay, but with a larger value for the decay constant than the actual one (since $\omega^* < \omega(t)$). But even this would mean that $v(t)$ is a velocity that decreases *exponentially* with time, i.e. the centre of mass of the disc would never stop after a *finite* time interval. In reality – because the decreasing $\omega(t)$ is still greater than ω^* – the velocity of the centre of mass decreases even more slowly than this. This (false) conclusion is a direct contradiction of our initial assumption that the motion is going to stop almost immediately. The conclusion from this contradiction is that the translational motion of the disc *cannot stop sooner* than its rotation does.

Using similar arguments, it can be proved that the angular velocity of rotation cannot decrease to zero while the linear velocity of the centre of mass is still significant. In fact, shortly before any such state could be reached, the net decelerating torque would have decreased to almost zero. This is because, although the net linear force would remain finite, the frictional forces acting on the 'paired' surface pieces are equal, and the torques they produce almost totally cancel each other out. In this state the rate of rotational deceleration would become very low, whereas, in stark contrast, the translational speed would be decreasing rapidly.

If neither the rotation nor the translation of the disc can end earlier than the other does, the only remaining possibility is that the two kinds of motion stop *simultaneously*.

> *Note.* Using numerical simulations and rather sophisticated calculations, it can be shown that, towards the end of the motion, the dimensionless ratio $R\omega/v$, characterising the relationship between the rotational and translational speeds, tends to the same value *whatever the initial conditions*: it always approaches the value 1.531! If we plot the trajectories of the disc in an $(R\omega, v)$ coordinate system for a variety of different initial conditions, as shown in Fig. 2, all the curves approach the origin from the direction of the dashed line, which has a slope of 1.531.

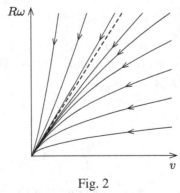

Fig. 2

S51 The key to the solution is realising that the small masses of the bearings and the absence of slipping together imply that the static frictional forces that act on the bearings have negligible components $F_\perp$ perpendicular to the rod. The argument producing this conclusion is as follows.

These forces are the only ones producing a torque about a bearing's axis, and so its wheel of mass m_b, radius r_b and moment of inertia I has an angular acceleration α, where

$$\alpha \propto \frac{F_\perp r_b}{I} \propto \frac{F_\perp r_b}{m_b r_b^2} = \frac{F_\perp}{m_b r_b}.$$

The condition that the bearings roll without slipping requires that the tangential accelerations of the ends of the rod (and hence of the bearings) are also proportional to $F_\perp/m_b$. This ratio can only be finite if $F_\perp$ is negligible, in the same sense as m_b is.

a) As it follows from the above argument that $F_\perp$ is negligible, the net torque acting on the rod is zero, i.e. its angular velocity is constant in time:

$$\omega(t) = \omega = \frac{|v_2 - v_1|}{L}.$$

For the same reason, the forces that *do* act on the rod are always parallel to its axis. However, they cannot affect the speed v of its centre of mass, as the rod cannot slip in a direction parallel to its length. The velocity v is therefore constant in magnitude, and given by

$$v = \frac{v_1 + v_2}{2},$$

though its direction constantly changes, being perpendicular to the rod's length at all times. In summary, the velocity of the rod's centre of mass rotates uniformly with angular velocity ω, i.e. the centre of the rod moves along a circular trajectory (*see* Fig. 1).

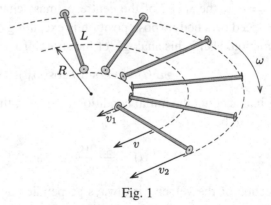

Fig. 1

The period of the circular motion is $2\pi/\omega$, and, if the radius of the circle is R, we have

$$\frac{2\pi}{\omega} v = 2\pi R,$$

from which it follows that

$$R = \frac{v}{\omega} = \frac{v_1 + v_2}{|v_2 - v_1|}\frac{L}{2}.$$

The two ends of the rod also move along circular paths, of radii $R + L/2$ and $R - L/2$. Fig. 1 illustrates a 'half-circuit' of each of these three circular trajectories.

b) The forces in the plane of the slope acting on the rod when it makes an angle φ relative to its initial orientation are shown in Fig. 2. As in part *a*), the net torque on the rod is zero, and its rotational angular velocity is ω_0 throughout.

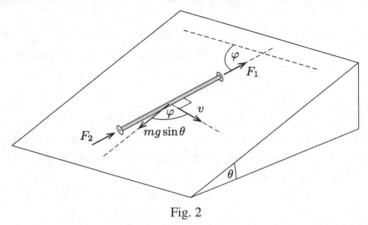

Fig. 2

The Newtonian equation of motion for the centre of mass, in the direction perpendicular to the rod's length, is

$$m\dot{v}(t) = mg\sin\theta\cos\varphi.$$

As $\varphi(t) = \omega_0 t$, the *speed* of the centre of mass changes in time in the same way as the speed of a body whose position is executing (one-dimensional) simple harmonic motion. Using this analogy, or the straightforward integration of

$$m\dot{v}(t) = mg\sin\theta\cos(\omega_0 t),$$

and taking into account the initial condition $v(0) = 0$, the time dependence of the speed is found to be

$$v(t) = \frac{g\sin\theta}{\omega_0}\sin\omega_0 t.$$

The direction of the velocity is always perpendicular to the rod, and so, using the coordinate system shown in Fig. 3, the x and y components of that velocity can be expressed in the form

$$\boldsymbol{v}(t) \equiv \left[\begin{array}{c} v_x(t) \\ v_y(t) \end{array}\right] = \frac{g\sin\theta}{\omega_0}\sin\omega_0 t \left[\begin{array}{c} \sin\omega_0 t \\ \cos\omega_0 t \end{array}\right].$$

Using trigonometrical identities, this can be rewritten as

$$v(t) = \frac{g\sin\theta}{2\omega_0}\begin{bmatrix} 1 - \cos 2\omega_0 t \\ \sin 2\omega_0 t \end{bmatrix}.$$

It can be seen that the tip of the velocity vector moves around a circle that is centred on $\dfrac{g\sin\theta}{2\omega_0}\begin{bmatrix} 1 \\ 0 \end{bmatrix}$ and has a radius of $v^* = \dfrac{g\sin\theta}{2\omega_0}$.

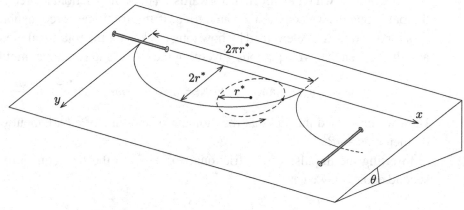

Fig. 3

So the centre of the rod moves in the same way as a circumferential point of a round cylinder rolling along the axis x with speed v^*, i.e. the trajectory is a common cycloid. The radius r^* of the 'rolling cylinder' can be found by equating its circumference to the x distance covered by the rod's centre during one period of the motion (π/ω_0):

$$r^* = \frac{1}{2\pi}\frac{\pi}{\omega_0}v^* = \frac{g\sin\theta}{4\omega_0^2}.$$

It follows that the maximum distance to which the centre of the rod 'drops down the incline' before returning to its initial elevation is $y = 2r^* = g\sin\theta/(2\omega^2)$. The trajectory of the centre of the rod and the 'rolling cylinder' can be seen in Fig. 3.

S52 *a*) Denote the width of the trough by 2ℓ, the initial speed of the balls by $v_0 = 3$ m s^{-1} and the radius of a ball by $R = 2.5$ cm. The centre of mass of each ball is at a height $r = \sqrt{R^2 - \ell^2}$ above the horizontal line joining the points on the trough on which it is resting at any particular moment. As it is instantaneously rotating about this line with angular velocity ω_0, the condition for it to roll without slipping is

$$v_0 = r\omega_0 = \sqrt{(R^2 - \ell^2)}\,\omega_0.$$

Following the first collision, because the balls slip on the edges of the trough, friction decelerates both their translational motions and their angular velocities. A second collision of the balls will occur if the absolute value of the consequent change in their speeds is larger than v_0. We first investigate the threshold case, in which this change of speed is exactly v_0.

A ball moving 'backwards' after the first collision (and rotating in the 'wrong sense' for the direction in which it is travelling) is decelerated by the kinetic frictional force, which is directed 'forwards'. The same frictional force decreases the ball's angular velocity at the same time. If the frictional force, denoted by F, acts for a period Δt (determined by how long it takes for the rotational motion to be annulled), then the two equations governing the linear and rotational motions are

$$F\Delta t = mv_0 \quad \text{and} \quad Fr\Delta t = I\omega_0 = \frac{2}{5}mR^2\omega_0 = \frac{2}{5}mR^2\frac{v_0}{r}.$$

Here we have used the fact that the moment of inertia I of a ball about one of its diameters is $\frac{2}{5}mR^2$.

Inserting the impulse of the frictional force (from the first equation) into the second equation, we get

$$mv_0 = \frac{2R^2}{5r^2}mv_0,$$

and after dividing through by mv_0, this yields

$$2R^2 = 5r^2 = 5(R^2 - \ell^2),$$

from which

$$\ell = \sqrt{\frac{3}{5}}R \approx 0.775R \approx 1.94 \text{ cm}.$$

So, the trough needs to be wider than 3.9 cm ($2\ell > 2\sqrt{\frac{3}{5}}R \approx 3.9$ cm) for the second collision to occur.

b) The given width of the trough is $2\ell = 4$ cm, and, as before, $v_0 = 3$ m s^{-1}; denote the speed of the balls just before the second collision[30] by v. Because of the symmetry of the situation, it is sufficient to investigate the behaviour of just one of the balls.

Although the ball was still slipping following the first collision, its speed must have changed by $v - (-v_0)$, and so the impulse it received due to friction was

$$F\Delta t = m(v_0 + v).$$

---

[30] Which will occur, because 2ℓ is greater than the threshold value of $2\sqrt{3/5}R$.

During the same interval, the angular velocity changed from ω_0 to $\omega = v/r$, but without changing sign! The torque arising from the frictional force caused the corresponding change in the angular momentum of the ball:

$$Fr\Delta t = I(\omega_0 - \omega) = \frac{2}{5}mR^2 \left(\frac{v_0}{r} - \frac{v}{r}\right),$$

where

$$r = \sqrt{R^2 - \ell^2} = 1.5 \text{ cm}.$$

Again, inserting the impulse of the frictional force from the first equation into the second, we get

$$r^2(v_0 + v) = \frac{2}{5}R^2(v_0 - v),$$

from which it follows that the speed of the balls before (and after) the second collision is

$$v = \frac{2R^2 - 5r^2}{2R^2 + 5r^2} v_0 = \frac{1}{19} v_0 \approx 0.16 \text{ m s}^{-1}.$$

Notes. 1. The above solution shows that the value of the coefficient of kinetic friction has no influence on the result. This fact is even more obvious if the problem is solved using the law of conservation of angular momentum. Following the first collision, the angular momentum of the ball about any point on the centreline of the horizontal plane containing the top edges of the trough does not change. This is because the sum of the torques of the frictional forces at the two edges, as well as the sum of the torques of the normal forces, are each zero for such points.

The total angular momentum consists of the ball's rotational angular momentum and that due to the linear motion of its centre of mass, which have opposite signs just after the first collision. If the total angular momentum is exactly zero after the first collision, then the ball will halt just as the slipping stops. This is the threshold case for the ball reversing its direction (and heading for a second collision):

$$\frac{2}{5}mR^2\frac{v_0}{r} - mrv_0 = 0,$$

from which we get $r = \sqrt{2/5}\,R$, in accord with the solution obtained in part *a*).

Part *b*) can be solved in the same way. In this case the right-hand side of the previous equation is not zero, but positive:

$$\frac{2}{5}mR^2\frac{v_0}{r} - mrv_0 = \frac{2}{5}mR^2\frac{v}{r} + mrv,$$

from which the previous result for v follows:

$$v = \frac{2R^2 - 5r^2}{2R^2 + 5r^2} v_0.$$

2. It can be shown more generally that the second collision occurs for the same range of trough sizes, even if the balls collide with a different combination of initial speeds. They can collide with opposing, or similarly directed, velocities, or with one of the balls at rest before the collision. In all cases, if the condition $\ell > \sqrt{3/5}\,R$ is obeyed, then the second collision takes place.

3. In principle, after the first collision, the number of subsequent collisions is infinite, but in reality the second collision happens at such a reduced speed, that, because of the previously neglected mechanical energy losses (air drag, rolling friction, etc.), collisions beyond the second cannot normally be observed directly. Videos made with slow-motion cameras show that more collisions do occur.

S53　*a*) In general, after receiving an impulse from the cue, the billiard ball both rolls and slips, and the instantaneous speed of its point of contact with the table is not zero. This 'slipping' continues until, as a result of kinetic frictional forces, the velocity of that point relative to the table decreases to zero; after that, the ball continues to roll but without slipping.

Consider the point P at which the ball touches the table before the shot is taken. Note that P denotes a *fixed* point *on the table*, and not the current contact point of the ball and table (which accelerates, or decelerates, during the stroke and the subsequent 'slipping'). The total angular momentum of the ball about this point is zero before the shot, as well as at the simultaneous end of the rolling and slipping motions (when the ball again becomes stationary).

During the motion that follows the cue stroke, the net torque about P of the forces acting on the ball is zero, because the gravitational force and the normal reaction of the table cancel each other, and the line of action of the frictional force always passes through P. The angular momentum of the ball about P can only remain at zero throughout (from before the stroke until after the final halt) if it does not receive any during the stroke itself. This requires that the line of action of the impulse, and hence that of the cue, must be directed through point P (*see* Fig. 1).

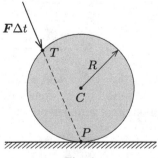

Fig. 1

b) If, immediately after the cue stroke, the ball starts a pure rolling motion without any slipping at all, there can be no static frictional force on the ball at the contact point with the table during the short time interval (Δt) of the impulse. As before, we must find a fixed point (in space) about which the angular momentum of the ball remains constant (zero) throughout the motion! We now do that.

During the short period of the impulse, the connection between the velocity v of the ball's centre of mass and its angular velocity ω around the centre of mass must be

$$v = R\omega \qquad (1)$$

at all times, where R is the radius of the ball. For motion to the right (as shown in Fig. 2) the rotation is clockwise.

From symmetry, the fixed point we seek must lie on the line PC, or on its extension; let us suppose that it is a height h above the table. The total angular momentum of the ball, of mass m, about that point is

$$J = \frac{2}{5}mR^2\omega - (h - R)mv. \qquad (2)$$

Bearing in mind condition (1), J will be zero if $h = \frac{7}{5}R$. The appropriate fixed point Q is shown in Fig. 2.

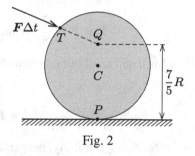

Fig. 2

So, for a pure rolling motion, the total angular momentum of the ball about Q must remain at zero throughout the stroke (and also thereafter). This will happen if the impulse is directed along the line TQ – the cue axis must follow the same line!

> *Notes.* 1. From the solution to part *a*) of the problem, it can be seen that, if the line of action of the impulse at point T is more vertical than segment TP, then after the 'slipping' finishes the ball moves 'backwards', otherwise it continues moving 'forwards'.
>
> 2. The coefficient of static friction between the ball and the billiard table depends on the coating of the ball and the quality of the cloth covering the table. If in part *b*) the table exerts a static frictional force on the ball during the cue stroke, then the line of action of the impulse can deviate from the line TQ to some extent.
>
> If – in contrast to what is shown in the figure accompanying the problem – the ball is struck at a point T that is below the level of Q, then the ball might

lose contact with the table for a short while[31] as a consequence of the impulse it receives.

S54   *a*) Denote the vector pointing from the centre C of the billiard ball to its lowest point (where it touches the table) by R, the mass of the ball by m, the velocity of its centre of mass by v, and its angular velocity by ω.

As noted in the problem, for a general *Coriolis-massé*, ω will not be perpendicular to v, and so the velocity of the lowest point of the ball,

$$v_P = v + \omega \times R,$$

will not be parallel to the velocity of the centre of the ball, even at the start of the motion. A similar connection holds between the corresponding accelerations and the angular acceleration:

$$\dot{v}_P = \dot{v} + \dot{\omega} \times R. \tag{1}$$

During the 'slipping' motion, the horizontal acceleration of the ball and its angular acceleration are both caused by the frictional force F, and so the dynamical equations for the translational and rotational motion can be written as follows:

$$F = m\dot{v},$$

$$R \times F = \frac{2}{5}mR^2\dot{\omega}.$$

Inserting expressions for $\dot{v}$ and $\dot{\omega}$, obtained from these two equations, into equation (1) gives

$$\dot{v}_P = \frac{1}{m}F + \frac{5}{2mR^2}(R \times F) \times R.$$

Now F and R are necessarily mutually perpendicular, and so using either the right-hand rule or the vector triple product identity, it follows that

$$(R \times F) \times R = R^2 F.$$

So finally we have that

$$\dot{v}_P = \frac{7}{2m}F. \tag{2}$$

The magnitude of the kinetic frictional force is μmg (where μ is the coefficient of friction), and its direction is opposed to that of the velocity of the lowest point of the ball:

---

[31] Such a hit is known as a 'jump shot' in billiards; many other factors, from chalk dust on the cloth or ball to static electricity, have been blamed for this phenomenon.

$$F = -\mu mg \frac{v_P}{|v_P|}. \tag{3}$$

Combining this with equation (2), we have

$$\dot{v}_P = -\frac{7}{2}\mu g \frac{v_P}{|v_P|}. \tag{4}$$

Equation (4) shows that the velocity of the ball's lowest point has a constant direction throughout the simultaneous rolling and slipping motion, and that its magnitude decreases uniformly to zero at a rate of $-\frac{7}{2}\mu g$. It then follows from (3) that not only the magnitude of the frictional force but also its direction are constant. As this direction does not coincide with that of the initial velocity of its centre of mass, the billiard ball moves along a *parabolic* (rather than a straight) trajectory (*see* figure).

When the velocity of the lowest point of the ball becomes zero (this happens at B in the figure), the ball continues to roll, but without any slipping, until air drag and rolling friction bring it to a halt. Its straight-line path is along the tangent to the parabola at point B.

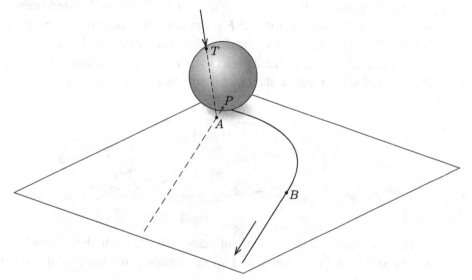

b) The final direction of the ball's motion can be found with the help of the law of conservation of angular momentum. We investigate the angular momentum of the ball about the line *PA*.

> *Note.* Angular momentum is a vector quantity, which is defined relative to a fixed (but arbitrarily chosen) *point* in space. But, it is also the case that a component of angular momentum in a given direction can be defined by an *axis* which lies in that direction. In this problem, for example (as will be shown later), the angular momentum of the ball relative to the *point P* is not conserved, but the component of angular momentum parallel to the line *PA* does remain constant.

Initially, the ball is at rest, so its angular momentum is zero. During the short time interval of the shot, the lines of action of the forces acting on the ball (the force of the shot exerted by the cue, the normal reaction force of the table, the frictional force and the gravitational force) all pass through various points on the line *PA*. So, just after the shot, the angular-momentum component defined by this line is also zero. This situation does not change as the ball moves along the parabolic arc *PB*, because the gravitational force and the normal reaction of the table cancel each other out, and the torque about this axis due to the frictional force is always zero (since the force and the axis lie in the same plane).

So, on the one hand, after finishing the 'slipping' section of the motion, the angular-momentum vector of the ball remains constant – it is horizontal, and perpendicular to the velocity of the centre of mass. But, on the other hand, as we have just shown, its component parallel to the line *PA* is zero. There is only one way to reconcile these two conclusions, and that is that the ball's path is *parallel* to the line *PA*.

S55 We use the notation shown in Fig. 1. Vector $\boldsymbol{R}$ points from the centre of the ball to its momentary contact point with the disc, $\boldsymbol{r}$ is directed from the centre of the disc to the same point, and $\boldsymbol{F}$ is the static frictional force acting on the ball. Since, relative to the centre of the disc, the centre of the ball is at $\boldsymbol{r} - \boldsymbol{R}$ and $\boldsymbol{R}$ is a constant vector, we have that its velocity is $\boldsymbol{v} = \dot{\boldsymbol{r}}$. The angular velocities are $\boldsymbol{\Omega}$ for the disc and $\boldsymbol{\omega}$ for the ball (the latter is not shown in the figure).

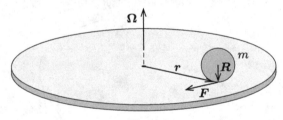

Fig. 1

The acceleration $\dot{\boldsymbol{v}}$ of the centre of mass of the ball, which has mass m, is caused by the static frictional force, and so the equation for horizontal motion can be written as

$$\boldsymbol{F} = m\dot{\boldsymbol{v}}. \tag{1}$$

The angular acceleration of the ball is also caused by the frictional force, and so the equation for the rotational motion takes the form

$$\boldsymbol{R} \times \boldsymbol{F} = \frac{2}{5}mR^2\dot{\boldsymbol{\omega}}, \tag{2}$$

since $\frac{2}{5}mR^2$ is the moment of inertia of the solid homogeneous sphere about one of its diameters.

As there is no slipping, the lowest point of the ball and the contact point on the disc move together and therefore

$$v + \omega \times R = \Omega \times r. \tag{3}$$

Taking the time derivatives of both sides of condition (3) gives a similar connection between the rates of change:

$$\dot{v} + \dot{\omega} \times R = \Omega \times v.$$

Using this and equations (1) and (2) yields

$$\dot{v} + \frac{5}{2mR^2}[R \times (m\dot{v})] \times R = \Omega \times v.$$

The direction of the vector triple product is the same as the direction of $\dot{v}$ and its magnitude[32] is $mR^2 \dot{v}$. After substituting this, and carrying out some simplification and rearrangement, we get

$$\dot{v} = \frac{2}{7}\Omega \times v,$$

which, using the notation $\Omega_0 = \frac{2}{7}\Omega$, can be written as

$$\dot{v} = \Omega_0 \times v. \tag{4}$$

That is, the acceleration is perpendicular to both the current velocity and the axis of rotation. Further, by taking the scalar product of equation (4) with v, we can deduce that the magnitude of v does not change with time:

$$\frac{d(v^2)}{dt} = \frac{d(v \cdot v)}{dt} = 2v \cdot \dot{v} = 2v \cdot (\Omega_0 \times v) = 0.$$

Uniform circular motion has exactly these properties!

Either by noting the implications of these observations, or after a straightforward time integration of (4), we can write the equation describing the rate of change of the position vector r in the form

$$\dot{r} = (\Omega_0 \times r) + v^*, \tag{5}$$

where v^* is a constant dependent on the initial conditions.

Now, for any (arbitrary) velocity vector v^*, we can always find a position vector r^* such that v^* can be written in the form $-\Omega_0 \times r^*$. For this reason, equation (5) can be transformed into

$$\dot{r} = \Omega_0 \times (r - r^*). \tag{5'}$$

---

[32] This can be proved by twice applying the right-hand rule for vector cross-products, or by using the identity for vector triple products to be found in the mathematics section of the Appendix.

It is straightforward to see that equation (5′) describes uniform motion, with angular velocity $\Omega_0 = \frac{2}{7}\Omega$, around a circle centred on r^*. With the help of this general result, answers can be given to the two specific questions asked.

a) For the ball to be moving along a circular path of radius r_0 and concentric with the centre of the disc, we need to set r^* in equation (5′) equal to the null vector. Then the initial velocity of the centre of mass is given by

$$v_0 = \Omega_0 \times r_0, \tag{6}$$

where r_0 gives the initial position of the ball, a distance r_0 from the axis of rotation. From this, the magnitude of the initial velocity has to be

$$v_0 = |\Omega_0 \times r_0| = r_0\Omega_0 = \frac{2}{7}r_0\Omega.$$

The initial angular velocity of the ball can be determined using the 'no-slipping' condition (3). Taking into account condition (6) and the relationship between Ω_0 and Ω, it yields

$$\omega_0 \times R = \frac{5}{2}v_0. \tag{7}$$

The magnitude and direction of the vector product on the left-hand side are not affected by the vertical component (parallel to R) of the ball's angular velocity vector ω_0, and so, in principle, this component can have any arbitrary value.

This conclusion assumes that the rubber ball is in contact with the disc only at a single point. In reality, the ball, as well as the disc, are deformed slightly, and so, in order to avoid effects arising from friction at the touching surfaces, the vertical component of the angular velocity should be chosen to be zero! In this case, the angular velocity vector is horizontal, and its magnitude can be found by taking the vector product of (7) with R:

$$R \times v_0 = \frac{2}{5}[R \times (\omega_0 \times R)] = \frac{2}{5}[R^2\omega_0 - (\omega_0 \cdot R)R] = \frac{2}{5}R^2\omega_0.$$

It follows that the required initial angular velocity (about a horizontal axis) is

$$\omega_0 = \frac{5}{2}\frac{|R \times v_0|}{R^2} = \frac{5}{2}\frac{v_0}{R} = \frac{5}{7}\frac{r_0\Omega}{R}.$$

b) This time, the initial position of the ball is still r_0, but its initial velocity is $-v_0$. Under these circumstances, equation (5′) reads

$$-v_0 = \Omega_0 \times (r_0 - r^*).$$

Comparing this with (6), it is clear that we must have $r^* = 2r_0$. So, in this case, the rubber ball still moves along a circular trajectory of radius r_0, but its centre is now at a distance $2r_0$ from the centre of the disc (*see* Fig. 2). The magnitude of the angular momentum of the ball can be found using the same line of argument as described in part *a*), and the result is $\omega_0 = 9r_0\Omega/(7R)$.

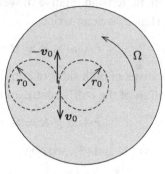

Fig. 2

S56 We use a Cartesian coordinate system fixed to the inclined plane (*see* figure) in which e_x and e_y are unit vectors in the directions of the corresponding axes. Denote the vector from the centre of the ball to its momentary contact point with the disc by R, the vector from the centre of the disc to that same point by r, the static frictional force acting on the ball by F, and the component of the gravitational force parallel to the inclined plane by G.

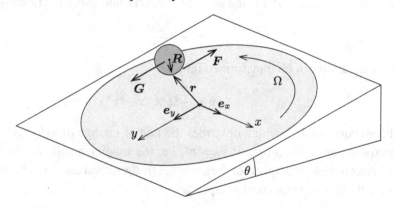

The last of these can be immediately expressed in terms of the given data:

$$G = mg \sin\theta\, e_y. \tag{1}$$

The equation for the translational motion of the centre of mass of the ball is

$$G + F = m\ddot{r}, \tag{2}$$

and that for the rotational motion around the centre of mass is

$$R \times F = \frac{2}{5}mR^2\dot{\omega}. \tag{3}$$

where ω is the angular velocity vector of the ball (which changes with time), and $\frac{2}{5}mR^2$ is the ball's moment of inertia about one of its diameters.

As the disc is rough, there is no relative movement between the ball and the disc at the contact point. Hence, we can write

$$\dot{r} + \omega \times R = \Omega \times r. \tag{4}$$

Taking the time derivatives of both sides of condition (4) produces a similar relation between the acceleration of the ball's centre and its angular acceleration:

$$\ddot{r} + \dot{\omega} \times R = \Omega \times \dot{r}.$$

From this and equations (2) and (3), we obtain

$$\ddot{r} + \frac{5}{2mR^2}[R \times (m\ddot{r} - F)] \times R = \Omega \times \dot{r}.$$

Using the standard identity for a triple vector product,[33] introducing the notation $v = \mathrm{d}(r - R)/\mathrm{d}t = \dot{r}$ for the velocity of the centre of mass, and, finally, taking into account equation (1) yields

$$\dot{v} = \frac{2}{7}(\Omega \times v) + \frac{5}{7}g \sin \theta \, e_y.$$

Noting that e_y can be written as $-(\Omega \times e_x)/\Omega$ and using the notation

$$v^* = \frac{5g \sin \theta}{2\Omega} e_x, \tag{5}$$

this equation can be transformed into

$$\dot{v} = \frac{2}{7}\Omega \times (v - v^*). \tag{6}$$

From equation (6), which describes the rate of change of velocity of the centre of mass, it can be seen that, if $v = v^*$, i.e. the magician rolls the rubber ball in the e_x direction with a speed of $5g \sin \theta/(2\Omega)$, then the centre of the ball travels in a straight line and at a constant speed.

> *Notes.* 1. Notice that the result does not depend upon the initial position of the ball, i.e. if the ball is rolled away from any point of the disc with velocity v^*, then the trajectory of its centre is a straight line parallel to the x-axis. But the required initial angular velocity of the ball does depend upon the position of release; its value can be found using equation (4).
>
> 2. The question may arise of how the ball moves if its initial velocity is not equal to v^*. It follows from equation (6) that if vector u is defined as $u = v - v^*$, then the rate of change of this new vector can be described by the following equation:

---

[33] See the mathematical Appendix at the end of the book. Note that R is orthogonal to both F and $\ddot{r}$.

$$\dot{u} = \frac{2}{7}\boldsymbol{\Omega} \times \boldsymbol{u}.$$

This means that vector $\boldsymbol{u}$ rotates uniformly with an angular velocity of $\frac{2}{7}\boldsymbol{\Omega}$ (as we saw in the solution to the previous problem). So the terminal point of vector $\boldsymbol{v} = \boldsymbol{v}^* + \boldsymbol{u}$ sweeps over a circle with centre $\boldsymbol{v}^*$ and radius $\boldsymbol{u}$. This means that, when $\boldsymbol{v} \neq \boldsymbol{v}^*$, the centre of the ball moves along a cycloid, which can be a curtate or prolate cycloid, or a trochoid (common cycloid), depending upon the initial speed.

S57 It can be shown, as we do below, that the ball – even though it is pulled down by the gravitational force – executes *simple harmonic motion* in the vertical direction. This surprising behaviour – an interesting 'interplay' between the ball's rotation and the motion of its centre of mass – is caused by the frictional force acting at the wall of the tube.

Denote the centre of the ball by C, and its temporary contact point with the wall of the tube by P. It is convenient to resolve the ball's internal angular velocity vector $\boldsymbol{\omega}$ into three mutually perpendicular components: the component ω_3 is vertical, ω_2 is directed along PC and component ω_1 is both horizontal and perpendicular to PC. During the motion, not only do the magnitudes of these components change, but, for ω_1 and ω_2, their directions also rotate – in the same way as segment PC does. We denote by Ω the magnitude of the ball's 'orbital' angular velocity around the axis of the tube.

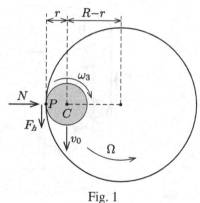

Fig. 1

Consider the top view of the ball and cylinder shown in Fig. 1, where both of the rotations around vertical axes are indicated, as are the horizontal forces acting on the ball. The latter are the normal reaction force N, which is perpendicular to the wall of the tube, and the horizontal component F_h of the static frictional force. The other forces acting on the ball (i.e. the vertical component of the static frictional force, and the gravitational force) are not shown in this figure.

By considering how C moves, we can write one of the conditions for a rolling motion without slipping as

$$(R - r)\Omega = r\omega_3, \qquad \text{that is} \qquad \omega_3 = \frac{R - r}{r}\Omega. \qquad (1)$$

It should be noted that the spin of the ball and the 'orbital' angular momentum of its centre of mass have *opposite* directions.

The only force acting on the ball that can change the angular velocity components Ω and ω_3 is the tangential static frictional force F_h. The torque of this force in the direction shown in Fig. 1 would increase Ω; but it would also decrease ω_3 at the same time. This is impossible, because according to condition (1) the ratio of the two angular velocity components is fixed!

The resolution of this apparent contradiction is that $F_h = 0$, and that neither Ω nor ω_3 can change during the motion. It follows that the horizontal line segment joining the centre of the ball to the tube axis, together with the segment PC, rotates *uniformly*. The magnitude of its (constant) angular velocity is determined by the initial velocity v_0 of the ball:

$$\Omega = \frac{v_0}{R - r} = \text{constant}.$$

For the description of the vertical motion of the ball, we choose a Cartesian coordinate system, whose z-axis coincides with the axis of symmetry of the tube. The vertical component of the ball's centre-of-mass velocity is denoted by v_z, and that of the static frictional force by F_v (see Fig. 2).

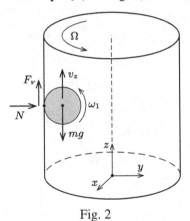

Fig. 2

The linear velocity component v_z and the angular velocity component ω_1 are not independent of each other; because there is no slipping, we must have, in addition to (1), that

$$v_z = r\omega_1. \qquad (2)$$

We now write Newton's equation for the vertical motion of a ball of mass m and moment of inertia $\frac{2}{5}mr^2$, together with the equations for rotational motion around the x- and y-axes. Because of the cylindrical symmetry present, we can, without loss of generality, take the x coordinate of the ball to be zero (in accord with Fig. 2) at the relevant moment, and then the following equations apply:

$$F_v - mg = m\dot{v}_z, \tag{3}$$
$$F_v r = -\frac{2}{5}mr^2\dot{\omega}_x, \tag{4}$$
$$0 = \dot{\omega}_y. \tag{5}$$

The quantities ω_x and ω_y in equations (4) and (5) are the components of the angular velocity of the ball along the x and y directions. At the specified moment, they have the following connections with the previously defined angular velocity components ω_1 and ω_2:

$$\omega_x = \omega_1, \qquad \omega_y = \omega_2. \tag{6}$$

Note. Looking at the first of equations (6), we could easily conclude that $\dot{\omega}_x = \dot{\omega}_1$. Then, using equations (2)–(4), we could deduce that the ball moves downwards (along the negative z-axis) with an acceleration of magnitude $\frac{5}{7}g$, i.e. the trajectory of the ball is a downwardly directed helix with an ever-increasing pitch. But this idea is *false!*

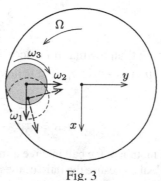

Fig. 3

The component of the ball's angular velocity in the x direction changes for two reasons (*see* Fig. 3): because of any change in magnitude of ω_1, and because the direction of the component ω_2 rotates with angular speed Ω,

$$\dot{\omega}_x = \dot{\omega}_1 - \Omega\omega_2. \tag{7}$$

A similar relation can be written for the rate of change of ω_y:

$$\dot{\omega}_y = \dot{\omega}_2 + \Omega\omega_1. \tag{8}$$

Two different expressions for the rate of change of force F_v can be obtained from equations (3) and (4); equating them, we get

$$m\ddot{v}_z = -\frac{2}{5}mr\ddot{\omega}_x.$$

Now, using the relationships (2), (5), (7) and (8), this can be transformed into a relationship containing v_z as the only variable:

$$\ddot{v}_z = -\frac{2}{7}\Omega^2 v_z(t).$$

Perhaps surprisingly, this shows that the ball's vertical speed oscillates with an angular frequency of $\sqrt{2/7}\,\Omega$, and so explains the periodic motion.

Taking into account the initial conditions of $v_z = 0$ and $z = z_0$, we have that

$$v_z(t) = -v_z^{\max} \sin\left(\sqrt{\frac{2}{7}}\,\Omega t\right)$$

and that the vertical position of the ball is given by

$$z(t) = z_0 - z_{\max}\left[1 - \cos\left(\sqrt{\frac{2}{7}}\,\Omega t\right)\right],$$

where $z_{\max}$ is the amplitude of the oscillation. The latter can be found by using the initial condition $\omega_2 = 0$; at that moment $\dot{\omega}_x = \dot{\omega}_1$. Inserting this into equation (4), and using (2) and (3), we get the following result:

$$(\dot{v}_z)_{t=0} = -\frac{5}{7}g.$$

But this is nothing more than the maximal acceleration of the ball's centre of mass, which can also be written as $-\frac{2}{7}\Omega^2 z_{\max}$. It follows that the amplitude of the vertical motion is

$$z_{\max} = \frac{5g}{2\Omega^2}.$$

> *Notes.* 1. The motion of the ball in three dimensions is not (*sensu stricto*) periodic, but quasi-periodic, because the ratio of the periods of the uniform circular motion (as seen from the top view) and the vertical harmonic oscillation (as seen from the side view) is not, in general, a rational number.
>
> 2. This phenomenon, which is perhaps astonishing at first sight, can be observed quite often in television broadcasts of golf competitions. The outside edge of a golf ball travelling too quickly towards a hole just catches the inside edge of the cylindrical cup; it appears that it should fall into the cup, but – after one period of the oscillation – it dances out again.

S58   If the tension in a spring varies throughout its length, then the total elongation $\Delta \ell$ of the spring cannot be calculated using the routine formula $T = k\Delta\ell$, where k is the relevant spring constant; we need a more sophisticated calculation.

The appropriate general method is to consider the spring as if it were divided into very small pieces, so that inside each piece the tension *can* be taken as constant. The separate extensions of these small pieces may then be calculated, and finally summed to give the overall elongation. To find the result in this way usually requires the integral calculus, but fortunately the current problem can also be solved using elementary methods.

When a spring, of mass m, is hung vertically, its fixed end carries its whole weight mg, whereas its free end carries no load, and so the average tension in the Slinky is $mg/2$; consequently, its overall extension is $mg/(2k)$, where k is its spring constant when it is uniformly stressed.

In the case of the 'sagging' spring (as shown in the figure included in the problem), each end of the spring must experience a vertically upward force of $mg/2$. In addition – because the Slinky's axis makes an initial angle of 45° with the vertical – each end must also experience a horizontal force of this same magnitude. Further, as there are no external horizontal forces acting on any of the individual segments of the spring, the horizontal component of its tension is constant throughout its length, and equal to $mg/2$. The vertical component of the tension changes from point to point because of the weight of the spring. It is zero at the midpoint, because the Slinky is horizontal there, and $mg/2$ at the end-points.

Clearly, the net tension (the square root of the sum of the squares of the horizontal and vertical components) is everywhere greater (except at the midpoint) than the horizontal component alone, i.e. than $mg/2$. So it can be stated – without the need for any more precise calculations – that the average tension is greater than $mg/2$, meaning that the elongation (namely the length of the spring) is *definitely larger* than $mg/(2k)$.

So, the elongated spring is shorter when hung vertically than it is when it is allowed to 'sag' and follow the arc described in the problem.

> *Note.* In the solution to the next problem, it is proved that the shape of a Slinky hung from both of its ends is a parabola.[34] If we denote by L_1 the overall length of the 'sagging' spring suspended so that its axis makes an initial angle of 45° with the vertical, then, using integral calculus, it can be shown that the ratio of L_1 to the overall length L_2 of a vertically hung spring is
>
> $$\frac{L_1}{L_2} = \frac{\sqrt{2} + \ln(1 + \sqrt{2})}{2} \approx 1.15.$$
>
> If this ratio L_1/L_2 is measured experimentally, smaller values, which may even be less than 1, are sometimes found. Some possible reasons for this anomaly are: that the spring 'constant' of the Slinky is not really constant when strong forces (corresponding to large extensions) are involved; that when the Slinky is hung from one of its ends, it is able to 'uncoil' (because of its weight, its lower end rotates through anything up to a few revolutions) – a spring attached at two ends cannot do this.

---

[34] This is to be contrasted with the case of a *uniform* rope, similarly suspended, for which the shape is a catenary.

S59   Take our coordinate system to be as shown in the figure, and denote the distance between the Slinky's fixed ends by d.

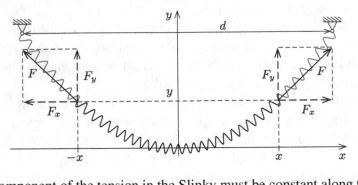

The x component of the tension in the Slinky must be constant along the spring, because any arbitrarily chosen piece of it does not accelerate in the horizontal direction, and no horizontal external force acts upon it. Now, the tension in the Slinky is parallel to its axis at any point, and so an implication of the constant horizontal component of the tension is that small equal-mass segments of the spring (with necessarily equal spring constants) also have equal horizontal projections (their various unstretched lengths all being negligible compared to their elongated lengths). Since both the segment lengths and the horizontal projections accumulate additively, it follows that the mass of any particular, but not necessarily short, piece of the Slinky is directly proportional to the length of its horizontal projection.

Next, consider the (symmetrical) piece of the Slinky indicated in the figure by the black line, and the forces of magnitude F that act upon its ends. Its horizontal length is $2x$, and so, by the previous argument, the mass of this piece is $2xm/d$, where m is the Slinky's total mass. The condition for the vertical equilibrium of this piece is

$$2F_y = \frac{2x}{d}mg.$$

Divide both sides by the (constant) horizontal component F_x of the tension:

$$\frac{F_y}{F_x} = \frac{x}{d}\frac{mg}{F_x}.$$

Now, as noted previously, the tension in the spring is always tangential, so the fraction on the left-hand side of this equation has the same value as the slope dy/dx of the Slinky at the position x:

$$\frac{dy}{dx} = \frac{x}{d}\frac{mg}{F_x}. \tag{1}$$

Since F_x is a constant, a straightforward integration now gives $y(x)$ as

$$y(x) = \frac{mg}{2F_x d} x^2 + C.$$

For the coordinate system chosen, the constant of integration C is equal to 0. We therefore have that $y \propto x^2$, and so the shape taken up by the Slinky is a parabola. The horizontal component of the tension can be expressed in terms of the spring constant of the Slinky ($F_x = kd$, as shown in the note appearing part way through the solution that starts on page 213), and the shape of the spring can be expressed solely in terms of the given values as

$$y(x) = \frac{mg}{2kd^2} x^2.$$

S60 *a*) Notionally divide the Slinky into $N \gg 1$ pieces of equal mass. The uniform mass of these small pieces is m/N, and their common spring constant k^* is N times larger than the spring constant k of the whole Slinky, i.e. $k^* = Nk$.

The tension in the nth piece from the bottom is

$$F_n = (n - 1)\frac{mg}{N}.$$

As the initial lengths of all the pieces are negligible compared to their final lengths (except, formally, for $n = 1$), their overall lengths can be approximated by their extensions. So the length ℓ_n of the nth piece is

$$\ell_n = \frac{F_n}{k^*} = (n - 1)\frac{mg}{N^2 k}.$$

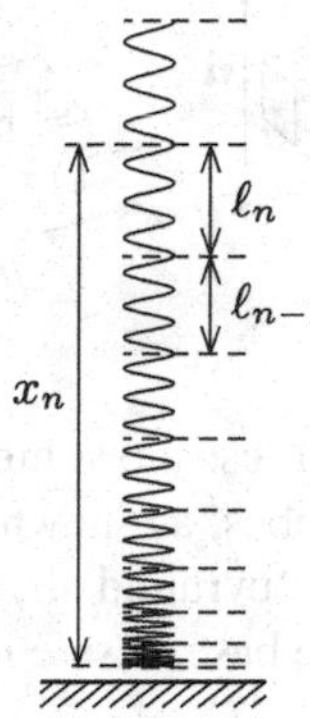

Fig. 1

The distance x_n from the top of the nth piece to the bottom of the Slinky (see Fig. 1) can be found by adding together the lengths of the pieces below it:

$$x_n = \sum_{j=1}^{n} \ell_j = \sum_{j=1}^{n} (j-1)\frac{mg}{N^2 k} = \frac{n(n-1)}{2} \frac{mg}{N^2 k} = \frac{n}{N}\left(\frac{n}{N} - \frac{1}{N}\right)\frac{mg}{2k}. \tag{1}$$

Replacing the general value n by the specific value N, and using the fact that $N \gg 1$, the overall length L of the Slinky can be expressed in terms of k and m:

$$L = \frac{mg}{2k}. \tag{2}$$

This is in line with the solution on page 204.

The work done during the lifting stage increases both the Slinky's stored elastic energy and its the gravitational potential energy, the latter because its centre of mass rises. Initially the elastic energy is zero, but in the final state it is the sum of the elastic energies of the individual pieces:

$$E_{\text{elastic}} = \sum_{j=1}^{N} \frac{1}{2} k^* \ell_j^2 = \sum_{j=1}^{N} \frac{1}{2} Nk \left[(j-1) \frac{mg}{N^2 k} \right]^2 = \frac{(mg)^2}{2N^3 k} \sum_{j=1}^{N} (j-1)^2. \tag{3}$$

Note. In the expression for the elastic energy, the sum of the squares of the first $(N-1)$ natural numbers appears. Similar summations occur in the solutions to several of the other problems, and so it is useful to show how such sums can be evaluated, either approximately or exactly. In the current situation we need the sum

$$s_n = \sum_{j=1}^{n} j^2$$

for large n.

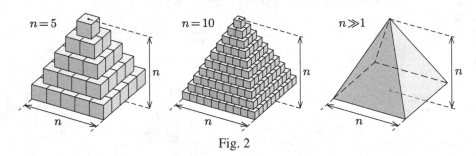

Fig. 2

This sum is more or less the volume of a square 'n-floor pyramid' built entirely out of uniform unit cubes, as shown in Fig. 2. For large n, a good approximation to the volume of the 'pyramid' is the volume of a straight-ridged geometrical pyramid with a square base of side n, and a height of n units. So we have that

$$s_n \approx n \cdot n \cdot \frac{n}{3} = \frac{n^3}{3} \tag{4}$$

as an approximation to the sum.[35]

---

[35] The precise expression for the sum, valid for all n, is $s_n = n(n+1)(2n+1)/6$. For a proof of this see, for example, K. F. Riley and M. P. Hobson, *Foundation Mathematics for the Physical Sciences* (Cambridge University Press, 2011), pp. 85 and 220. For large n, $s_n \approx n^3/3$.

Returning to the specific problem, expression (3) can be transformed with the help of the approximate formula (4) for the sum of the squares of the natural numbers:

$$E_{\text{elastic}} = \frac{1}{2} \frac{(mg)^2}{N^3 k} S_{(N-1)} \approx \frac{1}{2} \frac{(mg)^2}{N^3 k} \frac{(N-1)^3}{3} \approx \frac{1}{6} \frac{(mg)^2}{k} = \frac{1}{3} mgL.$$

In the final step, equation (2) was used.

To find the gravitational potential energy of the Slinky, we need h, the height of its centre of mass above the table-top at the end of the lifting process. Since the centre of gravity of the jth small piece is at its own centre, its distance from the table-top is $x_j - \frac{1}{2}\ell_j$. Consequently, h is given by

$$h = \frac{1}{m} \sum_{j=1}^{N} \frac{m}{N}(x_j - \tfrac{1}{2}\ell_j) = \frac{mg}{2N^3 k} \sum_{j=1}^{N} [j(j-1) - (j-1)] = \frac{mg}{2N^3 k} \sum_{j=1}^{N} (j-1)^2.$$

We note, in passing, that this expression has the same form (apart from a factor of mg) as that for the elastic energy. Making the corresponding approximation to this second sum, we have

$$h \approx \frac{mg}{2N^3 k} \frac{(N-1)^3}{3} \approx \frac{mg}{6k} = \frac{L}{3},$$

i.e. the centre of mass of the Slinky is one-third of the way up its stretched length. The gravitational potential energy relative to the plane of the table-top is thus

$$E_{\text{grav}} = mgh = \frac{1}{3} mgL.$$

So the total work done during the lift was

$$W = E_{\text{elastic}} + E_{\text{grav}} = \frac{2}{3} mgL.$$

Note. The same result can be obtained by integrating the work done by the continuously changing lifting force. When the first $n \gg 1$ pieces of the Slinky have been raised, the required force is equal to their weight,

$$F_n = \frac{n}{N} mg. \tag{5}$$

Equation (1) gives the length of the raised part of the spring as

$$x_n \approx \left(\frac{n}{N}\right)^2 \frac{mg}{2k}. \tag{6}$$

From equations (2), (5) and (6), it follows that the distance dependence of the force is $F(x) = mg\sqrt{x/L}$ and consequently that the total work done is

$$W = \int_0^L F(x)\, dx = \frac{mg}{\sqrt{L}} \int_0^L \sqrt{x}\, dx = \frac{2}{3} mgL.$$

b) After the Slinky is released, its centre of mass falls freely, so its speed just after its complete collapse is

$$v_0 = \sqrt{2gh} = \sqrt{\frac{2gL}{3}}.$$

c) The time taken to collapse is equal to the time of the free fall:

$$t = \sqrt{\frac{2h}{g}} = \sqrt{\frac{2L}{3g}}.$$

Note. Using the result of part *b)*, the kinetic energy of the Slinky can be found at the moment the collapse is complete:

$$E_{\text{kin}} = \frac{1}{2}mv_0^2 = \frac{1}{3}mgL.$$

But this is only *one-half* of the total (elastic and gravitational) energy of the Slinky before it was released! Where has the other half gone? The answer is that it has been dissipated as heat, generated by the (many) small inelastic collisions between adjacent turns of the coiled spring during its collapse.

S61   We start with a dimensional analysis of the situation, so as to 'get a feel' for what is happening. The length ℓ of the rotating spring should be a function of k, m, ω and r_0. In terms of the (relevant) base dimensions, length L, mass M and time T, the dimensions of these quantities are

$$[\ell] = \text{L}, \qquad [k] = \text{M T}^{-2}, \qquad [m] = \text{M}, \qquad [\omega] = \text{T}^{-1}, \qquad [r_0] = \text{L}.$$

It can be seen that the dimension of length appears only in r_0 (and not in k, m or ω), and so the length ℓ of the stretched spring must be proportional to r_0. The proportionality factor can only be a function of dimensionless combinations of the other three factors; here there is only one such combination, namely $\xi = \omega\sqrt{m/k}$, so

$$\ell = r_0 f(\xi). \tag{1}$$

The actual form of the function $f(\xi)$ can only be determined by detailed dynamical calculations, but, despite this, it can be stated that, in the limiting case of $r_0 \to 0$, the equilibrium length of the spring – if there is one – will be *zero*, meaning that the spring, even though it is rotating, *does not stretch* at all.

Consider next a simplified situation, in which a point-like body of mass m is attached to the end of a massless Slinky, as shown in Fig. 1.

If the length of the rotating spring is stable, it can be found from the equation for circular motion of the point-like body:

$$m(r_0 + \ell)\omega^2 = k\ell,$$

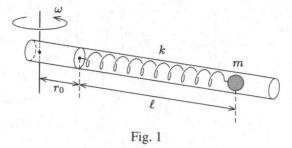

Fig. 1

from which

$$\ell = r_0 \frac{(m\omega^2/k)}{1 - (m\omega^2/k)} = r_0 \frac{\xi^2}{1 - \xi^2}.$$

Clearly, this solution is valid only if $\xi < 1$, i.e.

$$\omega < \sqrt{\frac{k}{m}} \equiv \omega_0.$$

For this range of angular velocities, if $r_0 \to 0$, then length ℓ becomes zero.

However, if $\omega > \omega_0$, then the tension in the spring cannot provide the centripetal force required to maintain the circular motion, whatever the value of ℓ. So, the 'ideal' Slinky can never be in equilibrium and, accordingly, it will be 'stretched to *infinity*', even for arbitrarily small values of r_0.

It is obvious that the conclusion drawn from equation (1) using dimensional analysis needs to be refined. If the function $f(\xi)$ becomes singular (infinitely large) somewhere, then for any arbitrary r_0 (including the limiting case $r_0 \to 0$), the length ℓ is calculated to be infinitely large. In reality, there is a natural limit to the elongation: the uncoiled length of the material forming the turns of the spring. Further, Hooke's law is unlikely to be obeyed long before this point is reached.

Let us now return to the original case of a spring with a continuous mass distribution. We specify any particular point on the spring by the mass m^* of that portion of the spring that lies between that point and the anchor point, and denote the distance between it and the rotational axis by $r(m^*)$; the tension in the spring at the same point is $F(m^*)$ (*see* Fig. 2). These two functions must satisfy the following boundary conditions (one at each end of the Slinky),

$$r(0) = r_0 \qquad \text{and} \qquad F(m) = 0. \tag{2}$$

Consider a small piece of the spring with mass Δm^* (and indicated by the heavy line in Fig. 2); its spring constant is $k(m/\Delta m^*)$. Consequently, the extension of this piece, caused by the spring tension at this point, is

$$r(m^* + \Delta m^*) - r(m^*) \equiv \Delta r = \frac{\Delta m^*}{mk} F(m^*).$$

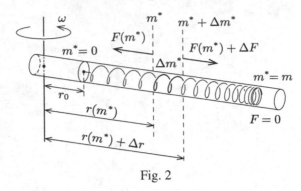

Fig. 2

The equation governing the circular motion of the small piece of spring is

$$F(m^* + \Delta m^*) - F(m^*) \equiv \Delta F = -\Delta m^* r(m^*) \omega^2.$$

These two equations can be transformed into differential equations as follows:

$$\frac{\Delta r}{\Delta m^*} \quad \rightarrow \quad \frac{dr}{dm^*} = \frac{1}{mk} F(m^*), \tag{3}$$

$$\frac{\Delta F}{\Delta m^*} \quad \rightarrow \quad \frac{dF}{dm^*} = -\omega^2 r(m^*). \tag{4}$$

We now note that the forms of (3) and (4) parallel the equations governing the simple harmonic motion of a point-like body. Indeed, using the correspondences

$$m^* \longleftrightarrow t \qquad \text{and} \qquad \frac{F}{mk} \longleftrightarrow v,$$

the equations describing the distribution of the spring's mass, and the tension in it, can be rewritten as follows:

$$\frac{dr(t)}{dt} = v(t), \tag{3'}$$

$$\frac{dv(t)}{dt} = -\left(\frac{\xi}{m}\right)^2 r(t), \tag{4'}$$

where $\xi = \omega\sqrt{m/k}$ is the dimensionless constant encountered earlier. These equations describe the harmonic oscillatory motion of a point-like body with 'angular frequency' ξ/m. Taking account of the boundary conditions set out in (2), $r(t = 0) = r(m^* = 0) = r_0$ and $v(t = m) = (1/mk)F(m^* = m) = 0$, the solution is

$$r(m^*) = \frac{r_0}{\cos \xi} \cos\left[\xi\left(1 - \frac{m^*}{m}\right)\right], \qquad F(m^*) = \frac{\xi k r_0}{\cos \xi} \sin\left[\xi\left(1 - \frac{m^*}{m}\right)\right].$$

Note. The same result can be found by taking the derivative of equation (3) and substituting (4) into it, i.e.

$$r''(m^*) = \frac{1}{mk}F'(m^*), \qquad \text{where} \qquad F'(m^*) = -\omega^2 r(m^*),$$

leading to

$$r''(m^*) = -\frac{\xi^2}{m^2}r(m^*),$$

which is a homogeneous second-order linear ordinary differential equation. Its solution is a linear combination of sine and cosine functions with the appropriate periodicity.

The total elongation of the spring so calculated is

$$\ell = r(m) - r(0) = r_0 \left(\frac{1}{\cos \xi} - 1 \right) = r_0 \left[\frac{1}{\cos(\omega\sqrt{m/k})} - 1 \right],$$

but this expression describes the actual state of affairs correctly only if $\xi < \pi/2$, i.e. if

$$\omega < \frac{\pi}{2}\sqrt{\frac{k}{m}} \equiv \omega_{\text{critical}}.$$

If the angular velocity is slowly increased, the extension of the spring becomes larger, and as the critical angular velocity is approached it becomes (theoretically) infinite. Beyond the critical value of the angular velocity, the spring does not have any stable rotational state, and its elongation is limited only by the extent to which its coils can be straightened; clearly, Hooke's law no longer applies.

It can be seen that the behaviour of a real Slinky, with a continuous mass distribution, is essentially the same as that of the spring in the simplified model (a massless spring with a point-like body attached); only the magnitude of the critical angular velocity differs – by a factor of $\pi/2$.

S62 We use the coordinate system shown in the figure, and find the shape of the spring (assumed to have already attained its stable configuration) in this frame. Denote the distance between the ends of the Slinky by d, its mass by m and its spring constant by k. The end-points of the Slinky are rotating in phase around two circles with identical radii r_0.

Consider the forces acting on an arbitrarily chosen small piece of the Slinky (indicated in the figure by the heavy line). The x component of the tension in the Slinky must be constant along the spring (equal to F_0, say), because the small piece does not accelerate in this direction:

$$F_{1,x} = F_{2,x} = F_0.$$

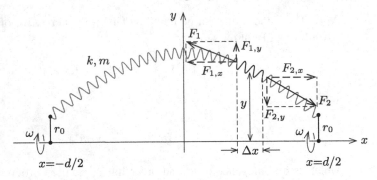

It follows that the *x*-directed projections of Slinky pieces with identical masses (and therefore with identical spring constants) are the same (the unstressed lengths of the pieces having been neglected). The *x*-directed projection of the chosen piece has a length of Δx, and so its mass can be expressed in terms of the total mass of the Slinky by $m\Delta x/d$.

In the *y* direction, the net force acting on the chosen small piece produces the centripetal acceleration $y\omega^2$ of that piece, where ω is the angular velocity of the twirling:

$$F_{2,y} - F_{1,y} = \Delta x \frac{m}{d} y\omega^2. \tag{1}$$

The direction of the tension force in the Slinky is always tangential, and so the ratio of the *y* and *x* components of the tension can be expressed in terms of the Slinky's slope at the given point:

$$\frac{F_{1,y}}{F_0} = -\left(\frac{\Delta y}{\Delta x}\right)_1, \qquad \frac{F_{2,y}}{F_0} = -\left(\frac{\Delta y}{\Delta x}\right)_2.$$

Dividing both sides of equation (1) by F_0, and using the above expressions for the slope, yields

$$\left(\frac{\Delta y}{\Delta x}\right)_2 - \left(\frac{\Delta y}{\Delta x}\right)_1 = -\Delta x \frac{m\omega^2}{F_0 d} y.$$

Now, dividing through by Δx, and taking the limit of $\Delta x \to 0$, produces the second derivative of $y(x)$ on the left-hand side and transforms the equation into a differential one:

$$\frac{\mathrm{d}^2 y}{\mathrm{d}x^2} = -\frac{m\omega^2}{F_0 d} y(x).$$

This equation is a spatial analogue of the equation governing one-dimensional simple harmonic motion and corresponds to a (spatial) angular frequency of $\Omega = \omega\sqrt{m/(F_0 d)}$. The analogous time-dependent equation is

$$\frac{d^2 r}{dt^2} = -\Omega_T^2 r(t),$$

where $r(t)$ is the position of a point-like body, measured from its equilibrium position, as a function of time. Its solution is

$$r(t) = A \cos(\Omega_T t + \varphi_0),$$

where A is the amplitude of the simple harmonic motion, and φ_0 is the initial phase. On this basis, the function describing the shape of the Slinky is

$$y(x) = A \cos\left(\sqrt{\frac{m}{F_0 d}}\, \omega x\right). \tag{2}$$

It was arranged, through the choice of coordinate system, that $y(x)$ would be an even function, and so the value of the 'initial phase' is zero.

> *Note.* The force F_0 can be related to the spring constant k of the Slinky and the distance d between its end-points by finding the elongation Δs of a small piece of the spring of mass Δm when it is subjected to a force F. As the spring constant of such a small piece is $m/\Delta m$ times the spring constant k of the whole spring, we have
>
> $$F = k \frac{m}{\Delta m} \Delta s.$$
>
> But, because the tension in the Slinky is tangential, it is also true that
>
> $$\frac{F}{F_0} = \frac{\Delta s}{\Delta x}.$$
>
> From these two equations, it follows that $F_0 \Delta m = mk \Delta x$, and then, from the summation of this result over the total length of the spring, we have $F_0 = kd$.

The 'angular frequency' of the function describing the shape of the spring can also be expressed in the form

$$\Omega = \omega \sqrt{\frac{m}{F_0 d}} = \omega \sqrt{\frac{m}{k}} \frac{1}{d} = \frac{\xi}{d},$$

where $\xi = \omega \sqrt{m/k}$ is a dimensionless constant – exactly the same one as appeared in the solution that starts on page 210.

The amplitude of the function $y(x) = A \cos(\xi x/d)$ describing the shape of the spring is determined by the boundary condition $y(d/2) = r_0$:

$$A = \frac{r_0}{\cos(\xi/2)}.$$

It follows that, if the ends of the 'skipping rope' are moving around circles of very small radius, i.e. $r_0 \approx 0$, then the spring does not move away significantly from the rotational axis ($y(x) \approx 0$), provided that $\xi < \pi$, i.e.

$$\omega < \pi \sqrt{\frac{k}{m}} < \omega_{\text{critical}}.$$

Approaching this critical angular velocity, the deflection of the spring increases, and at the critical value the amplitude would (theoretically) be infinite. But, in reality, the maximal amplitude of the spring remains finite, and close to the critical angular velocity Hooke's law is no longer obeyed.

> *Notes.* 1. The motion of the 'skipping rope' in weightless conditions is very similar to the behaviour of the spring in an earlier problem (one-dimensional elongation of the spring in a horizontal rotating tube). This is no coincidence! In fact, the equations describing the motion of the 'skipping rope' can be divided into two independent components: the equation in the direction of the rotational axis describes a uniformly stressed spring of length d; the equation for the extension perpendicular to the axis is essentially the same as that for the motion of the rotating spring in the tube. The difference between the critical angular velocities of a factor of 2 is due to the fact that it is only one-*half* of the 'skipping rope' that corresponds to the spring rotating in the tube, so its mass is $m/2$, and its spring constant is $2k$.
>
> 2. The 'half-wave' solution found above for a twirling 'skipping rope', with each end moving around a circle of small radius ($r_0 \approx 0$), is not the only one. All other cosine functions whose values are approximately zero at $x = \pm d/2$, but are non-zero for some other values of x, are possible solutions. This happens when $\cos(\xi/2) \approx 0$, that is, when $\xi \approx (2n + 1)\pi$, with n a positive integer. In this case the Slinky has $2n$ nodes – points at which there is never any movement perpendicular to the x-axis. If the ends of the spring are moved in antiphase (with a 'skipping rope', this is difficult to realise, but not impossible), then the shape of the spring is described by antisymmetric (sine) functions, and the number of nodes is odd.
>
> To see these phenomena, it is not necessary to travel to the International Space Station! If two people stand relatively far apart, so that a Slinky they are holding is significantly stretched, then – after some practice – with proper actuation of the end-points, the fundamental mode as well as some harmonics can be 'excited'.

S63 In everyday terms, the larger the bending moment applied to an elastic rod, the more the elastic rod curves. Strictly speaking, however, this observation should be stated as: At any given point on a bent beam the curvature is directly proportional to the *local* bending moment. For the tree branch, this means that the (small) vertical displacement y at position x along the branch (measured from the tree trunk) under a discrete load W at $x = a$ obeys an equation of the general form

$$\frac{d^2y}{dx^2} \propto W \times \text{(a linear function involving } x \text{ and } a\text{)}.$$

It follows from this that the displacement has a cubic dependence on the lengths involved. In particular, for a branch of length ℓ, if $a = \ell$, the displacement at the free end is $y(\ell) \propto W\ell^3$.

Direct application of this result shows that, if the displacement (sag) of the free end of a horizontal cantilevered rod of length ℓ, loaded vertically at its free end by a force F, is h, then a rod (of the same cross-section) that has length $\ell/2$, and is loaded at its free end with a force $4F$, sags by only $h/2$. This observation can be applied directly to the analysis of the two birds perching (separately) on the tree branch.

As the mass of the pigeon sitting at the midpoint of the branch is four times larger than that of the blackbird, if the *free end* of the branch moves down by h_1 for the blackbird, the *midpoint* of the branch is lowered by one-half of this, namely $h_1/2$, by the weight of the pigeon (*see* figure).

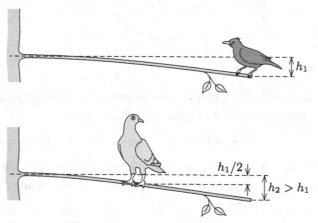

However, we are determining the sag of the *end* of the branch in both cases, and in that of the pigeon the branch is unloaded and straight from its midpoint to its free end. Because the curvature of this natural cantilever increases monotonically with distance from the tree trunk, the angle that the straight part of the branch makes with the horizontal (equal to that of the tangent at the phalf-way point) is greater than that made by the tangents at all other points between the fixed end and the midpoint; the latter produce an accumulated drop of $h_1/2$ over half the length of the branch. The straight-line part of the branch, which has the same length, will produce a further drop that is more than this. Thus the total drop will be greater than h_1, showing that the tree branch's free end will be depressed more by the pigeon at its midpoint than by the blackbird on its extremity.

> *Note.* The precise ratio of the displacements of the tree branch's free end can be found from a calculation of the shape of the loaded branch; the result is 5 : 4 in favour of the pigeon.

S64   Consider the elastic metal wire as divided into small *identical-length* segments. These small segments are curved even in an unloaded state, and, under loading, their curvatures either increase or decrease, depending on the direction of the flexural torque induced. The deformation of a small segment can be characterised by the angle ε, defined as the small difference in angle between the tangents to the wire at the two ends of the segment. As in the bending of an initially straight rod, the angle ε is directly proportional to the flexural torque in the small segment.

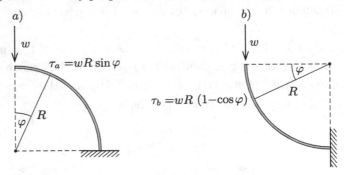

Fig. 1

The torques acting upon small segments of the two wires at corresponding positions characterised by the angle φ (*see* Fig. 1) are

$$\tau_a = wR\sin\varphi \qquad \text{and} \qquad \tau_b = wR(1 - \cos\varphi),$$

where w is the (vertical) load on the end of the wire. The small segments are deformed to the extent necessary to produce compensating (local) flexural torques. Now

$$\sin\varphi = 2\sin\frac{\varphi}{2}\cos\frac{\varphi}{2} = 2\sin^2\frac{\varphi}{2}\cot\frac{\varphi}{2} = (1 - \cos\varphi)\cot\frac{\varphi}{2},$$

and, in the range $0 \leq \varphi \leq \pi/2$, the factor $\cot(\varphi/2) \geq 1$. Thus $\sin\varphi \geq 1 - \cos\varphi$, with equality only at $\varphi = 0$ and $\varphi = \pi/2$, i.e. at the ends of the wire. It follows that the flexural torques at corresponding points cannot be smaller in case *a*) than in case *b*), i.e. except at two isolated points, where equality holds, $\tau_a(\varphi)$ is always larger than $\tau_b(\varphi)$.

We next determine the vertical displacements of the wires' end-points that are associated with the bending that occurs in equal small segments of the two wires, both segments corresponding to a particular angle φ. Denote the angular deflections (which are also the deflections of the arcs $\overset{\frown}{PA}$ and $\overset{\frown}{PB}$) of the segments at points P by ε_a in case *a*) and ε_b in case *b*), as shown in Fig. 2.

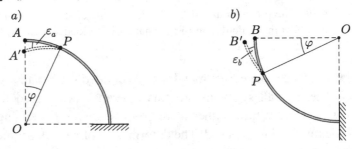

Fig. 2

First, we note that both deformations are proportional to the flexural torque at P, so

$$\frac{\varepsilon_b}{\varepsilon_a} = \frac{\tau_b}{\tau_a} < 1 \quad \text{for} \quad 0 < \varphi < \frac{\pi}{2}.$$

Next, these angular deflections have to be translated into the vertical components of the consequent movements of the tips of the wires. Because of the way that φ has been defined, the distances PA and PB are the same, each equal to $\ell = 2R \sin \varphi/2$, and so the movement produced by a deflection ε is $\varepsilon\ell$ in length, and perpendicular to PA or PB, as appropriate. Now, from the geometry of the equilateral triangles AOP and BOP, the angle between AP and the (downward) vertical is $\frac{1}{2}(\pi-\varphi)$, while that between BP and the vertical is $\frac{1}{2}\varphi$. The (downward) vertical components of the tips' movements are therefore:

$$\Delta h_a = \varepsilon_a\ell \sin \tfrac{1}{2}(\pi - \varphi) = \varepsilon_a\ell \cos \tfrac{1}{2}\varphi,$$

$$\Delta h_b = \varepsilon_b\ell \sin \tfrac{1}{2}\varphi.$$

Thus, the ratio of the droops Δh of the tips of the wires is

$$\frac{\Delta h_b}{\Delta h_a} = \frac{\varepsilon_b \sin(\varphi/2)}{\varepsilon_a \cos(\varphi/2)} = \frac{\varepsilon_b}{\varepsilon_a} \tan \frac{\varphi}{2} < \tan \frac{\varphi}{2} < 1 \quad \text{for} \quad 0 \le \varphi \le \frac{\pi}{2}.$$

So we have shown that (apart from at the end-points) the contributions that the corresponding stressed wire segments make to the lowering of their respective peg tips is always larger in case *a*) than in case *b*). As the total drop can be built up from the contributions of the individual small segments of the wire, it can be concluded that the lowering of the hat peg's tip is larger for design *a*) than for design *b*).

> *Note.* The numerical ratio of the two droops can be found using energy considerations and the integral calculus.
>
> If the end of the curved hat peg is slowly loaded up to a maximal weight of w, and as a consequence it sinks by Δh, then the work done is $W = \frac{1}{2}w\Delta h$ – the factor of $\frac{1}{2}$ arising because the average force is one-half of its maximal value. This work must be equal to the total elastic energy stored in the slightly deformed wire, which is the sum of the stored energies in the individual segments of the wire. The elastic stored energy of a segment – using the analogy of the energy formula of a stressed ordinary coil spring ($\frac{1}{2}kx^2 = T^2/2k$) – is proportional to the length $R\Delta\varphi$ of the segment and the square of the torque τ acting in it. Accordingly, the ratio of the droops of the two different hat pegs is
>
> $$\frac{\Delta h_a}{\Delta h_b} = \frac{W_a}{W_b} = \frac{\int \tau_a^2(\varphi)R\,d\varphi}{\int \tau_b^2(\varphi)R\,d\varphi} = \frac{\int_0^{\pi/2} \sin^2(\varphi)\,d\varphi}{\int_0^{\pi/2}(1 - \cos\varphi)^2\,d\varphi}$$
>
> $$= \frac{\frac{1}{2}\pi/2}{\pi/2 - 2 + \frac{1}{2}\pi/2} = \frac{\pi}{3\pi - 8} \approx 2.2.$$

The same result can also be obtained by integrating the 'elementary' contributions to the depression of the two peg tips, as shown in Fig. 2.

S65   Consider first a rod of length L, initially horizontal, whose end deflects by h when it is loaded with a weight w (*see* Fig. 1). We approximate the shape of the deflected rod by an arc of a circle of radius R.

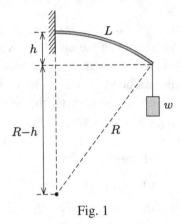

Fig. 1

Note. As is well known, the curvature (the reciprocal of the local radius of curvature) of any particular segment of a rod is proportional to the torque acting upon it. This is why, for the current configuration, the curvature is largest at the end where the rod is fixed, and decreases continuously to zero as the loaded end is approached.

Although the curvature cannot be constant, and the curved rod cannot be part of a true circle, for the sake of simplicity – and as we are only aiming for an estimate, and not an accurate result, for the critical load F – this rough approximation will be used.

Within this approximation, the radius of curvature can be found using Pythagoras's theorem on the right-angled triangle shown in the figure:

$$R^2 \approx (R - h)^2 + L^2.$$

Here, because the vertical deflection is very small, the length of the horizontal projection of the deflected rod has been taken as L.

Now since $h \ll L$, this approximate equation can be reduced to

$$R \approx \frac{L^2}{2h} \approx 50 \text{ m}. \tag{1}$$

We now investigate the energy implications of the situation. Imagine that the bent state of the rod was brought about by pushing the end of the rod down slowly (for this, a gradually increasing force was required, at any stage proportional to the then current deflection), until the given value of h was reached, and then the weight w was hooked onto the rod. As the exerted force was proportional to the displacement, the average force was one-half of its maximal value, and so the work done was

$$W = \frac{F_{\max}}{2}h = \frac{w}{2}h. \qquad (2)$$

This work was all stored as increased elastic energy in the rod (the gravitational change of the rod's own weight being negligible). It is well known (and also stated in the hint) that the stored elastic energy of a deformed rod is directly proportional to its length and inversely proportional to the square of its radius of curvature, i.e.

$$W = C\frac{L}{R^2}. \qquad (3)$$

Here C is a proportionality factor characterising the rigidity of the rod, and can be easily found from (1), (2) and (3):

$$C = \frac{w}{2}h\frac{R^2}{L} = \frac{wL^3}{8h} = 125 \text{ J m}.$$

Note. The same result can be found if the energy of the system, consisting of the rod and the hung weight w, is investigated as a function of the deflection, which varies along the rod. In equilibrium, the total energy (in which the elastic energy term can be approximated by a quadratic function of the deflection) is minimal. In this scenario, if the elastic energy of the rod is equated to the decrease in the gravitational potential energy of the load, we get a false result – wrong by a factor of 2. This is because such a calculation corresponds to letting the weight fall, and so the kinetic energy acquired by the load also needs to be taken into account if conservation of energy is to be applied correctly.

Consider now the loaded vertical rod – usually known as an Euler strut. Imagine that a weight is placed on the top of the rod. If this weight is smaller than the critical value of F, then the rod will be compressed to a small extent (as for a loaded supporting pillar), but it will not buckle. When the critical value of F is reached, the rod does buckle, and the weight moves down through a small distance x, as shown in Fig. 2. In this case, the elastic energy increase of the rod is the result of the potential energy decrease of the load.[36]

The shape of the buckled rod is again approximated by an arc of a circle; its radius is denoted by r, which bears no particular relation to the R used previously. The angle θ shown in Fig. 2 (and measured in radians) is

$$\theta = \frac{L}{2r}.$$

Since, when buckling starts, r is large, θ is small and the following approximation can be used, when needed:[37]

---

[36] If x is small, the loading force is practically constant during the buckling, and so equating the two kinds of energy change is correct; no error of a factor of 2 is generated. See the previous note.

[37] These are the first two terms of the standard Maclaurin series for $\sin\theta$, and can be checked numerically, even with a simple scientific calculator. See also the approximate formulae given in the Appendix.

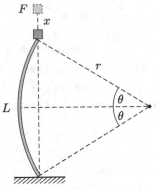

Fig. 2

$$\sin \theta \approx \theta - \frac{1}{6} \theta^3.$$

The increase of elastic energy in the rod is

$$\Delta E_1 = C \frac{L}{r^2},$$

and the decrease in potential energy of the weight is

$$-\Delta E_2 = Fx = F(2r\theta - 2r \sin \theta) \approx 2Fr \frac{\theta^3}{6} = \frac{1}{24} \frac{FL^3}{r^2}.$$

The condition for 'spontaneous' buckling of the rod is $|\Delta E_2| > \Delta E_1$, i.e.

$$\frac{1}{24} \frac{FL^3}{r^2} > C \frac{L}{r^2}.$$

This condition – which is independent of the value of r – will always be met if

$$F > F_{\text{critical}} = 24 \frac{C}{L^2} = 3w \frac{L}{h} \approx 3000 \text{ N}.$$

Notes. 1. The value of the critical force (in the approximation used above) is independent of r, i.e. the critical force does not depend on the extent of the buckling (provided r is not too small), and so in practice the vertical displacement x of the rod's uppermost point can be arbitrary. Realistically, this means that, when the vertical load acting on the supporting column reaches the critical value, the structure collapses or the rod becomes permanently bent.

2. The equations of linear elasticity (ones in which Hooke's law holds) show that a more accurate value of the critical loading force is

$$F_{\text{critical}} = \frac{\pi^2}{3} w \frac{L}{h} \approx 3300 \text{ N}.$$

Our calculation, which (wrongly) assumes a constant curvature, can, at best, be expected to give only an approximate result. Despite this, our estimation differs from the more accurate value only by a factor of $9/\pi^2$, i.e. the critical loading force has been found to within a 10 % error. It is perhaps surprising that the rough approximation used – which makes no use of any calculus – gives such a good estimate of the critical load.

S66   For the sake of simplicity, we will treat each cable strand as if it were long enough to be at the limit of its tensile strength, though of course, in reality, only slightly shorter cable segments could be connected to each other. In addition, the weight of elements used to connect the cables together will be neglected.

Let us start from the bottom of the composite cable, and move successively upwards. The very lowest segment, with a single cable, can be almost 1 km long. If we use two cables in the next segment, then their length can be 0.5 km, because then the tension reaches the critical value at the top of each of them. In the next segment, also 0.5 km long, four cables are required, and so on. The number of parallel cables in any given segment is always double that in the segment just below it. Accordingly, to reach a depth of 3 km, at least $1 + 1 + 2 + 4 + 8 = 16$ km of cable is necessary.

With increasing target depth, the required length of cable increases very rapidly. If the composite cable is lowered to a depth of h (measured in km), then the length of original cable needed (also in km units) is

$$L(h) = 1 + (1 + 2 + 4 + \cdots + 2^{2h-3}) = 2^{2(h-1)},$$

which is more than one million kilometres of cable for $h = 11$ km!

Note. A somewhat more economical procedure can be used, if the number of cables changes more frequently than at each 0.5 km, but it can be proved that the total length of cable used (measured in km) cannot be less than e^h. For a depth of 11 km, this length is approximately 60 000 km, which is still a pretty large drum of cable – and would go round the Equator one-and-a-half times!

S67   Whether the sausage is straight or torus-shaped, the split will occur along a line of maximal elastic stress (for a given internal pressure).

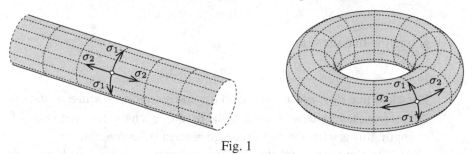

Fig. 1

As the thickness of the sausage skin is assumed uniform, instead of working with the elastic stress (with the dimensions of force/area), we can investigate the tensile forces acting on unit length of the sausage wall. Henceforth, this quantity will be called the *line stress*, and denoted by σ. Unlike the surface tension of liquids – which is a similar quantity – the line stress can depend upon the location of the line segment under investigation, and in general can be both anisotropic and inhomogeneous, as is illustrated in Fig. 1.

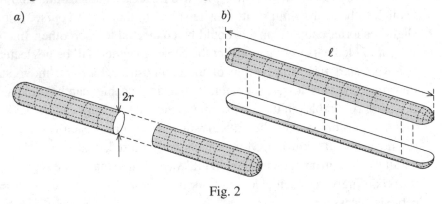

Fig. 2

Consider first the straight sausage. Let the length of the sausage be ℓ, its radius r and the overpressure (gauge pressure) within it Δp. If the sausage is imagined as being cut lengthwise into two equal pieces (*see* Fig. 2a), then, neglecting the effects of its semi-spherical ends, the two halves could be held together by a force of $2\ell\sigma_1$. However, the internal pressure tries to push apart the two halves of the sausage with a force of $2r\ell\,\Delta p$. As the sausage is in equilibrium:

$$2\ell\sigma_1 = 2r\ell\,\Delta p,$$

and so the sidewise line stress, perpendicular to the lengthwise cut, is

$$\sigma_1 = r\Delta p. \tag{1}$$

The same 'cut' method can be applied to find the lengthwise line stress acting on the sidewise segments (*see* Fig. 2a):

$$2\pi r\sigma_2 = r^2\pi\,\Delta p,$$

giving the line stress as

$$\sigma_2 = \tfrac{1}{2}r\Delta p. \tag{2}$$

It is clear that σ_2 is *always* less than σ_1, and so during boiling, as the overpressure Δp increases, the sidewise stress is the first to reach the limiting value of its tensile strength; this is why a straight sausage always splits *lengthwise*.

We now investigate whether the simple considerations used above can also be applied to a toroidal sausage. Let the major radius of the torus (the distance from the centre of the tube to the centre of the torus) be R, the minor radius (the radius of the tube) be r, and the internal overpressure again be denoted by Δp. Denote –

as for the straight sausage – the 'sideways' stress in the wall of the torus by σ_1, and the 'lengthwise' stress (perpendicular to the latter) by σ_2, as shown in Fig. 1.

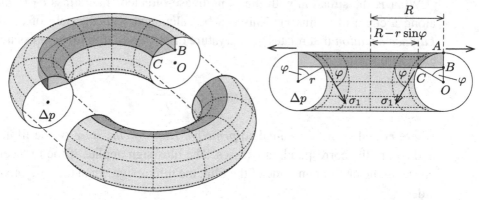

Fig. 3

As a thought experiment, let us imagine the torus cut by a plane that is parallel to (and above) the one that contains the central axis of the torus (assumed horizontal) – and consider the inner 'half' of the (upper) cut piece (i.e. all of its parts that are closer to the symmetry axis than R, and shown by dark grey shading in Fig. 3). If the vertical position of the cut is characterised by the tangential angle, φ, at the level where it meets the torus, then the equilibrium of the vertical components of the forces acting on the (dark grey) solid of revolution can be written as follows:

$$[\pi R^2 - \pi(R - r \sin \varphi)^2]\Delta p = 2\pi(R - r \sin \varphi)\sigma_1 \sin \varphi. \qquad (3)$$

To obtain this equation for the balance of *vertical* forces, we have equated (i) the vertical force that the internal overpressure exerts on the annulus of revolution produced by rotating (only) the segment CB about the axis and (ii) the vertical component of the line stress acting on a circular line passing through C. Because of the rotational symmetry, σ_1 is the same for all parts of the circle. At A the line stress has no vertical component. The weight of the solid of revolution has been ignored because, compared to the effect of the increasing overpressure Δp, it is negligible.

Using equation (3), the sidewise line stress can be found at an arbitrary angle φ:

$$\sigma_1 = r\Delta p \, \frac{1 - (r/2R) \sin \varphi}{1 - (r/R) \sin \varphi}. \qquad (4)$$

The sidewise stress on the 'outer' side of the torus can be calculated in a similar way; it is again given by expression (4), but for values of φ that lie in the range $\pi < \varphi < 2\pi$. It can be seen that the sidewise line stress depends on the ratio of r/R, and, in the case of $R \gg r$, expression (4) reproduces formula (1) for the straight frankfurter. Furthermore, σ_1 depends on position, through the variable φ, meaning that the stress is *inhomogeneous*. Its largest value occurs along the inner circle of the torus (at $\varphi = \pi/2$).

What is the situation with the 'lengthwise-directed' line stress σ_2 (which acts along a circle of radius r)? This can be calculated with the help of the *Young–Laplace equation* that relates the curvature of an elastic film to the pressure difference across it:

$$\Delta p = \frac{\sigma_1}{r_1} \pm \frac{\sigma_2}{r_2},$$

where r_1 and r_2 are the so-called *principal radii of curvature* of the film, and σ_1 and σ_2 are the corresponding stresses. The plus sign applies if the two centres of curvature lie on the same side of the film; the minus sign when they are on opposite sides.

> *Note.* Two particular applications of the Young–Laplace formula are to the pressure difference across an elastic cylindrical surface (*see* expression (1)) for which $r_1 = r$ and $r_2 = \infty$, and to films that form part or all of a sphere, for which $r_1 = r_2$ and $\sigma_1 = \sigma_2$ (examples include soap bubbles and capillary action in a tube).

The principal radii of curvature of surfaces of revolution can be calculated quite simply. If the curve c is rotated about the axis a (in Fig. 4 it results in a pear-like surface), one of the principal radii at a point P on the surface is simply the radius of the osculating circle, g, at P on the curve c. The other is the distance, in a 'slantwise' (perpendicular to the tangent to the curve) direction, from P to the rotational axis. As noted earlier, in the Young–Laplace formula the $\pm$ sign is positive if both curvatures 'pull' the surface in the same direction, and it is negative if the 'directions' of the two curvatures are opposing (as happens at the point marked P in Fig. 4).

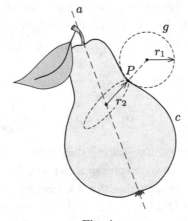

Fig. 4

At point C of the torus, the principal radii are:

$$r_1 = r \qquad \text{and} \qquad r_2 = \frac{R - r \sin \varphi}{\sin \varphi},$$

and the two curvatures are oppositely directed. According to the Young–Laplace formula

$$\sigma_2 = \frac{R - r \sin \varphi}{\sin \varphi} \left(\frac{\sigma_1}{r} - \Delta p \right),$$

which can be transformed by substituting expression (4) for the stress σ_1:

$$\sigma_2 = \frac{R - r \sin \varphi}{\sin \varphi} \left(\frac{1 - (r/2R) \sin \varphi}{1 - (r/R) \sin \varphi} - 1 \right) \Delta p = \frac{1}{2} r \Delta p. \tag{5}$$

It is strange that σ_2 does not depend upon φ (and so is the same everywhere on the torus). Further, it is independent of the ratio of r/R (which characterises the 'slenderness' of the torus), and, in particular, for $R \to \infty$, it equals the 'lengthwise' line stress of the straight sausage as given by (2).

From equations (4) and (5), it can be seen that for any location, and with any ratio of r/R, $\sigma_1 > \sigma_2$, and so the imaginary idealised toroidal sausage (with uniform skin thickness) would also split *lengthwise*. As σ_1 is largest *along the inner circular line* of the torus (at $\varphi = \pi/2$), the rupture would be expected to occur there.

S68 We will show that the depression of the rope is unaltered when the load and measurement points are interchanged – a result independent of their actual positions.

Let the tension in the rope, which has length ℓ, be F, and let us measure horizontal distances from the left-hand end of the rope. We assume that F is always the same throughout the tight-rope, though, in reality, the tensions to the left and right of the acrobat are slightly different; this small discrepancy can be neglected because of the small deflections.[38]

Suppose that, with the load at position x, the depression there is d, while the rope at y sinks by h. Contrariwise, under the same load at position y, the dip there is d', while the rope at x is lowered by h' (*see* figure).

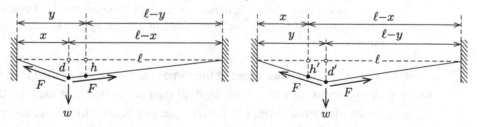

---

[38] With an obvious notation, the actual condition is $F_L \cos \theta_L = F_R \cos \theta_R$, but because both θ_L and θ_R are very small, their cosines are both ≈ 1, and so $F_L \approx F_R$. This approximation neglects terms of order θ^2; the terms in equation (2) are of order θ.

From the ratios of the sides of pairs of similar triangles in the two configurations, we have

$$\frac{d}{\ell - x} = \frac{h}{\ell - y} \quad \text{and} \quad \frac{d'}{y} = \frac{h'}{x}. \tag{1}$$

Since the sums of the vertical components of the tensions in the two parts of the tight-rope must each (separately) balance the acrobat's weight, we also have that

$$F\left(\frac{d}{x} + \frac{d}{\ell - x}\right) = F\left(\frac{d'}{y} + \frac{d'}{\ell - y}\right). \tag{2}$$

Here, the sines of the small angles involved have been approximated by their tangents.

From equation pair (1), the ratio d/d' is given by

$$\frac{d}{d'} = \frac{\ell - x}{\ell - y} \cdot \frac{x}{y} \cdot \frac{h}{h'},$$

while from (2) it follows that

$$\frac{d}{d'} = \frac{\dfrac{1}{y} + \dfrac{1}{\ell - y}}{\dfrac{1}{x} + \dfrac{1}{\ell - x}} = \frac{\ell - x}{\ell - y} \cdot \frac{x}{y}.$$

These results can only be consistent if $h' = h$.

This result is clearly general, but now to answer the specific question posed. When the acrobat is at T, then the depression of the rope at Q is just the same (5 cm) as that of point T when he was at Q.

> *Note.* The depression of the point on the rope specified by y, under a unit load acting at point x, is the so-called *impulse response function* $G(x, y)$ of the 'system'. The symmetry property, $G(x, y) = G(y, x)$, of this function of two variables plays an important role in mechanics, electrostatics and the description of waves, as well as in quantum theory. In honour of the British mathematical physicist George Green (1793–1841), the impulse response of a linear system is called its Green's function.

S69 Cut out a solid plane sheet of the same size and shape as the given triangle, and fix an eye-screw (a screw with a small ring as its head) at each of the three vertices of the triangular sheet. Fix the triangle in a horizontal plane, and pass thin threads of lengths L_1, L_2 and L_3 through the eye-screws. Knot together one end of each of the three threads, and attach a bob of mass m to each of the other ends (*see* figure).

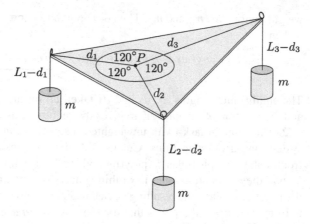

The mechanical energy of this system is equal to the sum of the gravitational potential energies of the three bobs, and, taking the zero level to be the plane of the triangle, this is

$$E_{\text{potential}} = -mg(L_1 - d_1) - mg(L_2 - d_2) - mg(L_3 - d_3).$$

This can be rearranged as

$$E_{\text{potential}} = \underbrace{mg}_{\text{constant}} \cdot (d_1 + d_2 + d_3) - \underbrace{mg(L_1 + L_2 + L_3)}_{\text{constant}}.$$

It can be seen that the total energy is minimal if the sum of the distances, $d_1 + d_2 + d_3$, is as small as possible. So, when the system is in equilibrium, the position of the knot gives the required point P on the triangle. Further, since each has the value mg, the tensions in the threads are all equal; consequently, the threads must meet at angles of 120°.

This latter observation shows how P could be constructed geometrically. If, for each side of the triangle, a circular arc on which that side subtends an angle of 120° is drawn within the triangle (by, for example, tracing the vertex of a series of arbitrary 120° triangles based on that side), then P is defined by the common intersection of all three arcs. However, there are several other, better-defined, purely geometrical procedures for determining the location of P, which is usually known as the *Fermat–Torricelli point* of the triangle.[39]

> *Notes.* 1. The mechanical device can be simply adapted to the more general question of seeking the point inside the triangle for which a *weighted* sum of the distances, $S = m_1 d_1 + m_2 d_2 + m_3 d_3$, is minimal. This is done by replacing the equal-mass bobs by ones whose masses are proportional to the

---

[39] For example, for a triangle $\triangle$ with all of its angles $\leq$ 120°: construct equilateral triangles on any two sides of $\triangle$; draw the lines joining each of the two new vertices to the corresponding opposite vertex of $\triangle$; the intersection of these two lines gives P.

weight factors m_1, m_2 and m_3. The corresponding expression for the total energy is then

$$E_{\text{potential}} = g(m_1 d_1 + m_2 d_2 + m_3 d_3) - \underbrace{g(m_1 L_1 + m_2 L_2 + m_3 L_3)}_{\text{constant}}.$$

The equilibrium lengths d_1, d_2 and d_3 taken up by the device thus minimise S, and the location of the knot again gives the required position on the triangle.

2. The original task (with unweighted distances) can also be tackled using a system based on soap films. Cut two identical copies of the original triangle from a sheet of plexiglas[40] (poly(methyl methacrylate)), and fix them together so that they are parallel and have thin spacers (of thickness t) between them at their vertices. When the resulting device is dipped into a soap solution, a structure is formed between the plates that consists of three rectangular films meeting in the form of a letter Y. Along the (short) segment of boundary common to all three films, the balance of three surface tension forces with identical magnitudes results in angles of 120° between pairs of soap films. The minimisation of the sum $S_0 = d_1 + d_2 + d_3$ by the equilibrium configuration taken up follows from the minimisation of the total surface energy of the films, which is proportional to $S_0 \times t$.

3. The previous analysis can only be applied to triangles that do not contain an angle greater than 120°. If the obtuse angle of a triangle is precisely 120°, then one of the terms in the sum of distances $d_1 + d_2 + d_3$ (the one corresponding to the 120° vertex) becomes zero; in this case P is simply the obtuse-angled vertex of the triangle. Even when the obtuse angle is larger than 120°, the obtuse-angled vertex provides the position yielding a minimal sum. This can be demonstrated directly using the mechanical model; the knot at the meeting point of the three threads comes to a halt in the eye-screw located at the obtuse-angled vertex. In the soap film model, this corresponds to the situation in which, instead of three, only two soap films are formed – along the sides adjacent to the obtuse angle.

S70 We show below that – however the sand is distributed (subject to the constraints stated in the problem) – the required work is always

$$W = Mg(s_2 \mu_1 + s_1 \mu_2).$$

In particular, if the centre of mass of the sack is at its centre, i.e. $s_1 = s_2 = \ell/2$, the work required is

$$W = \frac{\mu_1 + \mu_2}{2} Mg\ell.$$

Proof. Notionally divide the area of contact between the sack and the carpet into very narrow strips perpendicular to the direction of motion, so that the load on each can be taken as uniform. Denote the compressive force acting on the ith strip by $m_i g$, and its signed horizontal displacement from the centre of mass by x_i (this is

---

[40] Again, a word that has passed into common usage, though technically it is a registered trademark for Plexiglas.

positive for parts of the sack to the right of the centre of mass (CM) in the figure appearing in the problem).

During the transfer, the ith segment of the contact area moves through a distance $s_2 - x_i$ on the smoother surface, and through a distance $s_1 + x_i$ on the rougher one. It follows that its contribution to the work needed is

$$W_i = m_i g(s_2 - x_i)\mu_1 + m_i g(s_1 + x_i)\mu_2.$$

The total work required is therefore

$$W = \sum_i W_i = (s_2\mu_1 + s_1\mu_2)\left(\sum_i m_i\right)g + (\mu_2 - \mu_1)\left(\sum_i m_i x_i\right)g.$$

The first sum on the right-hand side of this equation is simply the mass M of the sack of sand, whereas, from the definition of centre of mass, the second sum is zero. QED[41]

> *Note.* It is important that the sack is towed at the bottom, because then the tractional and frictional forces cannot form a couple, and no deforming torque is generated. If a crate is similarly pulled onto the rough surface, and the horizontal traction is not applied at the level of the surface, then the result for the work required is not independent of the load distribution.

S71 *Solution 1.* Imagine the apex of the cone being depressed by Δh, and the radius of its base consequently increasing by Δr, i.e. the perimeter of the base circle[42] increases by $2\pi\Delta r$. Then the total work done by external forces would be

$$\Delta W = w\Delta h - F(2\pi\Delta r).$$

In accordance with D'Alembert's principle of virtual work, if the system is in equilibrium, this quantity must be zero; if it were not, the cone would move spontaneously to some new configuration. It follows that

$$F = \frac{w}{2\pi}\frac{\Delta h}{\Delta r}.$$

However, the changes in height and base circle radius are not independent, the connection between them being determined by the fixed length ℓ of one of the straight edges of the sheet. Using Pythagoras's theorem:

$$(r + \Delta r)^2 + (h - \Delta h)^2 = \ell^2 = r^2 + h^2,$$

from which we have

$$\frac{\Delta h}{\Delta r} = \frac{2r + \Delta r}{2h - \Delta h} \approx \frac{r}{h}.$$

---

[41] QED: *quod erat demonstrandum*, which is Latin for 'that which was to be demonstrated'.
[42] Technically, the base circle is the directrix of the cone, and either straight edge of the paper is a generatrix.

So the force in question has magnitude

$$F = \frac{r}{2\pi h}w.$$

Solution 2. Let us model the paper sheet, which has a continuous matter distribution, by a system consisting of many light thin rods. The lower ends of the rods are distributed evenly around the perimeter of a circle of radius r, while their upper ends are connected together (perhaps by a thin rope passing through eye-screws attached to the rods). If the lower ends of the rods are also connected by a thread, then the arrangement, resembling the frame of a Red Indian tent, can be loaded by a weight w at its apex, since the tension F in the thread will prevent the rods from slipping apart.

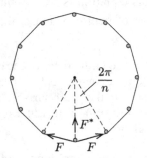
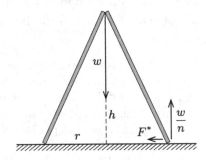

If the system consists of n rods in total ($n \gg 1$), then the angles between successive threads that join neighbouring rods are $2\pi/n$, and so the lower end of any particular rod is pulled towards the centre of the regular n-sided polygon by a horizontal force of magnitude

$$F^* = 2F \sin\left[\frac{\pi}{2} - \frac{1}{2}\left(\pi - \frac{2\pi}{n}\right)\right] = 2F \sin\frac{\pi}{n} \approx \frac{2\pi}{n}F.$$

In addition to this, the table-top pushes the bottom of the rod upwards with a vertical force of w/n.

Since the notional rods are light and we are ignoring friction, for equilibrium we must have that the moments of these two forces, taken about the top of the rod, cancel each other out. So,

$$\frac{w}{n}r = \frac{2\pi}{n}Fh, \qquad \text{that is} \qquad F = \frac{r}{2\pi h}w,$$

in accord with Solution 1.

S72 Let the sides of the triangle be of lengths a, b and c, and the linear density (mass per unit length) of the rods be λ. For the purpose of computing the centre of mass of the triangle, the rods can be replaced by point-like bodies with masses λa, λb and λc located at their corresponding midpoints, A', B' and C' (*see* figure).

The length of the line segment connecting any two of these points is half that of the iron triangle side that lies parallel to it.

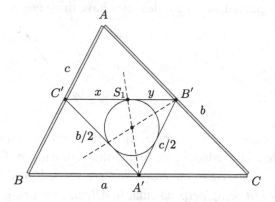

We next determine the centre of mass of the two point-like bodies on the line $B'C'$; they have masses λb and λc and are a distance $a/2$ apart. With x and y as shown in the figure, this is located at the point S_1 defined by

$$\lambda c \cdot x = \lambda b \cdot y.$$

Thus we have that

$$\frac{x}{y} = \frac{b}{c} = \frac{b/2}{c/2},$$

i.e. $B'C'$ is divided by the centre of mass of these two bodies in the ratio of the other two sides of the triangle $A'B'C'$.

It now follows from the converse of the angle bisector theorem that the point S_1 lies on the bisector of the angle $C'A'B'$. Further, the common centre of mass of point mass $\lambda(b + c)$ at S_1 and point mass λa at A' (i.e. the centre of mass of the whole iron triangular frame) lies on the bisector of the angle $C'A'B'$. As similar statements could be made starting from any other pair of points, the centre of mass of the iron triangle lies on all three angular bisectors of the triangle $A'B'C'$, which means that it coincides with the incentre of that triangle.

So Lisa was right; Frank's method is correct only for equilateral triangles.

S73 Any triangle must have at least one setting that is stable: the one in which the height of its centre of mass is (under the given conditions) the lowest of the three possibilities. It is also easy to find a – sufficiently asymmetric – obtuse-angled triangle that will tumble over from one of its edges. The much more difficult question is whether it is possible to make a triangle that has two edges on which it is unstable and *only one* on which it is stable. We now prove that such a triangle *does not exist*!

For any acute-angled triangle, however it is oriented, the foot of the vertical projection of its centre of mass (i.e. of its centroid) will always lie within the base edge of the triangle. All such triangles must have three edges on which they are stable.

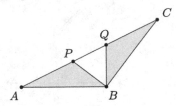

For an obtuse-angled triangle, the vertical projection of the centre of mass can fall outside a base edge. In the figure, this will happen if the centroid is inside *one of the two areas*, *ABP* and *BCQ*, that are shaded grey. Note that these areas are defined by *QB* being perpendicular to *AB*, and *PB* being perpendicular to *BC*. But if a triangle were able to tumble over from two of its edges, the centroid would need to be inside *both* grey areas, which is clearly not possible!

S74 When a tetrahedron, initially placed on one of its faces, falls over spontaneously, the altitude of its centre of mass must be reduced. A tetrahedron's centre of mass (coinciding with its centroid for a homogeneous body) is one-quarter of its height above its base. So, the volume of the body is the product of the area of one of its faces and 4/3 times the corresponding height of its centre of mass. As the volume is fixed, it follows that the larger the area of the face on which the tetrahedron is placed, the lower the altitude of its centre of mass. In particular, the face with the largest area has the lowest centre of gravity (among the four possibilities) and so must be stable.

We will prove in what follows that a tetrahedron that has three unstable faces *does not exist*. The essence of the proof is as follows. It will be shown that, if one of the faces is unstable, then there are at least two other faces that have larger areas than it has. From this, it follows that the largest and the second largest faces must both be stable.

Consider the face of tetrahedron *ABCD* that has the *second largest* area; let it be face *ABC* and suppose that the body falls over when placed on this face (*see* figure). This will happen if the projection *T′*, onto the plane *ABC*, of the tetrahedron's centre of mass *T* lies outside the triangle *ABC*. Note that the figure includes only points that lie in the plane of triangle *ABC* – in particular, *D* and *T* lie out of the plane of the figure and are not shown.

The centre of mass *T* of the homogeneous tetrahedron lies on the line segment joining the centroid *G* of the triangle *ABC* and the tetrahedron's fourth vertex *D*, and is located at the quarter point of the segment that is nearest to *G*. The same is also true for the projections *T′* and *D′* onto the plane *ABC* of the same two points: $GT' = \frac{1}{4}GD'$.

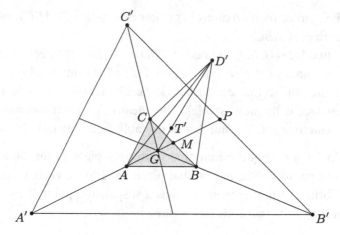

Next consider a four-fold magnification of the triangle ABC about its centroid G to form a larger, but similar, triangle $A'B'C'$. Combining the magnification with the properties of a centroid, and using the notation in the figure, it follows that $AM = 3GM = 4GM - GM = GP - GM = MP$, and from this it follows that the distance between the lines BC and $B'C'$ is equal to the distance between vertex A and the line BC. Similar connections are valid for the other two sides.

Since $GD' = 4GT'$ and, for the tetrahedron to topple as assumed, T' lies outside triangle ABC, it follows that the projection D' of vertex D lies outside the triangle $A'B'C'$. So, D' and G lie on opposite sides of edge $B'C'$ of the magnified triangle. Thus D' is further from line BC than A is. Consequently, the area $\mathcal{A}(BCD')$ of the triangle BCD' is greater than that of triangle BCA. Clearly, the area $\mathcal{A}(BCD)$ is greater than that of its projection $\mathcal{A}(BCD')$, and so for the face BCD we have that

$$\mathcal{A}(BCD) > \mathcal{A}(BCD') > \mathcal{A}(BCA).$$

This largely geometrical relationship also follows (more directly) from an alternative physics consideration, namely that the tetrahedron can only tumble from one of its unstable faces onto a face that corresponds to a smaller centre-of-mass height, and so inevitably has a larger face area (as discussed in the first paragraph above).

Now we show that *at least one* of the other two sides (ABD and ACD) of the tetrahedron has a larger face area than the area $\mathcal{A}(ABC)$ of the base face. As noted earlier, it is obvious that the areas of surface projections are smaller than the areas of the surfaces themselves, and so it is the case that

$$\mathcal{A}(ABD) + \mathcal{A}(ACD) > \mathcal{A}(ABD') + \mathcal{A}(ACD')$$
$$= \mathcal{A}(ABD'C)$$
$$= \mathcal{A}(ABC) + \mathcal{A}(BCD')$$
$$> 2\mathcal{A}(ABC).$$

As the sum of the two areas is greater than twice $\mathcal{A}(ABC)$, one of them must be larger than $\mathcal{A}(ABC)$.

So two faces (CBD, and one of ABD or ACD) have been found with larger areas than the area of the second largest face of the tetrahedron. This is an obvious contradiction, so the second largest face cannot be unstable (and, of course, the largest face is the most stable). This means that a tetrahedron with three unstable faces *cannot exist*. The number of unstable faces can only be zero, one or two.

S75   For the equilibrium of any arbitrary piece of the cable, both the net force and the net torque acting on that piece need to be zero. Consider, therefore, the equilibrium of the segment that has a suspension point P and the point of 'maximum sag' as its two ends (*see* figure).

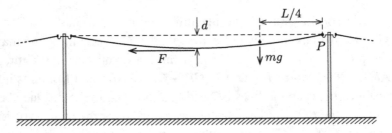

Because of the relatively small sag, the length of the selected piece can be taken as $L/2$ and its mass as $m = \lambda L/2$. Among the various forces acting on this line segment, only the horizontal tension F (exerted by the neglected half of the cable) and the gravitational force mg produce torques about the point P; the (almost) horizontal reaction force at P produces no torque about that point. The moment arm of force F is d, while that of the weight can be taken as $L/4$ (also because $d \ll L$), and consequently, as the net torque must be zero,

$$Fd - mg\frac{L}{4} = 0.$$

Inserting the expression for the mass into this equation gives the approximate magnitude of the tension in the cable as

$$F = \frac{\lambda L^2 g}{8d}.$$

At the same level as the other approximations already made, this can be taken as the tension throughout the cable.

> *Note.* In a Cartesian coordinate system with its origin midway between the two suspension points, the actual equation of the cable is the catenary
>
> $$y(x) = \mu^{-1}[\cosh(\mu x) - \cosh(\mu L/2)],$$

where $\mu = \lambda g/k$, and k is such that μ satisfies the equation $\sinh(\mu L/2) = \mu L_0/2$, in which L_0 is the true length of the full cable. The cable tension at x is $T(x) = k \cosh(\mu x)$.

The implications of these coupled equations are difficult to visualise, but a self-consistent set of approximations can be obtained when $\lambda g L \ll k$. Then μx is everywhere $\ll 1$ and the tension is nearly constant, with a value of $k = T(0) = F$. In retrospect we see that the condition $\lambda g L \ll k$ was equivalent to requiring that the total weight of the cable be much less than the tension in it. Further, we have

$$d = y(0) = \mu^{-1}[1 - \cosh(\mu L/2)]$$

$$= \mu^{-1}\left\{ 1 - \left[1 + \frac{1}{2!}\left(\frac{\mu L}{2}\right)^2 + \cdots \right] \right\}$$

$$\approx -\frac{\mu L^2}{8} = -\frac{\lambda g L^2}{8F},$$

in accord with our previous, much simpler, estimation of F in terms of d.

S76 Imagine that – in contrast to the situation described in the problem – the lower ends of the pole and the rope are each raised to the *same* height. This will require different horizontal forces in the two cases, of magnitudes F_1 and F_2 respectively, say (*see* figure).

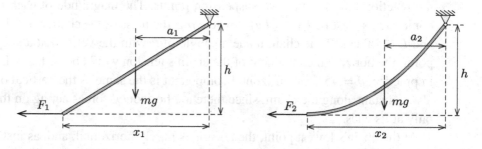

In this configuration, both bottom ends are a (common) vertical distance h below their suspension points. Further, it is clear that, because of its curvature, the rope will have a horizontal projection x_2 that is smaller than that, x_1, of the pole:

$$x_2 < x_1.$$

We now consider the moments about the two pivots of the various forces involved and note that:

(i) Both F_1 and F_2 have moment arms equal to h.

(ii) For the pole, the moment arm of its weight is clearly $a_1 = x_1/2$.

(iii) For the rope, the arm cannot be found quite so easily, but, as the pivot is approached, the gradient of the rope increases monotonically. It follows that, for rope segments with the same horizontal projection, the nearer they are to the pivot, the larger their mass. From this observation, we conclude that the moment arm of the rope's weight a_2 is less than $x_2/2$.

Combining (ii) and (iii) with our previous observation, we have:

$$a_2 < \frac{x_2}{2} < \frac{x_1}{2} = a_1.$$

The two torque balances about the relevant pivots are

$$F_1 h = mga_1 \quad \text{and} \quad F_2 h = mga_2,$$

where mg is the common weight of the pole and the rope. Since $a_2 < a_1$, we conclude that

$$F_2 < F_1,$$

and that to lift the two ends to the same height, the climbing pole requires a larger force than the rope does. If the force acting on the pole were slowly decreased to the value F_2 that is being applied to the rope, then its end would be lowered.

So, in equilibrium, and with identical horizontal pulling forces, the lower end of the climbing rope will be *higher* than the lowest point of the pole.

S77   a) The weight of the chain is balanced by the vertical components of the reaction forces at the two suspension points. The magnitude of each of these components must be $F_0 = \varrho Lg/2$, where ϱ is the mass of the chain per unit length, and $L = 40$ cm. If the chain made an angle of θ with the vertical at a suspension point, the horizontal component of the chain's tension would be $F_0 \tan \theta$. Because, in practice, $\theta = 45°$, the horizontal component is the same as the vertical one; it is also constant along the chain, since there are horizontal forces acting on the chain only at its ends.

At the chain's lowest point, the tension is purely horizontal, and, as just shown, its magnitude is F_0. So, around the lowest point, consider a small piece of the chain that subtends an angle 2φ at the centre of the osculating circle, whose radius is the value r_1 we seek. The length of the piece is $2\varphi r_1$ and it is pulled down by a gravitational force $2\varphi r_1 \varrho g$. This force is balanced by the upward net force of $2F_0 \sin \varphi$ due to the tension in the chain. In the limit of small angles, when $\sin \varphi \approx \varphi$, the equilibrium equation

$$2\varphi r_1 \varrho g = 2F_0 \sin \varphi \quad \text{leads to} \quad F_0 = \varrho g r_1.$$

Since $F_0 = \varrho Lg/2$, we have the simple result that $r_1 = L/2 = 20$ cm.

b) We use an approach similar to that used in part a), by again investigating the forces acting on a small piece of the osculating circle. At a suspension point, the chain tension is $\sqrt{2}F_0$, while the component of the gravitational force perpendicular to a small piece of the chain is $2\varphi r_2 \varrho g/\sqrt{2}$. Proceeding as before, we get

$$2\sqrt{2}F_0\varphi \approx 2\sqrt{2}F_0\sin\varphi = 2\sqrt{2}\frac{\varrho Lg}{2}\varphi = \frac{2\varphi r_2\varrho g}{\sqrt{2}},$$

from which r_2, the radius of curvature at each end of the chain, is $r_2 = L = 40$ cm.

S78 *Solution 1.* Any arbitrary piece of the rope is in equilibrium, and so the horizontal component of the tension in the rope between the pulleys is constant (equal to F_1, say). The vertical component changes from point to point; it is zero in the middle, and at the pulleys is equal to half the weight of the 'sagging' section, namely $\lambda(\ell/2)g$, where λ is the linear mass density of the rope.

Following the hint, as illustrated in Fig. 1, we imagine a small piece of the rope, of length $\Delta\ell$, cut out from the middle of the system, and re-inserted into the rope near one of the pulleys! What energy changes are involved?

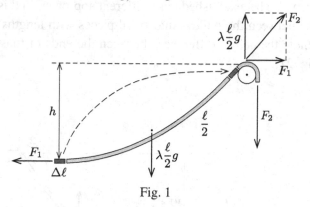

Fig. 1

Closing the gap in the middle of the rope requires work $F_1\Delta\ell$ to be done, while the work required to lift the small piece against gravity is $\lambda\Delta\ell\, gh$. Inserting it at the pulley actually allows some energy to be recovered, but formally the work required to do this is $-F_2\Delta\ell$.

As a result of these changes, we have done no more than return the rope to its initial state! It follows that the total work done must be zero, and so

$$F_1\Delta\ell + \lambda\Delta\ell\, gh - F_2\Delta\ell = 0,$$

showing that

$$F_2 - F_1 = \lambda gh. \tag{1}$$

The same thought experiment could have been carried out for any arbitrary pair of points on the rope, and so it is generally true that 'the difference in tension forces at two arbitrary points of the rope is directly proportional to the height difference between those points'.

As can be seen in Fig. 1, the connection between the rope tension F_2 (at the pulley) and its components is

$$F_2^2 = F_1^2 + \left(\lambda\frac{\ell}{2}g\right)^2. \tag{2}$$

A third equation, one that involves s, can be established by recognising that the hanging rope segments are each held in place by a force of strength F_2:

$$F_2 = \lambda gs. \tag{3}$$

From equations (1) and (3), we have that $F_1 = \lambda g(s - h)$. Substituting this into (2) gives the final result for the length of the hanging segments as

$$s = \frac{\ell^2 + 4h^2}{8h}.$$

Solution 2. Equation (1), which is generally valid for ropes and chains in equilibrium, can also be established via a different approach. In this, we divide the part of the rope between the pulleys into small pieces with lengths Δs_i ($i = 1, 2, \ldots, N$), and denote the height differences between the ends of these small pieces by Δy_i (*see* Fig. 2).

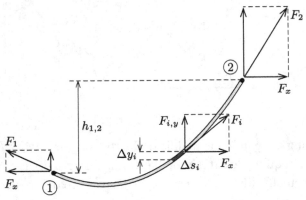

Fig. 2

If $F_{i,y}$ is the vertical component of the tension at the end (say, the right-hand end) of the ith piece, and F_x is the horizontal component, which does not depend on i, then we have, for the net tension F_i acting there, that

$$F_i^2 = F_{i,y}^2 + F_x^2. \tag{4}$$

Now, the net tensional force acting on a small piece of the rope is the difference between the vertical components of the tension at its two ends, and this must equal the weight of the piece:

$$F_{i,y} - F_{i-1,y} \equiv \Delta F_{i,y} = \lambda \Delta s_i g.$$

We can also use the fact that the tension force is always tangential, to express the slope of the rope in terms of the total tension and its components. This enables us to relate Δy_i to Δs_i:

$$\Delta y_i = \Delta s_i \frac{F_{i,y}}{F_i} = \frac{1}{\lambda g} \frac{F_{i,y} \Delta F_{i,y}}{F_i}. \tag{5}$$

We next consider small changes in both sides of equation (4), i.e. we subtract the equation for the $(i-1)$th piece of the rope from the equation for the ith piece. As F_x is constant along the rope,

$$F_i^2 - F_{i-1}^2 = F_{i,y}^2 - F_{i-1,y}^2,$$

from which we get

$$F_i \Delta F_i = F_{i,y} \Delta F_{i,y}. \tag{6}$$

To make the last step, we used the fact that, if an arbitrary quantity f has a small change Δf made to it, then the change in the square of f is

$$\Delta(f^2) \equiv (f + \Delta f)^2 - f^2 = 2f\Delta f + (\Delta f)^2 \approx 2f\Delta f.$$

Using equality (6), we can write (5) in the form

$$\Delta y_i = \frac{1}{\lambda g} \frac{F_i \Delta F_i}{F_i} = \frac{1}{\lambda g} \Delta F_i. \tag{7}$$

Summing both sides of equation (7) over i, between points 1 and 2 in the figure, we obtain

$$h_{1,2} = \frac{1}{\lambda g}(F_2 - F_1),$$

where $h_{1,2}$ is the (signed) height difference between the two points. Applying this expression to the rope's midpoint and its point of contact with one of the pulleys ($h_{1,2} = h$), we recover equation (1). From here on, the calculation proceeds as in Solution 1.

> *Note.* The 'sag' h of the rope, the rope length ℓ between the pulleys and the distance d between the pulleys are not independent quantities; specification of any two uniquely determines the third. Our result shows that, in the case of given h and ℓ (meaning also a given d), a value of s that keeps the system in equilibrium can always be found. However, a more interesting general result is also available.
> If, instead of the original problem, we try to answer the question of what is the equilibrium 'sag' when a rope of length $L = \ell + 2s$ is placed on the two pulleys a given distance d apart. Using calculus, it can be shown that, if $L/d < $ e (where e $\approx 2.718\ldots$ is Euler's number), the rope has no equilibrium position; but if $L/d > $ e, then it has two (*see* Fig. 3). One of them (with the smaller sag) is *stable* against small vertical displacements of the rope's midpoint, but the other, with the larger value of h, is *unstable*.

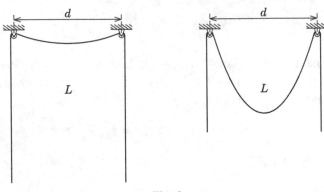

Fig. 3

S79   Imagine that the pearl necklace is pulled – tangentially to the cylinder, and at a height h above the lower end of the necklace – through a small distance Δx, as shown in the figure.[43]

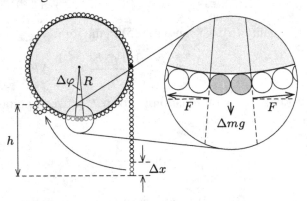

As a result, the free end of the necklace is raised by Δx. If the tension force in the necklace at the point where the loop is formed is $F(h)$, then the work required is $\Delta W = F(h)\Delta x$. As the surface is slippery, no work is done against friction and all of ΔW appears as an increase in the gravitational potential energy of the necklace; the latter is simply that of the missing (three-pearl) piece from the bottom of the necklace, after it has been raised through a vertical distance h:

$$F(h)\Delta x = \lambda \Delta x \cdot gh,$$

where λ is the mass per unit length of the necklace.

From this thought experiment, in which the point at which the loop was formed was chosen arbitrarily, we can conclude that the tension at any point depends only

---

[43] In the figure, the distance Δx is represented by three pearls, and so, in this thought experiment, what was originally a part of the cylinder's circumference occupied by two pearls has become the location of a loop of five pearls.

on the height h of that point above the free end of the necklace, where the tension is obviously zero:

$$F(h) = \lambda g h. \tag{1}$$

This result is the same as that found in the solution on page 239. The statement that 'the difference in the magnitudes of the tensions at two arbitrary points is proportional to the height difference between the points' is valid for all uniform ropes and chains touching frictionless surfaces.

If the segment of the necklace that hangs down is sufficiently long, then the equilibrium is stable. However, if the length of the dangling portion is decreased to below a certain critical value, then the pearl necklace drops away from the surface at the lowest point of the cylinder. When this happens, the normal reaction force exerted there, by the cylinder surface on the necklace, has decreased to zero.

Consider now a small piece of the necklace of length $R\Delta\varphi$ situated at the lowest point of the cylinder, where $\Delta\varphi$ is the angle subtended by this piece at the axis of the cylinder (*see* figure). The condition for its equilibrium, when there is no normal reaction from the cylinder surface, is

$$2F \sin(\Delta\varphi/2) = \lambda R\Delta\varphi \cdot g,$$

where F is the local tension in the necklace. Because the angle is arbitrarily small, the approximation $\sin(\Delta\varphi/2) \approx \Delta\varphi/2$ can be used, and so

$$F = \lambda g R.$$

This is the minimum tension needed at the bottom of the cylinder to keep the necklace in contact with it around its entire circumference. Comparing this with expression (1), we see that no part of the pearl necklace will leave the cylinder if the height difference h between the necklace's lower end and the lowest point of the cylinder is at least R, i.e. the length ℓ of the hanging segment is at least $2R$.

S80 *a)* The centre of mass of the rolled rug is not exactly above the point of support – in the figure in the problem, it is slightly to the left of it – and so the rug's weight produces a torque about that point, which then causes the rug to unroll.

b) When a length x of the rug has been rolled up, the mass $m(x)$ of the cylindrical rug roll is directly proportional to x:

$$m(x) = \frac{x}{L}M. \tag{1}$$

The cross-sectional area of the roll is also proportional to x, i.e. for the radius $r(x)$ of the roll we have:

$$r^2(x) = \frac{x}{L}R^2. \tag{2}$$

The horizontal force $F(x)$ needed to prevent the rug from unrolling can be found using D'Alembert's principle of virtual work. Imagine that a further length Δx ($\ll L$) is added to the roll (*see* figure), then

$$F(x)\Delta x = \Delta E_{\text{potential}}, \qquad (3)$$

where $E_{\text{potential}} = m(x)gr(x)$ is the gravitational potential energy of the rug.

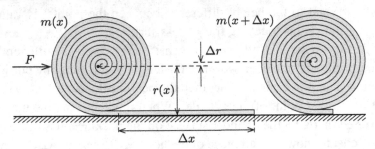

The change in the potential energy is

$$\Delta E_{\text{potential}} = mg \cdot \Delta r + \Delta m \cdot gr.$$

This can be transformed using formulae derived from (1) and (2), namely

$$\frac{\Delta m}{m} = \frac{\Delta x}{x} \qquad \text{and} \qquad 2\frac{\Delta r}{r} = \frac{\Delta x}{x},$$

to give

$$\Delta E_{\text{potential}} = \frac{3}{2}\frac{mgr}{x}\Delta x.$$

Substituting this into (3), and using (1) and (2), gives the force required to prevent further unrolling of the rug as a function of the length x of the coiled segment:

$$F(x) = \frac{3}{2}g\frac{m(x)\,r(x)}{x} = \frac{3MgR}{2L}\sqrt{\frac{x}{L}}.$$

When $x = L$, this expression provides the answer to the specific question asked:

$$F = F(L) = \frac{3R}{2L}Mg.$$

Note. We can check our result by calculating the total work required to roll up the whole rug. This is given by the integral with respect to x of the force $F(x)$:

$$\int_0^L F(x)\,dx = \int_0^L \frac{3}{2}\frac{MgR}{L}\sqrt{\frac{x}{L}}\,dx = MgR,$$

which is reassuringly equal to the gravitational potential energy of the fully rolled rug.

S81 (i) Denote the length of the closed loop on the novices' lasso by L, the position of the eyelet on the cone by P, the climber's position by A and the apex of the iceberg by O (*see* left-hand side of Fig. 1). The climber is lowest (i.e. the total energy of the system is minimal) if point P is as far from point O as possible; the lasso will take up the configuration that brings this about.

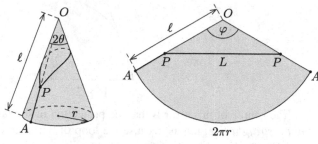

Fig. 1

Notionally, cut the lateral surface of the cone along the line (generatrix) OA, and spread it out into a plane.[44] Doing so, we obtain a circular sector with central angle φ, and with point P appearing on it at two positions (*see* right-hand side of Fig. 1).

The points P are as far from point O as possible if, in the figure, the rope segment of length L joining them is *straight*. Such a straight line can be drawn within the grey area of the spread-out lateral surface provided $\varphi < 180°$. This requires the slope length ℓ and the base radius r to satisfy the inequality

$$2\pi r = \ell\varphi < \ell\pi, \qquad \text{that is} \qquad r < \frac{\ell}{2}.$$

This condition means the acceptable range of original (spatial) ice-cone angles satisfies

$$\sin\theta = \frac{r}{\ell} < \frac{1}{2}, \qquad \text{that is} \qquad 2\theta < 60°.$$

The novices, using their two-piece lassos, can only tackle ice cones that are *sharper* than one with $2\theta = 60°$. If the cone angle is larger than this value, then the loop of the loaded lasso slips over the apex of the iceberg, and the climber falls.

> *Note.* Imagine the iceberg to be transparent, and consider the 'side view' of it when the novices' lasso is in place. In this view, the loop of the lasso appears as a single line, because one half of it is directly behind the other, and P is at one end of the line. At the other end, Q, the line is perpendicular to the generatrix of the cone passing through it. Naively, we might conclude from this that the angle of

---

[44] That this can be done follows from Minding's theorem and the fact that the so-called Gaussian curvature, the product of the two principal curvatures at any point, is zero *everywhere* on the cone's lateral surface. In cylindrical polar terminology: although one principal curvature is non-zero and varies as z is varied, the other, in any azimuthal plane, is *always* zero, and hence so is the product.

the cone could be increased up to $2\theta = 90°$ and still provide 'something for the loop to pull against' on the right-hand surface of the cone, when the loop is loaded at a point P on its left-hand surface. But – as shown above – this is not the case!

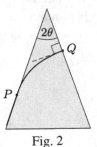

Fig. 2

The source of this error is that the projection of the lasso between P and Q is *not a straight line*; this is because the loop of the lasso cannot be a plane curve (*see* Fig. 2). If it were, then it would have to be a conic section, namely an ellipse. But an ellipse has a continuously varying tangent and cannot have a cusp at P; the rope of the loop, contrariwise, *must* – because of the loaded segment of the rope – have a cusp there.

Referring back to the plane sector shown in Fig. 1, as the angle of the cone θ is increased so is φ, and the centre of the straight line of fixed length L rises towards O. This central point corresponds to the point Q on the lateral surface of the cone, and it follows that, as 2θ increases, so Q moves higher and higher up the cone, reaching the apex O of the iceberg when $\varphi = \pi$ and $\theta = 60°$. Beyond that point lies only disaster!

(ii) But what about the experts' lasso? It consists of a single continuous frictionless rope, and so the tension in it must be the same everywhere. At the eyelet P, three equally tensioned ropes meet, and can only be in equilibrium if the angle between any two of them is $120°$ – in particular, the angle between the two segments of the loop has this value. In the plane-sector drawing of Fig. 1, this corresponds to each of the angles between line PP and the two lines marked OA being $60°$. This, in turn, requires φ to be $60°$, i.e.

$$2\pi r = \ell\frac{\pi}{3},$$

and consequently that

$$\sin\theta = \frac{r}{\ell} = \frac{1}{6}, \qquad \text{from which} \qquad 2\theta = 19.2°.$$

So equilibrium is only possible for this one particular value of 2θ (and not for a range of values). And, what is more, even this situation is *unstable*. If the apex angle of the iceberg is greater than $19.2°$, then the loaded lasso slips up to the vertex of the cone, passes over it, and falls back to Earth. If the angle is smaller than the critical value, then the noose simply slips down to the level of the frustrated climber. It is not by accident that this type of lasso is specified as only for experts!

S82 Consider the bike chain when it is at rest, influenced only by gravity. In this situation the tension (which is tangential at each point) varies continuously along the chain, in such a way that each link's weight is balanced by the two tension forces exerted on it by its neighbours. Denote this force by $F_1(s)$, a function of the arc length s measured from an arbitrary point (say, from the topmost point) of the chain. Necessarily, $F_1(0) = F_1(L)$, where L is the total length of the chain.

Consider now a chain of the same shape, operating in weightless conditions (say, inside the International Space Station), and with all of its links moving with the same constant tangential speed v. As the tangential acceleration of all the links is zero, the force stretching the chain, F_2, has to have the same magnitude everywhere, i.e. $F_2 = $ constant. In particular, the force is independent of the radius of curvature of the chain at any given point; this is proved in the following note.

> *Note.* If the radius of curvature of the chain, of mass per unit length λ, is R at some point (R can vary from place to place), then the mass of an arbitrarily small piece of length $R\Delta\alpha$ is $\Delta m = \lambda R\Delta\alpha$, while its acceleration is v^2/R, as shown in the figure.

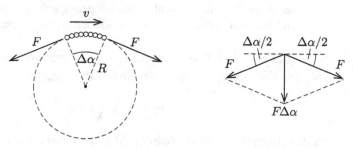

> The equation of motion of the piece is

$$\lambda R\Delta\alpha \cdot \frac{v^2}{R} = 2F \sin\frac{\Delta\alpha}{2},$$

> which leads to the relation

$$F = F_2(s) = \lambda v^2,$$

> when use is made of the fact that $\sin(\Delta\alpha/2) \to \Delta\alpha/2$ as $\Delta\alpha \to 0$. It should be noted, in particular, that F is independent of R.

So, whatever the shape of the chain, and whatever the chain's curvature at any particular point, the uniform tension provides the appropriate centripetal force for the chain to maintain that shape.

Finally, consider the situation described in the problem, a moving chain in a gravitational field. The forces acting are those obtained by superimposing the two simpler situations already discussed. If the tension varies as $F_1(s) + F_2(s)$ along the chain, then Newton's equation of motion is fulfilled for each chain link. The vector sum of $F_1(s)$ and the gravitational force is zero, and the term $F_2(s)$ (equal to λv^2

everywhere) provides the appropriate product of mass and centripetal acceleration for each link.

The bicycle chain's moving shape is *identical to its stationary one*! That shape is a 'catenary' for the hanging part of the chain, and, of course, a semicircle for the supported part.

> *Note.* The argument above can be illustrated by the following thought experiment. Let us 'switch off' the gravitational field of the Earth while the chain wheel is still at rest. The shape of the chain remains unaltered, but it no longer exerts a force on the wheel. As has been proved earlier (*see* previous note), if each chain link is set in motion with the same tangential speed, the chain maintains its original shape. Finally, we 'switch back on' the Earth's (uniform) gravitational field, which (as is well known) cannot deform the shapes of bodies. So the shape of the chain in the uniformly moving state remains the same as in the original rest state.
>
> During the increase of the rotational speed from zero, a stationary state cannot be formed immediately, and there must be some transient effects. These will be apparent as waves running round the chain. An investigation of the damping of the transients needs a more sophisticated analysis than we can provide here. (It seems unlikely in practical terms, but perhaps the transients will not be damped, and the chain will never settle down to a constant shape.)

S83   Let us follow Peter's calculations. The mass of the roll (which is continuously changing) expressed as a function of the distance already travelled is

$$m(x) = M \left(1 - \frac{x}{L} \right),$$

its radius (from the proportionality of the mass and cross-sectional area) is

$$r(x) = R \sqrt{1 - \frac{x}{L}},$$

and the horizontal component of its velocity (from the conservation of energy) is

$$v_x \equiv \frac{\mathrm{d}x}{\mathrm{d}t} = v_0 \frac{1}{\sqrt{1 - (x/L)}}.$$

The calculation of v_x used the fact that, because $R \ll L$, the vertical component of the velocity is very much smaller than the horizontal one, and so the vertical kinetic energy term could be neglected.

The vertical component of the velocity of the roll's centre is

$$v_y = \frac{\mathrm{d}r}{\mathrm{d}t} = \frac{\mathrm{d}r}{\mathrm{d}x} \frac{\mathrm{d}x}{\mathrm{d}t} = \frac{\mathrm{d}r}{\mathrm{d}x} v_x = -\frac{R v_0}{2L} \frac{1}{1 - (x/L)},$$

the negative sign appearing because, as x increases, the centre of the roll moves to a progressively lower position. From this, the vertical component of the linear momentum can be found:

$$p_y = m(x)v_y(x) = -\frac{MRv_0}{2L}.$$

As this quantity is constant in time, we have reached the, perhaps surprising, conclusion that the vertical component of the net force acting on the roll must be zero, i.e. $N - m(x)g = 0$. This means that the ground *does not exert* any extra vertical force on the roll beyond the (current) weight of the fire hose.

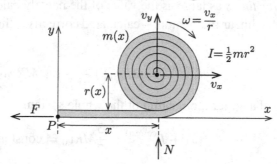

It follows that, according to Peter's analysis, the net torque produced by the external forces about P is zero; the line of action of F goes through P, and the normal force N and the weight of the roll cancel each other out. What about the angular momentum of the roll? If Peter's calculation is correct, then the total angular momentum must be constant in time.

The angular momentum about point P has two terms – the internal spin J_1 of the roll, and the orbital angular momentum J_2 due to the linear motion of the centre of mass. Taking the direction of the roll's rotation as positive, we note that, since there is no slipping, $\omega = v_x/r$. The internal spin J_1 can be expressed as follows:

$$J_1 = I\omega = \frac{1}{2}mr^2\omega = \frac{1}{2}mrv_x = \frac{1}{2}MRv_0\left(1 - \frac{x}{L}\right).$$

This quantity changes with time, since x does. The orbital angular momentum is

$$J_2 = p_x r - p_y x = mrv_x - xmv_y = MRv_0\left(1 - \frac{x}{L}\right) + MRv_0\frac{x}{2L},$$

and so the total angular momentum of the system is

$$J_1 + J_2 = MRv_0\left(\frac{3}{2} - \frac{x}{L}\right) \neq \text{constant}.$$

So, the angular-momentum theorem is not obeyed for the unrolling fire hose! – states Peter.[45]

But Pauline had other ideas! In her opinion, the vertical component of the linear momentum of the roll (and the system as a whole) should be calculated in a different way, because, she claimed, $p_y \neq m(x)v_y$.

---

[45] While fantasising about being the youngest Nobel Laureate ever!

And she was right. The linear momentum should be found by calculating the vertical component of the velocity of the centre of mass of the *complete* hose, and multiplying this by M. This prescription gives

$$p_y^{(\text{correct})} = \frac{\mathrm{d}}{\mathrm{d}t}\left[\frac{m(x)r(x)}{M}\right]M = MRv_x\frac{\mathrm{d}}{\mathrm{d}x}\left(1 - \frac{x}{L}\right)^{3/2} = -\frac{3}{2}\frac{MRv_0}{L}.$$

This is *three times* as large as the value of the naively calculated quantity mv_y!

Using this linear momentum – calculated correctly – the orbital angular momentum is

$$J_2^{(\text{correct})} = MRv_0\left(1 - \frac{x}{L}\right) + MRv_0\frac{3x}{2L},$$

and the total angular momentum of the whole system is

$$J_1 + J_2^{(\text{correct})} = \frac{3}{2}MRv_0 = \text{constant}.$$

Everything is now OK, because the angular-momentum theorem is still valid, even for a body with changing mass, provided the linear momentum and the orbital angular momentum are calculated correctly (albeit unusually).

> *Notes.* 1. The vertical linear momentum, calculated correctly, is also a constant in time, and so Peter's conclusion, that $N - m(x)g = 0$, was right.
>
> 2. Peter and Pauline have both used the same formula ($p_x = m(x)v_x(x)$) for the calculation of the horizontal momentum of the roll, and – as we have just seen – for systems with changing mass, this is not certain to give the right answer. Fortunately, this particular result *is* correct; this can be proved using Pauline's method (determining the time derivative of the horizontal position of the system's centre of mass taken as a whole).
>
> 3. When $x \to L$, some of the calculated quantities, namely v_x, v_y and $F(x)$, approach infinity, but it is obvious that this cannot happen physically. Our description and approximations cannot be valid at the very end of the motion.

S84 As the cylinder is rotating quite slowly, the sand in it is, to some extent, piled up against the wall of the cylinder that is moving upwards, and occupies a volume whose shape approximates that of a cylinder segment (*see* left-hand diagram in figure).

The change of temperature ΔT could be determined if we knew the mass m of the sand, its specific heat capacity c and the frictional work done: $\Delta T = W_{\text{friction}}/(mc)$. The mass of the sand is given ($m = 100$ kg), and its specific heat can be estimated from data books (by considering the data for similar materials, e.g. quartz glass or porcelain) as being in the range of 700–800 J kg^{-1} °C^{-1}.

A detailed description of the motion of the sand (and, from this, a calculation of the frictional loss of energy) would be a desperately hard (impossible) task. Fortunately, this is not necessary! It is sufficient to note that, sooner or later, the sand in the uniformly rotating cylinder assumes a settled configuration. Though the individual sand particles are moving, the body of sand as a whole has a stationary shape that does not change with time. Consequently the centre of mass of the sand is always at the same position, a horizontal distance a from the rotational axis of the cylinder (*see* right-hand diagram in figure).

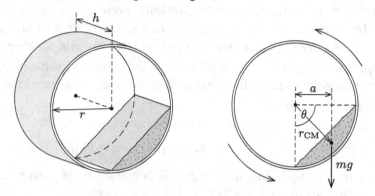

The increase in thermal energy of the moving sand (the energy dissipated by frictional forces within it) is equal to the mechanical work done in maintaining the rotation of the cylinder. This can be expressed as the product of the torque mga acting on the cylinder and the rotational angle $\Delta\varphi$ through which it turns:

$$W_{\text{friction}} = mga\Delta\varphi.$$

During a 10 minute interval the angle of rotation is

$$\Delta\varphi = 2\pi \times 0.5 \text{ s}^{-1} \times 600 \text{ s} = 1885 \text{ rad},$$

and the magnitude of the gravitational force is

$$mg = 100 \text{ kg} \times 9.81 \text{ m s}^{-2} = 981 \text{ N}.$$

But there is still one thing to be done – the calculation of the moment arm a of the gravitational force. We estimate first the distance r_{CM} between the centre of mass of the sand and the rotational axis. The volume of a cylinder segment of length h and radius r that subtends an opening angle θ at the cylinder's axis is

$$V = \tfrac{1}{2}hr^2(\theta - \sin\theta). \tag{1}$$

In our case, $h = 1$ m and $r = 0.5$ m. The volume of the sand in the cylinder can be determined from its mass and density. The density of sand ϱ depends on the constitution of the sand itself, how compacted it is and its moisture content, and varies over the range 1400–2000 kg m^{-3}. *The Engineering ToolBox* website gives a value of 1400 kg m^{-3} for dry uncompacted sand, and this is the figure we will use here. So

$$V = \frac{m}{\varrho} \approx 0.07 \text{ m}^3. \tag{2}$$

Inserting the estimated value (2) for the sand's volume into (1), and solving the transcendental equation for θ numerically, we get $\theta \approx 90°$ and $r_{CM} \approx 0.41$ m.[46]

The remaining question is: 'How is the plane surface of the sand-filled cylinder segment oriented, relative to the vertical?' From our everyday experience, or after a bit of deliberate experimentation, in a sand pit, or with an egg-timer or hourglass, we conclude that dry sand can be piled up with a 'scarp angle'[47] of up to approximately 45°. This, together with the estimated value of θ, implies that the highest point of the self-renewing sand avalanche is approximately level with the axis of the cylinder.

So, the length of the relevant moment arm is

$$a \approx r_{CM} \sin 45° \approx 0.29 \text{ m},$$

and the corresponding frictional work is about 540 kJ. Using this, and taking the specific heat of sand as 800 J kg^{-1} °C^{-1}, we get

$$\Delta T \approx 6.7 °\text{C}.$$

As the density and the specific heat of the sand are uncertain by up to 10 %, and the scarp angle assessment is only a rough one, the estimation of ΔT may have a 20 % error margin. Consequently, we give the temperature increase of the sand as being between 5 and 8 °C – distinctly ill defined, but definitely detectable.

S85 *a)* The density of iron is 7.8 times that of water, so the mass of the iron cube of volume 10^{-3} m^3, i.e. 1 litre, is the same as the mass of 7.8 litres of water. According to Archimedes' principle, the buoyancy force (upthrust) acting on the cube is the same as the weight of 1 litre of water. It follows that, for the cube to be in equilibrium, the tension in the cord must be equal to the weight of $7.8 - 1.0 = 6.8$ litres of water.

On the other side of the pulley, this tension must balance both the water's weight and the reaction (on the bottom of the bucket) to the upthrust experienced by the cube. The latter is, as already noted, the same as the weight of 1 litre of water, and so the amount of water in the bucket must be 5.8 litres. In reality, there will be a little less water than this in the bucket, because even a 'light plastic bucket' has some weight.

---

[46] Though it can hardly be justified – given all the other rough approximations elsewhere – r_{CM} has been calculated for the cylinder segment from the accurate formula

$$\frac{4r \sin^3(\theta/2)}{3(\theta - \sin\theta)}.$$

[47] See footnote in the hint, if necessary.

b) If the amount of water were increased, the bucket would sink and the cube would rise. Provided the additional water is not more than 2 litres, some part of the cube would remain under water and the cube would appear to float. If more than 2 litres were added, making the total mass of water more than 7.8 litres, the cube would be lifted clear of the water and continue to rise until it jams against the pulley.

c) If the amount of water in the bucket were to decrease, then the bucket would rise and the cube would sink to the bottom of the bucket. This equilibrium state (for the cube and bucket) would remain unaltered, even after all the water had evaporated. The tension in the cord would change during the evaporation, ending up at approximately half the weight of the iron cube. If the cord were 'light', the final tension in it would be one-half of the combined weights of the cube and bucket.

S86 Initially, the water pressure at the bottom of the container was 1.1 atm and, because the height of the cylinder was 10 m, at the top it was 1 atm less, i.e. 0.1 atm.

The key to the solution is the fact that water is virtually incompressible. Inside the *closed* container there was only the incompressible water and the air bubble – nothing else. We note that a, perhaps surprising, consequence of this is that the volume of the air bubble must have remained constant as it rose; the volumes of both the water and the container could not change, and so, consequently, neither could that of the air bubble.

Accordingly, after the air bubble had risen, its volume, temperature and amount of matter must have been as they were initially, and so its pressure could not have changed either. It follows that the water pressure at the top of the cylinder, which had to be equal to that in the bubble, had become 1.1 atm. Since the pressure at the bottom must have been 1 atm more than this (because of the hydrostatic pressure of 10 m of water), the answer to the problem is that the final water pressure at the bottom of the cylinder was 2.1 atm.

How is it possible that the pressure in the air bubble remains constant, but the pressure of the water increases? When we say that water is incompressible, we mean that, if the pressure on it is increased, the water's volume change is negligible, although, in theory, it is not exactly zero. In other words, a moderate increase in the pressure produces a volume decrease that is 'almost' zero. When the air bubble rises, its volume increases a little, and the decrease in water volume is the same – almost, but not quite, zero. However, it can correspond to, and cause, a significant increase in the pressure of the water.

S87 Let us suspend an easily deformed (low-spring-constant) spring from the roof of the *black box*, and hang a bucket on its lower end. The rubber hose coming into the box has a flexible extension that is filled with water and dips beneath the surface of water in the bucket (*see* figure). As the container and the bucket are connected, the equilibrium state is that their water levels (relative to the ground) are the same.

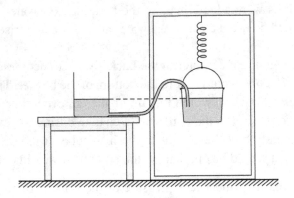

If more water is added to the container, some of it will flow into the bucket through the hose, and then, because of the bucket's increased weight, the spring will stretch further and the bucket will be lowered. If the spring is sufficiently weak and the lowering of the bucket is greater than the rise of the water level within it, then the level of the water surface in the bucket is lowered with respect to the ground. When equilibrium has been established, the same will have happened to the water level in the container. The process is reversible; if a little water is removed from the container, then the water level within it rises.

S88 Consider notionally covering the funnel with a cylinder that has a base that coincides with the rim of the funnel, and then pouring water into it up to the level of that in the funnel (*see* figure). In this situation the funnel is pushed upwards by the water inside it with the same force as the water outside it pushes it down.

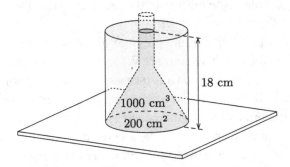

So, for the funnel to remain in place, its weight must be at least as large as that of the water outside it. The volume of this water is

$$18 \text{ cm} \times 200 \text{ cm}^2 - 1000 \text{ cm}^3 = 2600 \text{ cm}^3,$$

and so a funnel with a mass of 2600 g $= 2.6$ kg is just sufficient.[48]

S89 The external force acting on the hemispherical shell must balance two other forces, the weight of the shell and the force due to the hydrostatic pressure in the liquid. We need to find both the horizontal and vertical components of these forces.

The vertical forces are the weight mg of the hemispherical shell, and the weight Mg of the liquid, whose mass $M = \frac{2}{3}\pi R^3 \varrho$. The upward vertical component F_{vert} of the external force must balance the sum of these two forces:

$$F_{\text{vert}} = (m + M)g.$$

Horizontally, since the net force acting on the water must be zero, the reactions on it of the wall and shell – and hence also the forces the water exerts on them – must all be equal. That on the wall can easily be calculated, as the product of the average hydrostatic pressure $(\varrho g R)$ and the area of the circle that coincides with the rim of the shell. It follows that the horizontal component of the external force has the same value, and so it is $F_{\text{hor}} = \pi R^3 \varrho g$. This can be expressed in terms of the liquid mass as

$$F_{\text{hor}} = \tfrac{3}{2}Mg.$$

The magnitude of the required external force is therefore

$$F = \sqrt{F_{\text{hor}}^2 + F_{\text{ver}}^2} = Mg\sqrt{\frac{9}{4} + \left(1 + \frac{m}{M}\right)^2},$$

and its direction makes an angle

$$\theta = \arctan \frac{F_{\text{vert}}}{F_{\text{hor}}} = \arctan \frac{2(m + M)}{3M}$$

with the horizontal (*see* figure).

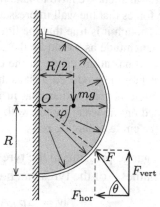

The appropriate point for the application of the external force can be determined by considering the necessary balance of the torques that act within the system.

For the hemispherical shell taken in isolation, the forces acting on it are shown in the figure: the horizontal and vertical components of the external force acting at the (as yet to be determined) appropriate point, the weight of the shell (whose centre of gravity is half-way along the radius that forms the axis of symmetry) and the forces caused by the hydrostatic pressure of the water in the shell (which are distributed along its the inner surface).

We note that, as the liquid exerts only a *radial* force on each piece of the surface, the net torque of these hydrostatic forces is zero if the centre O of the sphere is chosen as the reference point. As this greatly simplifies the problem, this is our obvious choice.

The required position of the force application point can be characterised by the angle φ shown in the figure. The remaining effective torques about O are then $mgR/2$ and $(3MGR \sin \varphi)/2$, both clockwise, and $(m + M)gR \cos \varphi$ anticlockwise. From their balance, it follows that

$$\left(1 + \frac{M}{m}\right) \cos \varphi - \frac{1}{2} - \frac{3M}{2m} \sin \varphi = 0.$$

This equation is quadratic in either $\sin \varphi$ or $\cos \varphi$, and can be solved if the ratio of the masses is known.

Notes. 1. If the hemispherical shell is much lighter than the liquid in it (i.e. $m \ll M$), then $F \approx 1.8Mg$ and $\theta \approx \varphi \approx 33.7°$. In the other limiting case ($m \gg M$), $F \approx mg$, $\theta \approx 90°$ and $\varphi \approx 60°$.

2. If the water and shell are considered as a single body, and we write the equation expressing the torque balance for it, then, of course, we should get the same result as previously for the positioning of the external force. In this case, in addition to the torques due to the applied force and the weights of the water and the hemispherical shell, we have to take into account the torque produced by the compressive forces that the wall impresses on the water.

Warning! Though it is true that the liquid presses on the wall with a net force of the same magnitude as would be the case if the pressure had the same value everywhere as it has at the centre of the circle, the same simplification does not apply to the net torque produced. Just as the hydrostatic pressure in the lower half of the base circle is greater than that in the upper half, so are the compressive forces imposed on the water by the wall, and their net torque about a horizontal diameter of the circle cannot be zero.

S90 Denote the gauge pressure in the tyres by Δp. If the weight of the car per tyre is w, then the 'flatness' of the tyres can be calculated from:

$$w = \Delta p \cdot A,$$

where A is the surface area touching the road (*see* figure).

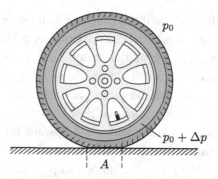

If the gauge pressure is a factor of $\frac{4}{3}$ times larger, then the surface area touching the road will be $\frac{3}{4}$ times smaller, because the weight $\Delta p \cdot A$ of the car (per tyre) does not change. For the sake of simplicity, let us suppose that the shape of the surface touching the road does not change, meaning its linear size decreases in both directions by a factor of $\sqrt{\frac{3}{4}}$.

When the steering wheel is rotated, the rubber surface in contact with the road slides, if the torque is sufficiently large. Notionally divide the dented rubber surface into small pieces, so that the total torque can be calculated as the sum of the torques produced by 'elementary' frictional forces acting on each of these pieces. After the increase in the tyre pressure, all of these pieces will be $\sqrt{\frac{3}{4}}$ times closer to the 'centre' of the reduced flat area in contact with the road, and so the moment arms of the elementary frictional forces will decrease in the same proportion.

As the normal forces, which press the pieces of the dented rubber surface onto the road, do not change (because their surface areas decrease at the same rate as Δp increases), the frictional forces acting on the pieces are unaltered. Consequently, the total torque needed for the relative movement of road and tyre will be $\sqrt{\frac{3}{4}} \approx 0.87$ times smaller than the original one. Correspondingly, the force required to rotate the steering wheel decreases by approximately 13 %.

In the above argument, it was assumed that the size of the rubber area that touches the road varies inversely with the gauge pressure, but maintains its shape. In reality, this is not strictly true! The construction of car tyres is such that the transverse size of the contact surface is about the same for 'soft' tyres as it is for those that have been inflated hard; the change in the surface area A is effected by a change of the lengthwise dimension, parallel to the length of the car. If such a tyre were very narrow, then an areal decrease by a factor of $\frac{3}{4}$ would require a lengthwise size change by the same factor, resulting in a 25 % decrease in the required steering force. The real situation is probably somewhere between these two extreme cases, and so the force decrease lies between 13 % and 25 %, say about 20 %.

S91 Initially, the gauge pressure, Δp, in the balloon is 10 cm of water. After the iron rod has been carefully placed on the top of the balloon, the water in the manometer will overflow if the gauge pressure exceeds $10 + 20 - (-20) = 50$ cm of water; or more scientifically, approximately 5 kPa. The weight of the cylindrical iron rod, of length 50 cm and diameter 2 cm, is $mg = 12$ N. The pressure on the bottom of the rod due to its own weight is $7.8 \times 10^3 \times 9.8 \times 0.50 \approx 40$ kPa; this is many times larger than the gauge pressure required to cause overflow. However, the manometer water *does not overflow*.

The reason for this is the additional upward force on the rod due to the deformation of the balloon's rubber wall (*see* Fig. 1) as it is placed on the balloon and allowed to sink, depressing the balloon's upper surface (but not so far that it touches its lower surface). Thus, the rod is held up, not only by the hydrostatic force from the gauge pressure Δp, but also by the elastic force F due to the tension in the rubber wall of the balloon.

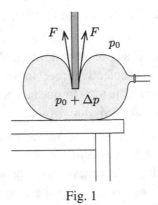

Fig. 1

It can be shown, as follows, that the pressure increase in the balloon is not very large, and that only its shape changes significantly. The new gauge pressure Δp can be estimated by considering the static equilibrium of a hypothetical cylindrical region within the balloon, as shown in Fig. 2, in which p_0 denotes the atmospheric pressure. The diameter of the volume is determined by the nearly circular contour on the balloon's upper surface on which F is horizontal. The position of the lower horizontal boundary A is somewhat arbitrary but must be below the bottom of the rod.

There are no vertical components of the elastic forces acting on this body (region), and the condition of equilibrium can be written as $mg = \Delta p A$, where the bottom area A is about 10–20 cm across and therefore of the order of 10^{-2} m^2 in size. This estimate gives the gauge pressure as

$$\Delta p \approx \frac{12 \text{ N}}{10^{-2} \text{ m}^2} = 1.2 \text{ kPa}.$$

This is equivalent to 12 cm of water, indicating that the (gauge) pressure increase is only 2 cm of water, far short of the 50 cm needed. The conclusion that the

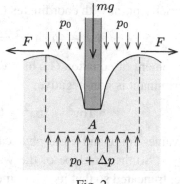

Fig. 2

water does not overflow is not at all sensitive to the estimation of area A. If we assume one-half, or even one-third, of our initial estimate for A, the additional gauge pressure is still not enough to cause overflowing. We note, in passing, that the effective A cannot be much larger than our assumed value, as this would lead to a value for Δp that is less than the original 10 cm of water – clearly a physically unacceptable conclusion.

> *Note.* The pressures involved in this problem are in accord with the fact that a balloon can be inflated fairly easily using the lungs, but the maximal *gauge* pressure in the human lung is clearly much less than atmospheric. If this were not the case, there would be an unbearably large force acting upon the chest.

S92 Consider a water jet emerging from a hole located a distance h below the surface level of the water (*see* figure). According to Torricelli's theorem, the speed of efflux is $\sqrt{2gh}$, and the formulae for horizontally launched projectiles give the equation of the trajectory of this jet (in the coordinate system shown in the figure) as

$$y(x) = \frac{x^2}{4h} + h. \tag{1}$$

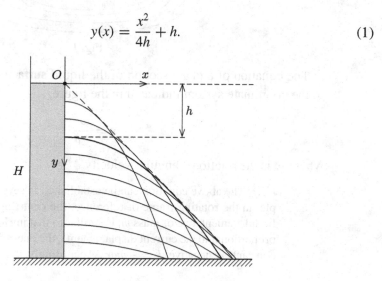

The water can reach a point with coordinates (x, y) only if the quadratic equation

$$4h^2 - 4hy + x^2 = 0, \qquad (2)$$

given by a rearrangement of equation (1), has real root(s) for h. The condition for this is that its discriminant is non-negative:

$$16y^2 - 16x^2 \geq 0.$$

As only positive values for x and y are physically possible, this is equivalent to the inequality $y \geq x$. So the envelope of the water jets is the lateral surface of a right-circular cone, truncated so that its top coincides with the water surface in the cylinder. The generatrix of the cone makes an angle of 45° with the horizontal.

> *Note.* If the measuring cylinder is placed on a horizontal table-top, then, to pro-
> duce the whole envelope, it is sufficient to bore holes only above the height $H/2$.
> The jets emerging from the holes at height $H/2$ are the ones that reach the table
> furthest from the cylinder, and they 'mark out' the perimeter of the truncated
> cone's base.

S93   If an upright cylindrical vessel containing liquid is steadily rotated in the Earth's homogeneous gravitational field, then, after a while, the surface of the liquid takes the shape of a paraboloid (a solid of revolution) (*see* Fig. 1).

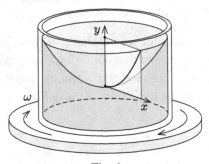

Fig. 1

The equation of a plane section of the liquid surface (the 'rotated parabola') is, in the coordinate system indicated in the figure,

$$y = \frac{\omega^2}{2g}x^2,$$

where ω is the rotational angular velocity.

> *Note.* The above equation can be established in several different ways. For exam-
> ple, in the rotating frame of reference, the centrifugal force acting on any small
> liquid element, one of mass m, is similar to that in Hooke's law (in that the force is
> proportional to the current displacement), the one difference being that the 'spring
> constant' is negative: $k = -m\omega^2$. Accordingly, the 'centrifugal potential energy'

is $-m\omega^2 x^2/2$, and when this is added to the gravitational potential energy, we get the total potential energy as

$$E_{\text{pot}} = -\frac{m\omega^2 x^2}{2} + mgy.$$

The free surface of the liquid must be an equipotential surface, with the same total potential energy everywhere. Setting E_{pot} equal to a constant yields the stated equation to within an unimportant shift in the origin of the y-axis.

With the given data, it can be shown that the surface of the rotating liquid can assume the appropriate paraboloidal shape, while still retaining all of the water within the flask's neck. The system behaves as just described, with the flask's neck as the rotating cylinder.

However, a further question arises (and this is the *key* question for the correct solution to this problem): If the paraboloid of revolution were notionally extended as far as the spherical part of the flask, would it, or would it not, 'cut' into the sphere? If the answer is 'yes', then it would be energetically favourable (i.e. a lower energy) for some of the water to reside in the sphere, provided it lay below the extended paraboloid surface. The gain in gravitational energy would be more than offset by the reduction in centrifugal energy brought about by its increased distance from the axis. How it might get there is another question (but one that we will tackle later); first we must decide if the possibility exists.

To do this, we take a plane section containing the rotational axis of the flask and determine the lowest point of the particular parabola (whose shape, but not vertical position, is predetermined by the given angular velocity ω) that just touches the circle in which the plane and the flask's sphere intersect.

Suppose that this point is a distance h below the centre O of the circle. Then, in a coordinate system whose origin is at O, the equation of the parabola is

$$y + h = \frac{\omega^2}{2g}x^2,$$

and the equation of the circle is

$$x^2 + y^2 = R^2,$$

where R is the radius of the flask's spherical part. Substituting for x^2 into the equation for the parabola, we get a quadratic equation for the y coordinates of points common to the circle and parabola:

$$y^2 + \frac{2g}{\omega^2}y + \frac{2gh}{\omega^2} - R^2 = 0.$$

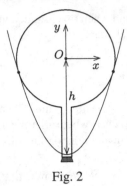

Fig. 2

When the parabola just touches the circle (as shown in Fig. 2), the above equation must have a double root and its discriminant must be zero; this gives a condition on height h as

$$h = \frac{g}{2\omega^2} + \frac{\omega^2}{2g}R^2.$$

Substituting the given data, $g = 9.81$ m s^{-2}, $\omega = 2\pi \times 3$ s^{-1} and $R = 0.1$ m, yields $h = 0.195$ m $= 19.5$ cm. The radius of the sphere is 10 cm, as is the length of the flask's neck, and together they are 0.5 cm longer than the calculated value for h of 19.5 cm. Thus the lowest point of this critical paraboloid lies half a centimetre above the bottom of the neck of the inverted flask.

Any water remaining in the flask's neck behaves as if it were in a rotating cylinder, and the difference in height between the rim and the lowest point of the free surface is $(\omega^2/2g)r^2$, where r is the radius of the neck. Numerically, this difference is about 2 mm. Clearly, within the neck, there is not enough volume 'below the critical surface' to accommodate all of the original water. The 'missing' water is inside the sphere in a strip running around it below the critical paraboloid (*see* Fig. 3). As already discussed, when it is there, the water has a lower energy than it could have if it stayed in the neck.

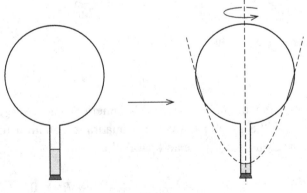

Fig. 3

How did the water get there, we may wonder? Perhaps some water splashed up during the accelerating phase of the flask's rotation. But this is not the proper answer to the question; rather, it is that water evaporates from the neck of the flask and then condenses on its wall at the relevant place. The driving force for this thermodynamic process is simply the tiny pressure difference in the flask produced by the combined effect of the gravitational force and the rotation of the flask.

In the rotational reference frame, the water is in equilibrium along any paraboloid of revolution that has the appropriate proportionality constant, $\omega^2/(2g)$. If two different paraboloid surfaces are compared, the specific (i.e. per unit mass) potential energy of water is larger along the higher surface than along the lower one. In equilibrium, the surface of the water must follow the same paraboloid in the neck of the flask as it does in the sphere. If this were not so, the concurrent processes of evaporation and condensation would 'move the system' to one with lower total energy.

After a (very) long time, the arrangement of the water in the flask is as sketched in Fig. 3.

> *Note.* This unusual phenomenon is quite unbelievable at first sight, and we might think that it would not happen in reality. We could, for example, argue that: 'It cannot happen – just as water does not climb out of a water glass onto a supporting table-top, even though the water would lower its energy by doing so.'[49] The interesting fact is that the water can climb out of the glass, even if the thermal equilibrium is perfect. The reason it can do so is the tiny barometric pressure difference between the level of the water surface in the glass and the level of the table-top. Experimentally, this can be shown by covering the glass of water on the table-top with a large bell jar, and then waiting for a (very) long time – that is the most difficult part of the experiment! The water 'climbs out' of the glass, with or without the bell jar, but because of the large volume of air in a room, it cannot condense anywhere, and remains in the air as unsaturated water vapour.

S94 *Solution 1.* When the razor-blade is floating on the surface of the water, and the level of the water in the glass is at a height h_1, then the compressive force acting on the bottom of the glass is

$$F_1 = \varrho g h_1 A,$$

where A is the inner cross-sectional area of the bottom of the glass, and ϱ is the density of water (*see* Fig. 1).

When the blade, of mass m, sinks, the height of the water level changes to h_2, and now the force acting on the bottom of the glass, is given by:

$$F_2 = \varrho g h_2 A + mg \left(1 - \frac{\varrho}{\varrho_{\text{blade}}} \right).$$

---

[49] But do not mention superfluid liquid helium, which is well known for doing just that!

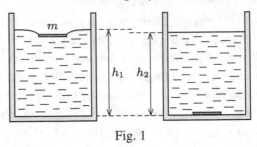

Fig. 1

The negative term in the brackets represents the buoyancy force that acts on the blade when it has sunk to the bottom of the glass. We have assumed here that there is a thin water layer between the blade and the bottom of the glass. If this is not so, then the same negative term would arise from the slight pressure difference between the bottom and the top surfaces of the blade – at the top the hydrostatic pressure is a little less than at the bottom.

However, F_1 and F_2 must be the equal, because each of them is equal to the total weight of the water and blade. Measurements on a set of scales would show exactly this weight difference between the glass with water and blade, and the empty glass, whether or not the blade is floating.

It follows that $h_1 > h_2$, and so, when the razor-blade sinks, the height of the water *decreases* a little. The amount by which the water level is lowered is given by the height of a water layer, spread uniformly across the glass, that has the same weight as the razor-blade, less the upthrust the blade experiences.

> *Note.* For the sake of simplicity, we have implicitly assumed that the material of the glass is such that the contact angle between the water and the glass is 90°, i.e. the water level remains horizontal right up to the wall. If this is not so, then the surface tension force exerted on the glass by the water surface's rim must be taken into account. But this effect would add the same terms to the expressions for both F_1 and F_2 and would not change the conclusion reached.

Solution 2. Consider the water displaced by the razor-blade, i.e. the region $ABCD$ of 'missing' water denoted by grey shading in Fig. 2.

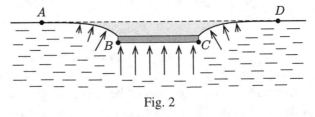

Fig. 2

In this region, besides the razor-blade (dark grey), there is also air (light grey), though the mass of the latter is negligible. So, the razor-blade maintains its static equilibrium because the gravitational force acting on it is balanced by the net (upward) force exerted by the liquid (within which the pressure is greater than atmospheric). This same net force could equally well have balanced the weight of the water that is missing from the same volume *ABCD*.

It is therefore clear that the weight *mg* of the razor-blade is just equal to the weight of the displaced water, as was recognised by Archimedes many years ago. The interesting thing is that the 'displaced water' in Archimedes' principle is not only the missing water from the volume occupied by the razor-blade, but also that from the additional region affected by surface tension. So, the volume of displaced water is greater than the blade's own volume, and it follows that the water level must *drop* when the blade sinks.

> *Note.* For a floating body of greater density than water, its weight is balanced by the sum of the compressive force of the water below it and the vertical component of the surface tension forces acting along its edges. The former is equal to the weight of 'missing' water directly above the body; its magnitude is proportional to the *area* of the body's horizontal projection. The latter, the surface tension term, is equal to the weight of the remaining part of the displaced water; this contribution is proportional to the *perimeter* of the body's horizontal projection.
>
> For the flotation of relatively large bodies, the hydrostatic pressure makes the dominant contribution, and the surface tension only stabilises the balance. For bodies with smaller cross-sectional areas (e.g. the legs of water boatmen and pond skaters, or floating metal paper clips), the situation is just the opposite: their weights are mainly balanced by surface tension forces, and the effect of hydrostatic pressure is not important. In the case of the razor-blade, the forces associated with the two different effects have the same order of magnitude, and so both of them are significant.

S95 In the central region of horizontal radial flow, the water moves rapidly until it reaches the jump. The whole surface of this flowing water is in contact with air at atmospheric pressure; accordingly, the pressure in it is everywhere equal to 1 atm (neglecting small effects due to the viscosity of water). From Bernoulli's law, it follows that the speed of the water in this horizontal plane segment must be uniform. The continuity equation then implies that the depth of the fast-flowing layer at any point must be inversely proportional to the distance of that point from the centre of the disc. In reality, although the fast flow is laminar, because of viscosity, the decrease in depth of this layer does not exactly follow this simple 'inverse proportionality' rule.

The vertical water jet pushes the disc downwards with quite a large force when the former is abruptly halted by the depression in the latter (this is called hydrodynamic pressure), and so the disc should sink. The gravitational force acting on the disc is also directed downwards (as is the weight of the water that flows quickly across the top of the disc, but this is negligible). So the key problem is that of finding the origin of the upwardly directed force that balances the downward ones. In this phenomenon, the role of the hydrostatic pressure is paramount (as it is in the solution referred to in the original problem). The disc does not float exactly at the top of the water, but in a slightly lower position. In this situation, the hydrostatic pressure of the water under the disc can provide the upward force required to maintain it in equilibrium.

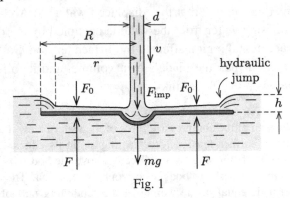

Fig. 1

The lowering of the disc can be estimated by considering the force balance in the system. The forces acting on the disc, as well as the geometrical nomenclature, are shown in Fig. 1. The force F_{imp} of the hydrodynamic pressure can be calculated from the vertical momentum lost when the water jet impacts on the disc:

$$F_{imp} = \frac{d(mv)}{dt} = \varrho \frac{dV}{dt} v = \varrho Q v,$$

where v is the impact speed of the jet, ϱ is the density of water and $Q = dV/dt$ is the flow rate of the water jet; the latter can be measured by collecting water from the jet over a measured time. From the flow rate, the impact speed of the water can be found if we also measure the diameter d of the jet:

$$v = \frac{Q}{\pi(d^2/4)} = \frac{4Q}{\pi d^2},$$

yielding

$$F_{imp} = \frac{4\varrho Q^2}{\pi d^2}.$$

At the top of the disc, the pressure is roughly the atmospheric pressure p_0, whereas at the bottom it is $p_b = p_0 + \varrho g h$, where h is the height of the water level outside the hydraulic jump relative to the bottom of the disc (*see* Fig. 1). Neglecting the (relatively small) cross-sectional area of the jet at the centre of the disc, where the impact takes place, the net upward force ΔF_{press} due to the pressure difference between the bottom and top of the disc is

$$\Delta F_{press} = F_b - F_t = p_b(\pi R^2) - p_0(\pi r^2) = \pi(R^2 - r^2)p_0 + \pi\varrho g h R^2.$$

The force balance of the disc can be expressed as

$$mg + \frac{4\varrho Q^2}{\pi d^2} = mg + F_{imp} = \Delta F_{press} = \pi(R^2 - r^2)p_0 + \pi\varrho g h R^2.$$

As mentioned in the next paragraph, r increases if Q does, and if the flow rate is sufficiently large, then r approaches R, and the first term on the right-hand side of the equation becomes zero. In an experiment we carried out – using a copper disc of diameter 10 cm, thickness 0.20 mm and weight $mg = 0.137$ N – this happened when the flow rate was $Q \approx 4.3 \times 10^{-5}$ m³ s⁻¹ and the diameter of the jet was $d \approx 5$ mm. Calculating the height h from our measured data, we obtained $h \approx 3$ mm, which is a reasonable result, and in accord with our visual observations. It was of some interest that the hydrodynamic pressure produced a force of only $F_{imp} = (4\varrho Q^2)/(\pi d^2) = 0.094$ N, which was less than the weight of the disc.

When the flow rate is reduced, the impact of the water jet is smaller, so the upwardly directed force due to the pressure difference ΔF_{press} must also be smaller. When doing the experiments, we observed that, as expected, a weaker flow rate results in a smaller value for r, i.e. the hydraulic jump recedes towards the centre of the disc. According to a more sophisticated analysis, the radius r of the 'hydraulic jump circle' is proportional to $Q^{2/3}$, confirming that a reduced flow rate causes the faster-flowing area around the centre of the disc to decrease in size. The consequent widening of the high-water ring spreading inwards from the disc's perimeter increases the downward force acting on the disc. If the flow rate decreases below a certain critical value, then the disc sinks.

Notes. 1. If the flow rate is increased, the system remains stable, but the radial location of the jump does not increase beyond $r = R$; only h increases, meaning that the disc lies deeper in the water. It takes extremely strong jets to push the disc beneath the water surface. It is rather surprising how stable the system is – the disc is almost unsinkable!

2. If you would like to try this experiment, then a disposable CD (or DVD) is recommended as a readily available alternative to a metal disc with an engineered depression. Attach one or two quite heavy coins[50] to the CD with adhesive tape, so as to cover the central hole symmetrically. The opposite side of the hole should also be covered with tape, but carefully, because a small hollow (the 'depression') is important for the stability of the system. This latter tape smoothes off the sharp rim around the hole, and greatly reduces the splashing produced when the water jet hits the depression. When the CD is placed on the water, make sure that the coin is underneath (*see* Fig. 2). A kitchen sink with a suitable tap is ideal for this demonstration, which will surely produce hours – well, maybe a few minutes – of enjoyment.

---

[50] A UK £2 coin is one possibility.

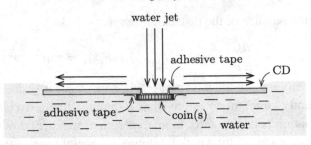

Fig. 2

S96 The soap film between the two bubbles is a spherical cap of radius ϱ (to be found). The line joining the centres, A and B, of the two spheres is a normal to the film. Fig. 1 shows a plane section of the spheres that contains this line; we have assumed that $R > r$.

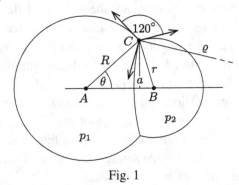

Fig. 1

Inside the two-sided spherical soap bubble of radius R, the gauge pressure is

$$p_1 - p_0 = \frac{4\gamma}{R}, \tag{1}$$

where p_0 is the atmospheric pressure, and γ is the surface tension of the liquid (water containing soap or some detergent). Similarly, the gauge pressure in the smaller bubble is

$$p_2 - p_0 = \frac{4\gamma}{r}, \tag{2}$$

while for the separating soap film, the pressure difference between its two sides is

$$p_2 - p_1 = \frac{4\gamma}{\varrho}. \tag{3}$$

Subtracting (1) from (2), and inserting the difference into (3), the radius of curvature can be expressed in the form

$$\varrho = \frac{Rr}{R - r}.$$

For the record, we note that the film's centre of curvature lies on the same side of it as that of the smaller bubble.

The radius a of the perimeter circle can be determined by the following argument. A small segment of the circle going through point C in Fig. 1 is in equilibrium under the action of three forces with identical magnitudes – determined by the common value of the surface tension – and so the angles between the forces must be 120°.

Now, a tangent to a circle is perpendicular to the corresponding radius, and so the angle between radius AC and the tangent to the film is 30°, as is the angle between the latter and radius BC. It follows that angle $\angle ACB = 60°$.

Denoting the angle $\angle BAC$ by θ, we now apply the cosine and sine laws to the triangle ABC:

$$(AB)^2 = R^2 + r^2 - 2Rr \cos 60°$$

and

$$\frac{\sin \theta}{\sin 60°} = \frac{r}{AB}.$$

From these, the radius a of the perimeter circle can be written as

$$a = R \sin \theta = \frac{\sqrt{3}}{2} \frac{Rr}{\sqrt{R^2 + r^2 - Rr}}.$$

Note. Three special cases have been sketched in Fig. 2:

a) If $R \to r$, then $\varrho \to \infty$ (the separating surface becomes a plane), and $a = r\sqrt{3}/2$.
b) If $R \gg r$, then $\varrho = r$, and $a = r\sqrt{3}/2$.
c) If $R = 2r$, then $\varrho = R$, and $a = r$.

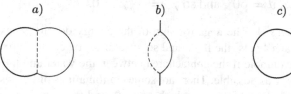

Fig. 2

S97 The shape of the puddle is such that the sum of the surface energy and the gravitational potential energy is minimal. The surface energy has two terms: the energy of the water–air boundary, and the corresponding energy at the water–floor interface.

A water drop on the floor is in contact with both the air and the floor. The energy per unit area of the water–air boundary is γ, usually referred to simply as the surface tension of water (without specifically mentioning the air).[51] But,

---
[51] Surface tension γ is measured in J m^{-2} or N m^{-1}, according to context.

in addition, the water has contact with the floor over some particular area. If this contact area changes (e.g. the water spreads out and as a result the contact area between the floor and air decreases), then there is an energy change proportional to the change in area. Denote this 'surface energy density' by γ'.

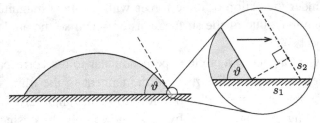

Fig. 1

The three quantities γ, γ' and the contact angle ϑ, which characterises the equilibrium at the edge of the body of liquid, are not independent of each other. For if, notionally, the contact line at the edge of the water drop (*see* Fig. 1) were shifted from its equilibrium position by a small distance s_1, then – by the principle of virtual work – the sum of the first-order changes in the surface energies would be zero. Along a contact line of unit length,

$$\gamma' s_1 + \gamma s_2 = 0,$$

where s_2 is the resultant lengthening of the water–air interface. Using the geometrical connection $s_2 = s_1 \cos \vartheta$ (as shown in Fig. 1), we have

$$\gamma' = -\gamma \cos \vartheta. \tag{1}$$

In our case, $\vartheta = 60°$, and so $\gamma' = -\frac{1}{2}\gamma < 0$.

> *Notes.* 1. The negative sign of the energy density γ' expresses the fact that the water 'wets' the floor, and so, insofar as lowering energy is concerned, it is more favourable if the contact area between the water and the solid floor material is as large as possible. There are some combinations of liquids and solids (e.g. mercury and clean glass) for which $\gamma' > 0$, and the contact angle is then $\vartheta > 90°$. For such combinations 'wetting' does not take place, and the tendency is to decrease the area of contact, because this provides a lower energy.
>
> 2. The quantity γ' can be written as the difference between two surface energy densities, γ_{SL} and γ_{SG} (the so-called interfacial surface tensions). The first of these is the energy per unit area of the liquid–solid interface, and the second that of the gas in contact with the solid. But this separation into two terms is *purely formal*, as in formulae only their difference γ' ever arises; the separate energy densities of a solid (with respect to liquids and gases) have no roles in themselves.

The surface area of a puddle containing $V = 5$ litres of water is relatively large (*see* Fig. 2), and so the areas of the water–air and water–floor interfaces can, to a good approximation, be taken as having a common value A. So, it is good enough for our purposes to write the total energy of the puddle as

$$E = \gamma A + \gamma' A + mg\frac{h}{2},$$

where m is the mass of the puddle and $h = V/A$ is its height. Using result (1), and expressing m in terms of the density ϱ of water, E can be written as a function of A:

$$E(A) = \gamma(1 - \cos\vartheta)A + \frac{\varrho g V^2}{2A}. \tag{2}$$

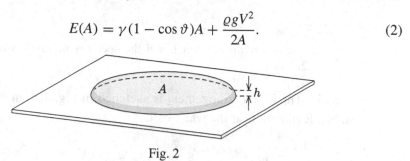

Fig. 2

Since $E(A)$ is of the form $\alpha A + \beta/A$, it is minimal when $dE/dA = \alpha - \beta/A^2 = 0$, i.e. $A = \sqrt{\beta/\alpha}$. Using this result, the energy-minimising area of the puddle can be found directly (without the need to actually evaluate $E_{\min}(A)$):

$$A = \sqrt{\frac{\varrho g V^2}{2\gamma(1 - \cos\vartheta)}} \approx 1.8 \text{ m}^2.$$

As he had knocked the bucket over on several previous occasions, the student knew that, almost irrespective of the volume of water spilt, the depth of the resulting puddle is always the same, about 2.7 mm (*see* note below). He was therefore able to produce the above answer within seconds – but only after getting out his pocket calculator!

Note. From the surface area of the puddle, its depth can also be calculated:

$$h = \frac{V}{A} = \sqrt{\frac{2\gamma}{\varrho g}(1 - \cos\vartheta)} \approx 2.7 \text{ mm}.$$

This result can also be obtained from the force balance of an appropriately chosen segment of the puddle.[52] The dimensional quantity that appears in this formula (i.e. ignoring dimensionless constants) is the so-called *capillary length* $\lambda_{\text{cap}} = \sqrt{\gamma/\varrho g}$; it has the dimensions of length and, for most liquids, is of the order of millimetres in magnitude. It gives the characteristic scale size when the effects of surface phenomena and gravity are comparable. Liquid drops that are much smaller than λ_{cap} are spherical to a good approximation; their shape is determined

---

[52] A similar calculation can be found in the predecessor of this book: see 'Problem 130' in P. Gnädig, G. Honyek & K. F. Riley, *200 Puzzling Physics Problems* (Cambridge University Press, 2001).

by the pressure produced when surface tension acts within a curved surface. The shape of 'drops' with typical sizes much larger than λ_{cap} is mostly influenced by the gravitational field, and so they spread out and form puddles.

These considerations are not confined to static phenomena. For example, the speed v of surface waves of wavelength λ, on a liquid of density ϱ and surface tension γ, is given by

$$v^2 = \frac{g\lambda}{2\pi} + \frac{2\pi\gamma}{\varrho\lambda}.$$

The two terms are equal, and the speed is minimal, when $\lambda = 2\pi\sqrt{\gamma/\varrho g} = 2\pi\lambda_{cap}$.

S98 The initial arrangement is sketched in Fig. 1*a*), in which the glass rod is shown levitating near the water ball.

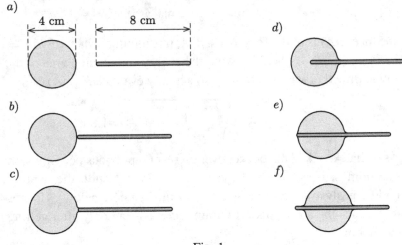

Fig. 1

The process starts when one end of the rod, of radius r, gently touches the water drop (Fig. 1*b*)). The water wets the glass, and so a little of it 'flows' onto the blunt end of the rod (Fig. 1*c*)). But the process cannot stop there, because the net force now acting on the glass rod *is not zero*.

The pressure inside the water drop, of radius R and surface tension γ, is a little higher than the outside air pressure ($\Delta p = 2\gamma/R$), and this means that a force of $\pi r^2 \Delta p$ tends to push the rod out again. But the attractive force $2\pi r\gamma$ exerted by the water film on the perimeter of the rod is *much larger* than this. So the rod penetrates further into the water drop, and an intermediate situation is shown in Fig. 1*d*).

But the forces acting on the rod are still not balanced, and so there is no reason for it to stop. It continues moving forwards until the situation is as shown in Fig. 1*e*). Here, the rod has reached the left-hand edge of the water drop; actually, a little beyond it, with the water surface pushed out slightly. A force balance on the rod can only be achieved if the left-hand end of the rod fully emerges from the water drop, as shown in Fig. 1*f*). In this position, the shape of the water surface surrounding the rod is the same at both of its ends; the situation is symmetric and there is no net force acting on the rod.

We still need to consider whether or not the water drop spreads itself over the whole rod. The total energy of the system is the sum of the surface energy of the water in contact with the air and the surface energy of the glass in contact with water (and maybe with air); for an equilibrium state, this total energy needs to be as small as possible. As the rod is *thin*, the total surface area of the glass is negligible compared to that of the water ball. Thus, the equilibrium of the system is dominantly controlled by the requirement for a minimal surface area of the water, and, for any given volume of water, this is going to happen when its shape is that of a sphere.

So, in the final state, the water drop is almost spherical, the glass rod forms a diameter of that sphere and 'sticks out' of it at both ends.

S99 *a*) For the water, which 'perfectly' wets the glass, it is energetically favourable for it to be in contact with the wall of the container over as large an area as possible. As gravity imposes no limitations, the water will cover the total inner surface of the shell, and an air bubble will be formed somewhere inside it. Its shape will be governed by the condition that its surface – for any given fixed volume – should be as small as possible. The volume of the air *is* fixed, because the pressure of curvature produced by the surface tension is negligible compared to normal atmospheric pressure. Consequently, both the pressure and temperature of the gas are independent of the bubble's shape, and so therefore is the volume it occupies.

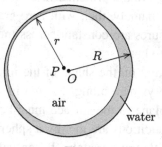

Fig. 1

Of all the solids having a given volume, the sphere is the one with the smallest surface area.[53] So, the shape of an air bubble, enclosed in a spherical shell that has

---

[53] Even in ancient times, this was well known.

radius $R = 4$ cm and centre O, is a smaller sphere (*see* Fig. 1). The radius r of the latter can be calculated from its volume:

$$\frac{4\pi r^3}{3} = \frac{2}{3} \times \frac{4\pi R^3}{3},$$

from which

$$r = \sqrt[3]{\frac{2}{3}}R = 3.5 \text{ cm.}$$

The position of the centre P of the air bubble is somewhere within 0.5 cm of O, but is otherwise arbitrary; within this constraint, the equilibrium is *indifferent* as to the position of P. The air bubble can move in the hollow shell as the consequence of the smallest of effects (e.g. a slight acceleration of the space station), but when it reaches the shell wall, it bounces back and does not 'stick' to it.

> *Note.* At every point of a spherical surface, the curvature is the same, and, as a result, so is the pressure of curvature produced by surface tension at the wall of the air bubble. That is how it must be, because, under weightless conditions, neither the pressure of the gas inside, nor the pressure of the water outside, can vary along the interface; their (constant) difference is everywhere equal to the pressure of curvature.

b) The air bubble enclosed within a silver spherical shell assumes a shape that is such that its mean curvature does not vary over the water–air interface; as a formula, this is

$$\frac{1}{2}\left(\frac{1}{r_{max}} + \frac{1}{r_{min}}\right) = \text{constant},$$

where r_{max} and r_{min} are the principal radii of curvature. The reason for this is the same as in part *a*), namely, that, with zero gravity and STP conditions, both the air and water pressures are constant across the whole surface – and so the mean curvature also has to be.

A further constraint on the shape of the interface is that, because the angle of contact is 90°, everywhere along the line in which the water surface meets the wall, the tangent plane to the surface must be perpendicular to that of the inner wall. These two conditions are met by a spherical cap (a portion of a sphere 'cut off' by a plane) with appropriately chosen values for its size and the position of the sphere's centre; this is therefore a *possible* solution. But the solution to this formally mathematical problem is – for physical reasons – unambiguous; it is *the solution*! The shape of the water in the silver shell is that of two spherical caps, joined to each other by their common plane face (*see* Fig. 2). The radius of one cap is obviously R; that of the other is calculated below.

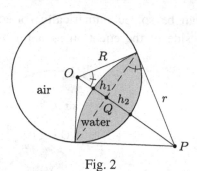

Fig. 2

Note. In a more general situation, the liquid–air interface is neither a spherical surface, nor a segment of one, but takes on a more complex shape (though still one with a constant mean curvature). For example, if the hollow within the levitating vessel were not a sphere, but an irregularly shaped cavity, the liquid–air interface could not be a spherical surface. This is because, if it were, then its tangent planes along the line of intersection of the wall and liquid surface could *not all* make the requisite angle with the wall (an angle that is predetermined, and equal to the contact angle between the liquid and the shell material). In our problem, the required conditions can be satisfied by quite a simple surface, i.e. by a spherical cap, because along any line of intersection of two spheres, the angle between the tangent planes is the same everywhere.

The radius of curvature r of the interface and the distance of its centre P from O are jointly determined by the volume of the water and the contact angle. Since the contact angle is 90°, a number of right-angled triangles can be used to ease the calculation; we also note that the two marked angles in Fig. 2 are equal (to α, say).

Using Pythagoras's theorem and the ratio properties of similar triangles, we have

$$OP = \sqrt{R^2 + r^2}, \quad OQ = R\cos\alpha = \frac{R^2}{\sqrt{R^2 + r^2}}, \quad PQ = r\sin\alpha = \frac{r^2}{\sqrt{R^2 + r^2}}.$$

The heights of the two spherical caps, shaded in Fig. 2, can now be found as

$$h_1 = r - PQ = r\left(1 - \frac{r}{\sqrt{R^2 + r^2}}\right), \quad h_2 = R - OQ = R\left(1 - \frac{R}{\sqrt{R^2 + r^2}}\right).$$

The volume of a spherical cap with radius ϱ and height h is $\frac{1}{3}\pi h^2(3\varrho - h)$, and so, equating the two expressions for the given volume of water:

$$\frac{\pi}{3}h_1^2(3r - h_1) + \frac{\pi}{3}h_2^2(3R - h_2) = \frac{1}{3} \times \frac{4\pi}{3}R^3.$$

From this and the values found for h_1 and h_2, we can obtain an equation for the radius ratio $x = r/R$:

$$x^3\left(1 - \frac{x}{\sqrt{1 + x^2}}\right)^2\left(2 + \frac{x}{\sqrt{1 + x^2}}\right) + \left(1 - \frac{1}{\sqrt{1 + x^2}}\right)^2\left(2 + \frac{1}{\sqrt{1 + x^2}}\right) = \frac{4}{3}.$$

This equation can be solved numerically (or graphically by plotting a graph of the left-hand side of the equation as a function of x) and yields the result $x = 3.206 \approx 3.2$.

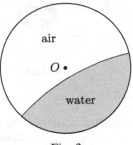

Fig. 3

So, the interface between the air and water enclosed in the levitating hollow silver sphere is the surface of a spherical cap of radius $r = 3.2R = 12.8$ cm; it is drawn to scale in Fig. 3. Because of the weightless conditions, the location of the 'water lens' is arbitrary, though it must be 'up against the wall'.

In principle, it is possible for the air bubble to take the form of a sphere as in Fig. 1, and float somewhere inside the spherical shell; but this state is unstable. If the bubble – as the result of any tiny disturbance – approaches the wall of the shell, it does not bounce back (as it did in case a)), but sticks to the wall. It can be shown that the total surface energy is much smaller for the arrangement shown in Fig. 3 than for that illustrated in Fig. 1. The corresponding energy comparison is much easier if the bubble sticks to a plane surface. Then the sphere becomes a hemisphere with the same volume, and its new surface area will be $1/\sqrt[3]{2} \approx 0.8$ times smaller than the original one.

S100   Suppose that we construct a container with two adjacent but separate compartments; all the outer walls are to be insulating, but the common connecting one is to be heat-conducting. One compartment is for tap water and the other for the distilled water, the only interaction between them being thermal conduction through the intervening wall.

For our analysis, we assume that the specific heats of distilled and tap water are equal, and so the ultimate equilibrium temperature for any particular stage is the mass-weighted arithmetic mean of the temperatures of the components when the stage begins. For example, if distilled water of mass M and temperature T_0 (measured in degrees Celsius) and tap water of mass m and temperature $100\,°C$ are poured into the compartments, then the common equilibrium temperature will be

$$T_1 = \frac{MT_0 + m \times 100\,°C}{M + m}.$$

For $M = m = 1$ kg, the common temperature would be $50\,°C$, which is not good enough!

By dividing the hot tap water into a number, say six, of smaller equal portions, we can do better. If all the cold distilled water is put into one of the compartments of the heat exchanger, and $\frac{1}{6}$ kg of tap water at $100\,°C$ is placed into the other, then, after equalisation, the common temperature will, from the above formula, be

$$T_1 = \frac{1 \times 0 + \frac{1}{6} \times 100}{\frac{7}{6}} = 14.3\,°C.$$

We now pour away the cooled tap water, and refill its compartment with a further $\frac{1}{6}$ kg of water at $100\,°C$. After the second heat exchange, the common temperature will be

$$T_2 = \frac{1 \times 14.3 + \frac{1}{6} \times 100}{\frac{7}{6}} = 26.5\,°C.$$

Continuing in this way, after the sixth and final heat exchange, the temperature of the distilled water will be $T_6 = 60.3\,°C$ – and the task set in the problem has been accomplished!

> *Notes.* 1. The problem can also be discussed in a more general way. Suppose that the 'hot water' at temperature T_{hot} is divided, not into six, but into n portions, and the individual portions are brought, one by one, into thermal contact with the distilled water, whose initial temperature is T_0.
>
> Denoting by T_m its temperature before exposure to the $(m+1)$th portion of hot tap water, we can write the heat balance equation as
>
> $$T_m + \frac{1}{n}T_{hot} = \left(1 + \frac{1}{n}\right)T_{m+1},$$
>
> which can be rearranged as the recurrence relation
>
> $$T_{m+1} = \alpha + \beta T_m \quad \text{with} \quad \alpha = \frac{T_{hot}}{n+1} \quad \text{and} \quad \beta = \frac{n}{n+1}.$$
>
> This recurrence relation has the general solution[54]
>
> $$T_m = T_0 \beta^m + \alpha \frac{1 - \beta^m}{1 - \beta}, \qquad m = 0, 1, 2, \ldots, n.$$
>
> This gives the temperature of the distilled water at each stage, and, in particular, its final temperature, T_n:
>
> $$T_n = \left[1 - \left(\frac{n}{n+1}\right)^n\right]T_{hot} + \left(\frac{n}{n+1}\right)^n T_0.$$
>
> For $n = 6$ we recover our earlier result of $60.3\,°C$, but for $n = 5$ the final temperature is only $59.8\,°C$, and short of our target.

---

[54] See, for example, equation (15.25) on page 498 of K. F. Riley, M. P. Hobson & S. J. Bence, *Mathematical Methods for Physics and Engineering*, 3rd edn. (Cambridge University Press, 2006). Its validity can be verified by direct substitution in the recurrence relation; the boundary value T_0 can be checked by setting $m = 0$.

As we have done in other problems, we can use the fact that

$$\lim_{n \to \infty} \left(1 + \frac{x}{n}\right)^n = e^x,$$

and that therefore $\lim(n/(n+1))^n = 1/\lim(1 + 1/n)^n = 1/e$, to show that, even if the hot water is divided into an arbitrarily large number of portions ($n \to \infty$), the final temperature cannot exceed

$$T_\infty = \left(1 - \frac{1}{e}\right) T_{hot} + \frac{1}{e} T_0.$$

This temperature limit for the data given in the problem is $63.2\,°C$, which is only slightly larger than $60\,°C$, a value that was exceeded for n as small as 6.

2. Up until now the distilled water has been treated as 'indivisible'. If its partitioning is allowed, then the warming process can be made much more effective.

First, divide both the distilled water (DW) and the hot tap water (TW) into two identical portions. The equilibrium temperature of half a litre of TW at $100\,°C$ and half a litre of DW at $0\,°C$ is $50\,°C$. Next, equalise the temperatures of the remaining half-litre of hot water and the half-litre of DW previously warmed to $50\,°C$; this results in half a litre of DW at $75\,°C$.

But do not be satisfied with this! Equalise the temperatures of the TW that has been cooled to $50\,°C$ and the so-far unused half-litre of DW at $0\,°C$, to give a common temperature of $25\,°C$. Next we put this half-litre of DW at $25\,°C$ in contact with the half-litre of TW that is at $75\,°C$. We have now produced, in addition to the half-litre of $75\,°C$ DW prepared previously, a further half-litre of DW at $50\,°C$. Finally, we pour together the two half-litres of DW, to obtain one litre of distilled water at a temperature of $62.5\,°C$.

If both the hot water and the distilled water are each divided into three identical portions, then, using the method described above, we can get a litre of DW at a temperature of $68.75\,°C$; this is substantially higher than the maximum temperature obtainable using undivided DW. It is perhaps surprising that, by refining the division even further, we can, in principle, warm up the DW to a temperature arbitrarily close to $100\,°C$.

This principle outlined in the previous paragraph is used in *counter-flow heat exchangers*, where the fluids enter the exchanger from opposite ends, and flow in opposite directions along long tubes. For example, this is the typical method of air ventilation in so-called passive houses, where the cold fresh air moves close to, but in the opposite direction to, the used warm air. In practice, because of various losses, 100 % efficiency cannot be obtained, though 90 % is a realistic goal.

In Nature, some animals, living in cold climates, are regionally *heterothermic* and are able to allow their less insulated extremities to cool to temperatures much lower than their core temperature – nearly to $0\,°C$. They can do this because their smallest blood vessels, the network of capillaries, act as a heat exchanger working with very good efficiency. Returning blood flowing in their veins is warmed by the warmer blood flowing outwards in nearby arteries. Consequently, close to the surface of the skin, the blood has cooled down nearly to the temperature of its surroundings, while the core temperatures of the animals remain high compared to that of the environment.

S101 At first sight, we might think that in a stationary state the outer and inner pressures are equal; but this is *not* true! More explicitly, the system is not in thermodynamic equilibrium with its surroundings; because of the small hole, the furnace cannot be considered as isolated, and the temperature inside it is higher than that of its surroundings. However, considered on its own, it is (after a suitable interval) in a stationary state. This implies that, during any given period, the number of gas molecules leaving the furnace is equal to the number entering it, with the hole as the only means of entry or exit.

The number of molecules passing through a hole of cross-sectional area A during a time interval t is $AvNt$, where v is the component, perpendicular to the wall, of the average molecular velocity, and N is the number density (the number of particles in unit volume). In dynamic equilibrium,

$$A v_{out} N_{out} t = A v_{in} N_{in} t, \qquad \text{that is,} \qquad \frac{v_{out}}{v_{in}} = \frac{N_{in}}{N_{out}}.$$

The square of the speed of the molecules (a measure of the internal energy of the gas) is proportional to the gas temperature T, and, from the ideal gas equation, the number density is proportional to the quotient of the pressure p and the temperature, i.e. $N \propto p/T$. Collecting these various equalities together, we have

$$\frac{v_{out}}{v_{in}} = \sqrt{\frac{T_{out}}{T_{in}}} \qquad \text{and} \qquad \frac{N_{out}}{N_{in}} = \frac{T_{in}}{T_{out}} \frac{p_{out}}{p_{in}}.$$

From these it follows that

$$p_{in} = p_{out} \sqrt{\frac{T_{in}}{T_{out}}} = 100 \text{ kPa} \times \sqrt{\frac{330 \text{ K}}{273 \text{ K}}} \approx 110 \text{ kPa}.$$

> *Note.* In the solution, it was assumed that gas molecules moving towards the hole, from either direction, can pass through it unimpeded. This is the case only if the depth of the hole is much smaller than the average distance travelled by a molecule between successive collisions, the so-called *mean free path*. For normal air the mean free path is about 10^{-7} m – a very small distance, and certainly much less than any realistic thickness for a furnace wall. For the approximation to be valid for a real wall, the pressures involved, both inside and outside the furnace, have to be several orders of magnitude less than atmospheric.

S102 *a)* Let us consider this case first, in which the system is heat-insulated. Let the initial pressure and volume of the gas be p_1 and V_1, and denote the corresponding final values by p_2 and V_2. Because of the small hole, the gas seeps slowly from the upper to the lower part, with the pistons moving uniformly, and with no net force acting on them. Consequently, in each 'chamber' the pressure remains constant; further, these constant values must be p_1 and p_2. We can be even more specific. As the process is quasi-stationary, $p_1 = p_0 + (mg)/A$ and $p_2 = p_0 - (mg)/A$, where p_0 is the atmospheric pressure, m is the mass of a piston and A is the cross-sectional area of the cylinder. These pressures are constant, and totally independent of whether the system is heat-insulated or not.

During the gas transfer, the work done by the upper piston is $p_1 V_1$ and that done by the lower one is $-p_2 V_2$. No heat is added to or taken from the gas, and so, in accordance with the first law of thermodynamics, the total work done is equal to the change in the internal energy of the gas:

$$p_1 V_1 - p_2 V_2 = \frac{f}{2}nRT_2 - \frac{f}{2}nRT_1 = \frac{f}{2}p_2 V_2 - \frac{f}{2}p_1 V_1.$$

Here n is the number of moles of gas present, and f is the number of degrees of freedom it has (neither matters from the point of view of the problem); we have also used the ideal gas equation to make the final step. The overall equality reduces to $p_1 V_1 = p_2 V_2$. It follows that $T_2 = T_1$, and so in the final state the internal energy of the gas is the same as it was initially (with the common temperature equal to that of the outside air).

b) In this case, we can say immediately that, because the gas is always in thermal equilibrium with its surroundings, the final temperature is equal to the initial one.

So, in the two cases, *a*) and *b*), the final temperatures and pressures are equal; this means that the final volumes must also be the same. So the answer to the question in the problem: the lower piston will stop *at the same position* in both cases.

Notes. 1. We can easily be led to a false result if we try to apply the equation $pV^\gamma = $ constant, which is normally appropriate for adiabatic processes (here γ is the heat capacity ratio of the gas). This equation is only obeyed in *reversible* processes in which no heat enters or leaves the gas and its entropy is constant.

However, the process in this problem is inherently *irreversible*, as the gas, squeezed through the hole, gains some kinetic energy, which is then dissipated by gas viscosity and transformed into thermal energy. Although $Q = 0$, the entropy of the gas increases. As the entropy is a function of state, the final state of the gas cannot be that which would be found using an *incorrect* application of the formula $pV^\gamma = $ constant.

Another well-known example in which the heat gain is zero, but the quantity pV^γ is not constant, is the free expansion of an ideal gas into a vacuum. This process is also irreversible (the gas does not climb back into the open container), and so the entropy of the gas increases. The mechanical work done is zero, and in the sense of the first law of thermodynamics, its internal energy, and consequently its temperature, do not change.

2. In both cases, *a*) and *b*), the change in the internal energy of the gas, the mechanical work done, and the heat gain or loss, are all zero. We may wonder whether the decreases in gravitational potential energy of the pistons should appear somewhere in the reckoning. They do not, because it can be shown that each potential energy loss is exactly equal to the mechanical work done against atmospheric pressure.

S103 At an altitude of h, relative to the ground, let the atmospheric pressure be $p(h)$. At a marginally higher altitude $h + \Delta h$, the pressure is lower by an amount

$$\Delta p = -\varrho(h)g\Delta h,$$

where $\varrho(h)$ is the density of air, and g is the gravitational acceleration. Using the ideal gas law, the density can be expressed in terms of the pressure and temperature of the air:

$$\frac{\Delta p}{\Delta h} = -\frac{Mg}{R}\frac{p(h)}{T(h)}, \tag{1}$$

where $M = 29$ g mol^{-1} is the average molar mass of air, and R is the gas constant. Both p and T are functions of h, and so we need one more relationship between them in order to find $T = T(h)$.

Imagine that a small volume of air at a height h in the troposphere, which is assumed to have a time-independent temperature distribution $T(h)$, *suddenly* rises by Δh to a slightly higher altitude. For the reasons given in the problem, it cannot receive heat, and so it expands adiabatically, and gets colder.

If the final temperature T' of the raised air is greater than the temperature $T(h + \Delta h)$ of the surrounding air, then the air mass, which is less dense than its surroundings, rises still further, and the atmosphere becomes unstable. But if the final temperature of the expanded air is lower than that of its surroundings, it will sink back to its original position, and the temperature distribution will be a stable one. We now investigate the (critical) threshold situation that separates these two cases, namely when $T' = T(h + \Delta h)$.

In an adiabatic expansion, the quantity $p^{1-\gamma}T^\gamma$ remains constant; here $\gamma \approx 7/5$ is the heat capacity ratio of air. So, in the hypothetical situation leading to the adiabatic expansion of the air, we have

$$\frac{p(h + \Delta h)}{p(h)} = \left[\frac{T(h + \Delta h)}{T(h)}\right]^{\gamma/(\gamma-1)}. \tag{2}$$

The small changes Δp and ΔT in the pressure and temperature are

$$\Delta p = p(h + \Delta h) - p(h), \qquad \Delta T = T(h + \Delta h) - T(h).$$

Using these, equation (2) can be written in the form

$$1 + \frac{\Delta p}{p(h)} = \left[1 + \frac{\Delta T}{T(h)}\right]^{\gamma/(\gamma-1)}.$$

The quotient $\Delta T/T(h) \equiv \varepsilon$, on the right-hand side, is arbitrarily small, and so we can use the approximation $(1 + \varepsilon)^n \approx 1 + n\varepsilon$ to obtain:

$$\frac{\Delta p}{p(h)} \approx \frac{\gamma}{\gamma - 1}\frac{\Delta T}{T(h)}. \tag{3}$$

With the help of equations (1) and (3), an expression for the critical temperature gradient, characteristic of the troposphere, can be found:

$$\frac{\Delta T}{\Delta h} = -\frac{\gamma - 1}{\gamma}\frac{Mg}{R}.$$

Inserting the known data ($\gamma \approx 1.4$ and $M = 29$ g mol^{-1}), we get the value -9.8×10^{-3} °C m^{-1}; this is (perhaps fortuitously) very close to the observed temperature decrease of 1 °C over 100 m.

Because of the constant temperature gradient, the temperature is a linear function of the altitude:

$$T(h) = T_0\left(1 - \frac{\gamma - 1}{\gamma}\frac{h}{h_0}\right), \tag{4}$$

where the notation $h_0 = RT_0/(Mg)$, called the *scale height*, has been introduced. Taking the 'sea-level' air temperature as $T_0 = 15$ °C, then $h_0 \approx 8.4$ km.

> *Notes.* 1. The altitude dependence of the air pressure can be determined from equation (4), when it is used in conjunction with the adiabatic equation $p^{1-\gamma}T^\gamma = $ constant:
>
> $$p(h) = p_0\left(1 - \frac{\gamma - 1}{\gamma}\frac{h}{h_0}\right)^{\gamma/(\gamma-1)}, \tag{5}$$
>
> where p_0 is the air pressure at sea level. This expression (which is often called the adiabatic barometric formula) gives values very similar to those derived from the isothermal barometric formula $p(h) = p_0\,e^{-h/h_0}$ when $h < h_0$. But there are two significant differences between the two equations: (i) for the derivation of the isothermal barometric formula, a constant atmospheric temperature is assumed; and (ii) equation (5) predicts zero pressure at an altitude of $h = h_0\gamma/(\gamma - 1) \approx$ 30 km, whereas in the isothermal barometric formulation the atmosphere has no sharp border. In reality, equations (4) and (5) become inaccurate at altitudes much lower than 30 km, namely, at the diffuse border between the troposphere and the stratosphere (10–12 km).
>
> 2. With the help of the adiabatic temperature–altitude formula (4), the air temperature at the flight altitude of commercial aircraft can be calculated. The result is about 100 °C lower than the ground-level temperature, i.e. about -80 °C. Actually, the temperature at such heights is a little higher, ≈ -60 °C. The difference is caused by the water vapour in the atmosphere. During the cooling of a rising air mass, some of the water vapour in it condenses, resulting in (latent) heat being released. This heat reduces the rate of temperature decrease with height, as compared to that predicted by equation (4).
>
> The higher temperature produced by condensation in a rising humid air mass results in a lower air density and the air reaching higher altitudes, while the condensation causes strong cloud formation to take place at the same time. In the tropics (where the humidity is very high), the rate of atmospheric temperature decrease with altitude is much smaller than that predicted by the adiabatic formula (4), though it *does* gives correct values in dry regions of the Earth.

S104 When the helium gas is close to its minimal volume, to first order there is no change in the volume for a finite change in the temperature. This means that, in this region, the process is approximately isochoric, and so the molar heat capacity of the monatomic helium gas must be $C_V = \frac{3}{2}R$. Equating this to the given form for the molar heat capacity,

$$\frac{3}{2}R = \frac{3RT}{4T_0},$$

shows that the temperature at the end of the process (the point at which the volume is minimal) is $T_f = 2T_0$.

We now apply the first law of thermodynamics, using the given molar heat capacity,

$$dU = dQ + dW,$$

i.e.

$$\frac{3}{2}nR\,dT = \frac{3nRT}{4T_0}\,dT + dW,$$

where the change in internal energy dU is expressed in terms of the molar heat capacity ($C_V = \frac{3}{2}R$) of helium gas at constant volume. The total work done on the system, as it goes from the initial to the final temperature, can be expressed as the following integral:

$$W = \frac{3}{2}nR \int_{T_0}^{2T_0} \left(1 - \frac{T}{2T_0}\right) dT = \frac{3}{8}nRT_0.$$

Notes. 1. The same result can be obtained without explicitly using integral calculus, if we take the average molar heat capacity over the relevant temperature range:[55]

$$C_{av} = \frac{C_0 + C_V}{2} = \frac{\frac{3}{4}R + \frac{3}{2}R}{2} = \frac{9}{8}R.$$

The total heat transferred is then $Q = nC_{av}(T_f - T_0) = \frac{9}{8}nRT_0$, while the total change of the internal energy is $\Delta U = nC_V(T_f - T_0) = \frac{3}{2}nRT_0$. It follows that the work that must have been done on the helium gas is $W = \Delta U - Q = (\frac{3}{2} - \frac{9}{8})nRT_0 = \frac{3}{8}nRT_0$.

2. Using the ideal gas equation, we can express the gas pressure as $p = (nRT)/V$, and insert it into the equation stating the first law of thermodynamics, provided we write the work done in the form $dW = -p\,dV$:

$$\frac{3}{2}nR\,dT = \frac{3nR}{4T_0}T\,dT - \frac{nRT}{V}\,dV.$$

---

[55] This is justified only because the molar heat capacity is directly proportional to the temperature, and so its average × its range is equal to its integral.

This straightforward differential equation can be integrated, and from it we can find the volume of the gas as a function of temperature in the particular (unspecified!) process of the problem:

$$V = V_0 \left(\frac{T_0}{T} \right)^{3/2} \exp\left[\frac{3(T - T_0)}{4T_0} \right],$$

where V_0 is the initial volume of the helium gas.

S105 *Solution 1.* The role of the easily expanded bag is the same as that of a frictionless piston that encloses gas in a horizontal cylinder. And so, from here on, we will use phraseology appropriate to the piston analogy.

If the outside pressure is decreased slowly, then the piston moves slowly, and the interim states of the gas can be considered as intrinsically equilibrium states. In this scenario the process is adiabatic, and so the decrease in the internal energy of the gas is just equal to the work it does during the expansion.

In the second case, in which the outside pressure decreases rapidly, there is a finite pressure difference across the piston, and the net force acting on it is not zero. Consequently, the piston initially accelerates and its speed increases, but later, when it has overshot the final equilibrium position, it decelerates and stops, after which it starts moving backwards.

If there were no friction (neither between the piston and the wall of the cylinder, nor inside the gas as viscous 'inner friction'), then the piston would never stop moving, and the volume, the pressure and the temperature of the gas would all change periodically; it would make no sense to seek 'the final temperature of the gas'. But, in reality, this does not happen, the oscillation stops sooner or later, because the (ordered) kinetic energy of the piston's motion is dissipated, and results in an increase in the internal energy of the gas, i.e. in the kinetic energy associated with the disordered microscopic motion of its constituent particles.

The final (equilibrium) volume of the gas – as well as its internal energy and hence its temperature – cannot be the same in this second case as they were in the first (quasi-static) process. They have to be larger, because the decrease in internal energy is smaller than the work done by the gas during the expansion, and the energy difference (or a part of it) 'gets back' to the gas-plus-piston system as a result of friction.

So, the temperature drop of the gas in the container (in the bag) is greater if the outside pressure is decreased slowly.

Solution 2. When the outside pressure is decreased slowly, the successive states of the gas are equilibrium states throughout the expansion; so the process is *reversible*. Contrariwise, if the decrease of the outside pressure occurs suddenly, then the interim states of the gas cannot be considered as equilibrium states.

The bag (or the piston) is going to oscillate quickly in a motion that is damped by the dissipative processes; the change of state of the system is an *irreversible* one. We can use this distinction between the two scenarios to draw a conclusion about the relative final temperatures of the gas in the two cases.

According to the second law of thermodynamics, in reversible processes, the change of entropy, ΔS, of a system is the sum (or integral) of quantities of the form $\Delta Q/T$, where ΔQ is a heat input or output and T is the absolute temperature at which the heat is gained or lost. In an irreversible process, the gain in entropy is always larger than this.

In the first of our cases, the process is reversible, and (because of the thermal insulation) there is no heat transfer ($\Delta Q = 0$), so $\Delta S_{rev} = 0$. In the second case (involving a sudden pressure decrease), the process is irreversible, and the heat transfer into the gas is either zero (if the damping of the oscillation is caused by inner friction of the gas) or positive (if the friction of the bag/piston contributes significantly to the dissipative forces). Whatever the cause of the damping, $\Delta S_{irrev} > 0$.

The change of entropy of a given amount of gas undergoing a thermodynamic change is independent of the details of the process involved – it is determined unequivocally by the initial and final states. In other words, entropy is a state function, and, as such, can be expressed in terms of other state functions, such as pressure and temperature.

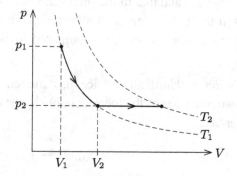

Denote the initial temperature of the gas by T_1, and its final one by T_2. To calculate the change of entropy between the initial and final states, consider a process in which, first the pressure is decreased from p_1 to p_2 at constant temperature T_1, and then the gas is warmed up (or cooled down) to the temperature T_2 at constant pressure p_2 (*see* figure). Our formulation will be for general values of T_1 and T_2, though it will (formally) take p_1 to be greater than p_2. In the figure, it is (again formally) assumed that $T_2 > T_1$, despite the fact that for our particular application the reverse is true; the calculated expressions apply to both heated and cooled final situations.

The change of entropy in the first, isothermal section is

$$\Delta S_1 = \frac{Q}{T_1} = \frac{W_{\text{gas}}}{T_1} = \frac{1}{T_1} \int_{V_1}^{V_2} p \, dV$$

$$= \frac{1}{T_1} \int_{V_1}^{V_2} \frac{p_1 V_1}{V} \, dV = \frac{p_1 V_1}{T_1} \ln \frac{V_2}{V_1} = -nR \ln \frac{p_2}{p_1}.$$

And in the second, isobaric process, it is

$$\Delta S_2 = \int \frac{dQ}{T} = \int_{T_1}^{T_2} nC_p \frac{dT}{T} = nC_p \ln \frac{T_2}{T_1},$$

where C_p is the molar heat capacity of the gas at constant pressure. The total change of entropy of the gas from state (T_1, p_1) to state (T_2, p_2) is

$$\Delta S = \Delta S_1 + \Delta S_2 = -nR \ln \frac{p_2}{p_1} + nC_p \ln \frac{T_2}{T_1}.$$

It should be noted that for both of our processes $p_2 < p_1$ and $T_2 < T_1$, and so ΔS_1 is positive while ΔS_2 is negative.

Now, whether the expansion is fast or slow, the ratio of initial to final pressure is the same, and so, therefore, is the ΔS_1 term. As the initial temperature is also the same in both cases, the difference between the reversible and irreversible entropy changes is related only to the final temperatures. As $\Delta S_{\text{irrev}} > \Delta S_{\text{rev}}$, it follows that $(\Delta S_2)_{\text{fast}} > (\Delta S_2)_{\text{slow}}$, and that in the final state $T_{\text{fast}} > T_{\text{slow}}$.

So the gas in the bag suffers a smaller temperature drop when there is a fast decrease in the outside pressure, as compared to when the pressure decreases slowly.

S106 *Solution 1.* Until all the ice has melted, the temperature difference between the two junctions of the thermocouple is constant, and so, therefore, is the potential difference:

$$V_{AB} = S_{AB}(T_A - T_B),$$

where S_{AB} is the Seebeck coefficient of the materials of the thermocouple. This potential difference drives a current $I = V_{AB}/R$ round the circuit, and so the rate of Joule heating in the resistor is

$$P_{\text{Joule}} = \frac{V_{AB}^2}{R} = \frac{S_{AB}^2 (T_A - T_B)^2}{R}.$$

The Peltier effect in the circuit causes heat release (output) at junction B and heat absorption (input) at junction A. The rates of heat release and absorption are

$$P_A^{\text{input}} = \Pi_A I = S_{AB} T_A I$$

and

$$P_B^{\text{output}} = \Pi_B I = S_{AB} T_B I = S_{AB} T_B (V_{AB}/R) = S_{AB}^2 T_B (T_A - T_B)/R,$$

where we have used the connection between the Peltier and Seebeck coefficients.

> *Note.* It might be questioned how we can be sure of the direction of the current in the circuit, and so of the sense in which heat is transferred at each of the junctions. To answer this, we note that the above expressions for the power in terms of S_{AB} and T show that the rate of heat input or output is larger at the higher-temperature junction. Consequently, an alternative process in which heat would be released at A, and B would further cool its surroundings – to say nothing of the Joule heat production in the resistor – would violate the law of conservation of energy.

It follows from the above that the ratio of the Joule heat warming up the water, to the output heat given to the ice, is

$$\frac{Q_{\text{Joule}}}{Q_B^{\text{output}}} = \frac{cm\Delta T_{\text{water}}}{Lm} = \frac{P_{\text{Joule}}}{P_B^{\text{output}}} = \frac{T_A - T_B}{T_B},$$

where c is the specific heat of water, L is the latent heat of fusion of ice and m is the common mass of ice and water. So the temperature increase of the water by the time that the last of the ice has melted is

$$\Delta T_{\text{water}} = \frac{T_A - T_B}{T_B} \frac{L}{c} \approx 7.9\,^{\circ}\text{C}.$$

Solution 2. The system can be considered as a heat engine. The hot reservoir is the room-temperature air, the cold reservoir is the ice at $0\,^{\circ}\text{C}$, and the net work done by the engine is not mechanical work but the Joule heat released in the resistor:

$$W = Q_{\text{Joule}}.$$

In accord with the conservation of energy, the heat dissipated in the resistor is equal to the difference between the input heat at point A and the output heat at point B:

$$Q_{\text{Joule}} = Q_A^{\text{input}} - Q_B^{\text{output}}.$$

Although the heat dissipation in the resistor is irreversible, the operation of the thermocouple itself *is* reversible. To see this, we note that the net electrical energy produced by the cell could have been used to charge a battery, which could be later connected to the resistor to warm up the water. However, we might decide to use the stored energy in the battery, not for warming up the water, but to drive a heat pump with the ice and the room-temperature air as the two reservoirs; in this case the ice would freeze and the air would be warmed up, i.e. the direction of the whole process would be reversed.

It follows from the reversibility of the process that the entropy of the thermo-couple is constant, i.e. the entropy increase at point A because of heat input equals the entropy decrease at point B where heat is released:

$$\frac{Q_A^{\text{input}}}{T_A} = \frac{Q_B^{\text{output}}}{T_B}.$$

Using this and the previous equation, we can find the connection between the Joule heat generated in the resistor and the heat absorbed by the ice:

$$Q_{\text{Joule}} = \left(\frac{T_A}{T_B} - 1\right) Q_B^{\text{output}},$$

in agreement with Solution 1. After this, we may find the temperature increase of the water in the same way as previously.

> *Note.* Solution 2 shows that the thermoelectric generator works as an unusual continuously functioning Carnot engine; heat input and heat output take place not as separate phases of a cycle, but with continuous and simultaneous heat transfer to and from the two heat reservoirs.

S107   The surface temperature T_{Sun} of the Sun, its radius R and the Sun–Earth distance D are all well-known data; denote the radius of the lens by r and its focal length by f. The total amount of energy emitted by the Sun per unit time (its luminosity) is

$$L_{\text{Sun}} = 4\pi R^2 \sigma T_{\text{Sun}}^4,$$

where σ is the Stefan–Boltzmann constant.[56] Of this emitted power, a fraction $\pi r^2/(4\pi D^2)$ reaches the lens, and so the image of the Sun produced by the lens receives a total power of

$$P = \left(\frac{R}{D}\right)^2 \pi\sigma T_{\text{Sun}}^4 r^2.$$

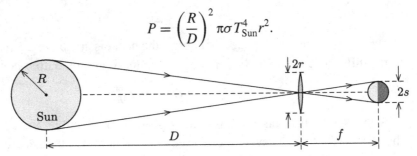

From geometrical considerations (*see* figure, which is clearly not to scale), the radius s of the Sun's image is

---

[56] The Sun's actual spectrum is not that of a true black body, to which the formula strictly applies. The effective surface temperature $T_{\text{Sun}} \approx 5800$ K is that which a true black body would need to have to match the observed total radiation from unit area of the Sun's surface.

$$s = \frac{R}{D}f, \qquad \text{and so} \qquad P = \pi \left(\frac{rs}{f}\right)^2 \sigma T_{\text{Sun}}^4.$$

Each point of the image has the same brightness, and so the radiative power incident on unit area of the image is

$$w = \left(\frac{r}{f}\right)^2 \sigma T_{\text{Sun}}^4 = \frac{1}{16} \sigma T_{\text{Sun}}^4.$$

An object illuminated by this power will warm up to a limiting temperature T, such that the absorbed power will be equal to the power of the black-body radiation, $A\sigma T^4$, emitted by the object. As the latter depends on the surface area A of the body, the maximal temperature T reached depends upon the shape of the object. If the radius of the sphere in the problem is $\varrho = s$ (as shown in the figure), the absorbed power is still P, but the emitted one is $4\pi s^2 \sigma T^4$. It follows that

$$T = (1/64)^{1/4} T_{\text{Sun}} = (1/64)^{1/4} \times 5800 \text{ K} \approx 2040 \text{ K} \approx 1770\,°\text{C}.$$

If the body were a thin disc, rather than a sphere, and had a radius $\varrho \leq s$, then the input power would be $\pi\varrho^2 w$, and the output $2\pi\varrho^2 \sigma T^4$; the maximal temperature would then be $T = (1/32)^{1/4} T_{\text{Sun}} \approx 2430 \text{ K} \approx 2160\,°\text{C}$.

Returning to the original problem, if the radius of the sphere were smaller than the radius of the Sun's image ($\varrho < s$), then the sphere would still warm up to the same value. Suppose, for example, that the radius of the sphere is half that of the image. Then the solar energy reaching the sphere in unit time is one-quarter of that striking the lens, but as its (re-radiation) surface area is only one-quarter of that of a sphere of radius s, it attains the same temperature.

If the sphere's radius is larger than the radius of the Sun's image, and the material of which it is made is not a perfect heat insulator, then its maximal temperature will be smaller than the value calculated above.

> *Notes.* 1. Our result, that it is impossible to warm up a small body to any arbitrary high temperature using a lens, may be quite surprising. The reason for it is that, even with perfect optics, the Sun's image is not a point-like dot, but a small disc. The Sun is far from us, that is true, but clearly its size (its angular diameter or visual angle) is finite, and so its image has to be. The illumination (the incident energy per unit area in unit time) of this image determines the maximal temperature of the heated body. This temperature cannot exceed the surface temperature of the Sun.
>
> In reality, $T < T_{\text{Sun}}$, and the maximum temperature of the warmed body is reduced relative to that of the Sun by a factor that depends on the fraction r/f. Thus the fraction r/f, which depends only on the physical properties of the lens, and not on its use, is a quantity that characterises its 'light strength'. It is very similar to the so-called *f-number* or *focal ratio* of the lens, defined as $N = f/d$, where f is the focal length and d is the effective diameter of the lens aperture

(perhaps deliberately reduced by a shutter). Note that the f-number is essentially one-half of the reciprocal of r/f.

2. In principle, the ratio r/f could be larger than 1 – even $r \gg f$ is imaginable. In such cases, the derived formula would indicate that a temperature $T > T_{Sun}$ is possible; in reality, this is not so. In the derivation, we tacitly supposed that the light rays contributing to the image travelled close to the optical axis. When this is not the case, detailed calculations show that, for both large-diameter convex lenses and large-aperture parabolic mirrors, we always get a temperature $T < T_{Sun}$. It can be proved more generally that, for linear optical systems (i.e. ones in which the principle of superposition holds) using only solar energy, it is impossible to exceed the surface temperature of the Sun. This theoretical limit is a consequence of the second law of thermodynamics.

3. If a small body is heated by solar energy, but not by using (linear) optical devices, then the theoretical limit mentioned above no longer applies. As a simple 'nonlinear device', we could use solar cells to collect electrical energy during daylight, and store this energy in batteries. Later we could produce an electric arc (a plasma discharge) using the stored energy. Clearly, the electric arc does not 'remember' the temperature of the body (in our case, the Sun) from which the stored energy came.

4. In our calculations, we assumed that the Sun's radiation reaches the Earth's surface unattenuated. In reality, the Earth's atmosphere reflects a considerable part of the incident radiation; further, in the atmosphere there is also some absorption. Both of these effects will reduce the maximum realisable temperature.

S108   As we all know, in calm weather, and usually overnight, water vapour condenses on surfaces that are colder than $0\,°C$ in the form of hoar frost. This frost contains tiny ice crystals, and, strictly speaking, this phenomenon is called vapour deposition. We have to explain how the windscreen of the car can be colder than the air around the car.

The windscreen glass at temperature T is in thermal contact with the surrounding air at temperature T_0, and, in addition, it can gain or lose energy in the form of heat radiation. A stable situation can be maintained if the heat gain and the heat loss due to these two processes cancel each other out.

The heat flux Φ (the amount of energy that flows through unit area in unit time) that is taken in via thermal conduction through the stationary air adjacent to the screen is given by

$$\Phi = -\alpha(T - T_0).$$

Here α is the so-called *conductance*, which has a value of about $20\ \mathrm{J\ m^{-2}\ s^{-1}\ K^{-1}}$ for stationary air. In the stable situation under consideration, $T < T_0$, and so $\Phi > 0$.

Now for the radiation aspect. According to Kirchhoff's law of thermal radiation, all bodies at the same temperature, when receiving or emitting thermal radiation, have the same ratio of emissivity to absorptivity at any particular wavelength.

We now consider the various sources (and sinks) of radiative energy that might affect the windscreen.

The air can hold very little heat, and so the radiation received from the air is negligible. The windscreen is not affected by radiation from the surface of the Earth, and, if the sky is cloudless, then it too will contribute nothing. In brief, there is no significant incoming radiation to the windscreen. The outgoing heat flux Φ' from the glass can be calculated from the Stefan–Boltzmann law:

$$\Phi' = \varepsilon \sigma T^4,$$

where ε is the emissivity of the glass, and σ is the Stefan–Boltzmann constant.

For a stationary state, $\Phi' = \Phi$, and, since $T_0 \gg T_0 - T$, the difference between the general air temperature and the temperature of the windscreen is

$$\Delta T = T_0 - T \approx \frac{\varepsilon \sigma T_0^4}{\alpha}.$$

According to tabulated data and other measurements, the relevant value of the emissivity is $\varepsilon \approx 0.2$, and substituting this into the approximate formula, we find that a temperature difference of as much as $3\,°\mathrm{C}$ is possible.

To protect against hoar frost, it is sufficient to place a heat-reflecting (to stop the windscreen radiating effectively) or heat-radiating (to provide a source of incoming radiation) surface near the windscreen. As an illustration, it is common experience that, for a car parked next to a tree or house, those of its windows that 'look at' the tree or house do not frost up.

> *Notes.* 1. In the building industry, the conductance is called the *U*-factor or *U*-value, and is an overall heat transfer coefficient that describes how well (or badly, for preference!) a building element conducts heat. It is calculated as the rate of heat transfer (in watts) through $1\ \mathrm{m}^2$ of a structure divided by the difference in temperature across the structure. Conductance α for a particular material is closely connected to its thermal conductivity k: $\alpha = U = k/L$, where L is the material's thickness. In our problem, we assumed that the thickness of the stationary air layer covering the windshield, was about 1 cm, and that the heat conduction takes place through this layer. In windy weather, this layer is thinner, and so the conductance is larger. The heat conduction can then compensate for the cooling effect of the radiation, and results in the windscreen being held at a temperature closer to that of the surrounding air and consequently protected from frost.
>
> 2. The cooling of the windscreen cannot be explained by the evaporation of the dew on it, because, in calm weather, the latent heat released when the water vapour condenses from the increasingly colder surrounding air produces a larger (and opposing) effect than the subsequent evaporation; the net effect of humidity is to increase the temperature of the glass.
>
> 3. In more modern cars with an electronic display on the dashboard, a pictogram (a snowflake), or some other visible indication, is displayed if the road

could become slippery because of potential icing. The numerical result of the problem explains why the computer system switches on the snowflake if the air temperature outside drops to a predetermined value of about 3 °C.

S109 Denote the temperature at the centre of the triangle by T_0, and the heat flows into the three vertices by I_k ($k = 1, 2, 3$). In a stationary state, we must have

$$I_1 + I_2 + I_3 = 0.$$

Now imagine the two new arrangements produced by rotating the system by 120 and 240 degrees, and superimpose them on the original, as illustrated in the figure. In the resulting configuration, the temperature is the same at all three vertices of the triangle, and equal to $T_1 + T_2 + T_3$. As the net heat flow ($I_1 + I_2 + I_3$) is zero at all three vertices, the temperature distribution must be homogeneous.

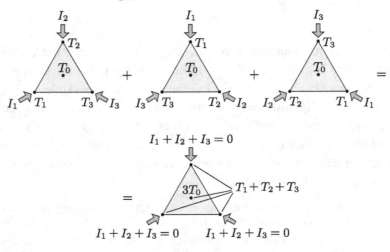

At the centre of the superposed triangles, the (superposed) temperature is $3T_0$, and, since the temperature is the same everywhere, we must have $3T_0 = T_1 + T_2 + T_3$, i.e. the temperature at the centre of the original equilateral triangular plate is the arithmetic mean of the temperatures at its vertices:

$$T_0 = \frac{T_1 + T_2 + T_3}{3}.$$

Note. The problem can be generalised to regular polygon-shaped plates with n vertices, and to Platonic solids (regular, convex polyhedrons). For example, the temperature at the centre of a homogeneous dodecahedron is

$$T_0 = \frac{1}{20} \sum_{k=1}^{20} T_k.$$

S110 The ice cubes float on the surface of the liquid in both vessels. In the first beaker, the density of the melted water from the ice cube, at a temperature of

0 °C, is greater than that of the tap water at room temperature, and so it sinks to the bottom of the vessel and is replaced by room-temperature water. The convection currents arising in this way assist the melting.

In the second beaker, the density of the brine is greater than that of the cold melt water, and so the latter does not sink, but remains around the ice cube. In this case, convection does not set in, and the melting of the ice cube is much slower than it was in the tap water.

> *Notes.* 1. Our explanation can be strengthened by the following '*experimentum crucis*'. A small amount of lead shot is frozen into the ice cubes, so that they sink in both the tap water and the brine. No convection currents are set up around the cube melting in tap water at room temperature, because the latter has a density less than that of the cold melt water. The cold water produced in the depths of the brine rises to the surface, convection currents are set up, and the cube melts more quickly than its freshwater counterpart.
>
> 2. The phenomenon is quite sensitive to the initial temperature of the water in the vessels. If this initial (room) temperature is relatively low – and consequently the density of the tap water is very close to that of the water from the melting cube – then, experimentally, we have found differences between the melting times of the two floating ice cubes of as much as a factor of 10.
>
> The convection currents in the water are readily apparent if the ice cubes are made of tap water dyed with food colouring. Both experiments are relatively simple to set up, and each provides a reassuring insight into the obviously different melting processes in the two vessels.
>
> 3. The melting of icebergs floating in the oceans can take many years, and the main reason for this is that these 'giant ice cubes', made of fresh water, are floating in their own 'fresh water puddles' and convection currents, which would speed their melting, do not get set up.

S111 Suppose that we first freeze the 1 kg of supercooled water at -10 °C, and after this warm up the ice so formed to 0 °C. During the freezing stage, the energy released is simply the heat Q that is to be determined, and warming up the ice by 10 °C requires 21 kJ of energy, as the specific heat of ice is about 2.1 kJ kg^{-1} K^{-1}.

However, there is another way in which the same final state could be reached, starting from the given initial one. We could first carefully warm up the supercooled water from -10 °C to 0 °C (being careful not to initiate freezing – more easily done in this theoretical exercise than in practice!), and then freeze the water at its normal freezing point. In the first part of this alternative process, the heat that has to be supplied is 42 kJ – assuming that the specific heat of supercooled water does not differ considerably from the value 4.2 kJ kg^{-1} K^{-1} for water at room temperature; during the freezing stage the heat released is 334 kJ.

Because, in the two processes, the initial and final states are the same, and the same amount of work is done on the surroundings, the two amounts of heat energy supplied must be equal:

$$-Q + 21 \text{ kJ} = 42 \text{ kJ} - 334 \text{ kJ}.$$

From this, $Q = 313$ kJ, and so the latent heat (enthalpy) of fusion – or, more simply, the freezing heat – of supercooled water at a temperature of $-10\,^\circ$C is about 313 kJ kg^{-1}.

> *Note.* If the supercooled water at $-10\,^\circ$C suddenly starts freezing, then, after a short time, a mixture of water and ice, both at $0\,^\circ$C, is formed; the energy needed to warm the water–ice mixture is that released during the freezing. This same final state could be created from pure water at $0\,^\circ$C by removing about 42 kJ of heat, corresponding to the formation of about 42 kJ $\div$ 334 kJ kg^{-1} $\approx$ 0.13 kg of ice. This implies that, during the sudden freezing of supercooled water at $-10\,^\circ$C, *only* about 13 % of the total mass will freeze.

S112   In our investigation of what happens to the water vapour when its volume is changed, we need consider only its own saturated vapour pressure (SVP), because – according to Dalton's law of partial pressures – the presence of other gases has no influence on it.

To a good approximation (unless it is in the vicinity of its critical point), the saturated water vapour follows the ideal gas equation:

$$pV = NkT.$$

If the volume of the saturated water vapour is decreased *slowly*, then its temperature remains constant; since the SVP depends only on the vapour's temperature, it too cannot change. So, from the gas law, it follows that the number of particles N must decrease. This, in turn, means that some of the slowly compressed vapour *must* condense. A slow expansion would cause the reverse process to take place; some of any liquid water present would evaporate.

If an increase in the volume is so *fast* that there is no time for significant heat transfer through the walls separating the mixture of air and water vapour from the outside surroundings, then the mixture will undergo an adiabatic expansion and its temperature will fall. Because SVP increases rapidly with temperature, even a modest fall in the latter produces a significant drop in the former; this more than compensates for the increased volume, and some of the water vapour condenses.

So, condensation of some of the water vapour can be brought about by either a *slow decrease* or a *fast increase* in the volume.

> *Notes.* 1. If we decide to follow the change of state of the water vapour in a more precise way than through the ideal gas approximation, then we could use, for example, the van der Waals equation of state, or we could investigate the matter empirically, on the basis of measured results.
>
>   In Fig. 1 a real isotherm (and the various water phases) are shown in a p–V diagram. (The gas and supercritical phases, situated above the isotherm going through the critical point C, are not plotted.)

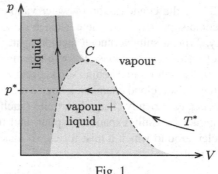

Fig. 1

During the isothermal compression, the (unsaturated) water vapour pressure increases until it reaches the value p^*, the SVP corresponding to the given temperature T^*. After this, the pressure does not increase any further, but condensation of the vapour does begin. As the volume continues to decrease, the amount of condensed liquid increases. The pressure remains constant until the whole of the vapour has condensed. In this way, vapours can be liquefied at *constant temperature* by *decreasing* their volumes.

If the aim is to liquefy the vapour via an adiabatic process, then somehow the temperature needs to be decreased. As there is no heat transfer ($Q = 0$), the first law of thermodynamics indicates that, in order to reduce its internal energy, and hence its temperature, the vapour has to do some external work, $\Delta U = -W_{\text{vapour}}$. The work done by the vapour is positive if the *volume increases*.

As previously noted, the condensation of the vapour can happen in a variety of ways, two examples being an isothermal decrease of the volume, and an adiabatic increase in it.

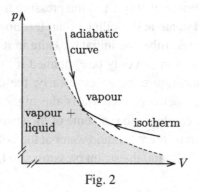

Fig. 2

In the p–V diagram, the curve that separates the vapour phase from the two-phase states can be described by a function of the form $p \sim 1/V^{1+x}$ over quite a wide range (*see* Fig. 2). The parameter x is usually quite a small positive number; in the case of water vapour it is about $1/16$. This means that the limiting curve is 'steeper' than the isotherm (for which $x = 0$ in the ideal gas model), but less steep than an adiabatic curve (for which $x = 2/f$ for an ideal gas with f thermodynamic degrees of freedom).

2. For the condensation of water vapour that has been expanded rapidly and become supersaturated, the existence of so-called condensation nuclei is necessary. Typical suitable nuclei include dust, ionised atoms and soot nanoparticles. Soot particles are responsible for the formation of the condensation trails (referred to briefly as contrails) behind some aircraft. The role of ionised atoms is important in the Wilson cloud chamber, a device used for detecting ionising radiation. Fast-moving, electrically charged particles (such as protons or electrons) ionise atoms in an adiabatically expanded vapour, and the resulting ions act as condensation nuclei, around which a mist forms, so making the particle trajectories visible as mist trails.

S113   As the wording of the question suggests, in the final stationary state, the pressure is the same all along the tube. If it were not so, then net forces would act on some or all parts of the steam and cause them to accelerate and move. We can ignore the very small hydrostatic pressure difference between the top and bottom of the test tube, because it is negligible compared to the saturated vapour pressure.

Initially, the test tube is open at the top and water is boiling in it. So, at the bottom, there is water at 100 °C, and above that there is saturated water vapour (steam) at 100 °C; the pressure of the steam must be equal to the atmospheric pressure (≈ 101 kPa).

When the test tube is sealed, there is still a little water at the bottom. As, finally, the temperature is 200 °C at the top of the test tube, and 100 °C at the bottom, the average temperature of the steam has increased. So its pressure cannot have decreased.

But, at the bottom of the test tube, the remaining water and the steam just above it are in equilibrium at 100 °C; so the pressure there must be 101 kPa, the pressure of the saturated steam at its boiling point. It follows that, since the pressure is constant all along the test tube, the initial pressure in it cannot have increased (anywhere).

This conclusion can only be explained if the steam's local density decreases as the local temperature increases; clearly, the decrease will be greatest at the top, where the temperature increase is the highest. The density decrease is brought about by the condensation of some of the steam to liquid water.

So, ultimately, there is liquid water at the bottom of the tube (more than when the tube was sealed), and the steam pressure still has its initial value, that for saturated steam at 100 °C.

So the pressure of the steam in the test tube in the final state is still 101 kPa!

> *Note.* The density of the steam decreases from the bottom to the top of the test tube, and so only the steam just above the meniscus is saturated – further up the tube it is unsaturated.

S114   *a*) At a temperature of 77.4 K (i.e. at the boiling point of nitrogen), the pressure of saturated nitrogen vapour is 1 atm = 101.3 kPa, while the saturated pressure of oxygen becomes 1 atm at a higher temperature, namely at 90.2 K.

On Earth, the molar ratio of oxygen and nitrogen is approximately 1 : 4. The ratio of the partial pressures of the two components will also be very close to this figure, because, until the start of liquefaction, the behaviour of each gas constituent is very close to that of an ideal gas. So, when, on the Earth, the liquefaction of oxygen begins at the stated pressure of 113 kPa, one-fifth of this, that is, 22.6 kPa, is due to the partial pressure of the oxygen; this pressure is, at the same time, the saturated vapour pressure of oxygen at a temperature of 77.4 K.

It also follows that, under these conditions, the pressure of nitrogen is 113 kPa − 22.6 kPa = 90.4 kPa. This is less than the saturated vapour pressure of nitrogen at this temperature, which, since 77.4 K is nitrogen's boiling point, has a value of 101 kPa. Consequently, the nitrogen *does not* liquefy at this pressure.

During the subsequent compression, the partial pressure of the oxygen, already in two phases, does not change, while the nitrogen pressure increases from 90.4 kPa to 101 kPa. This latter pressure will be reached when the volume has been reduced by a factor of $(90.4/101) \approx 0.9$. After that, the total pressure remains constant (at 22.6 kPa + 101 kPa $\approx$ 124 kPa) until the liquefaction is complete.

On Exo-Earth, the liquefaction of the nitrogen, at a constant temperature of 77.4 K, begins at a pressure of 113 kPa. So, in this state, the partial pressure of nitrogen is 101 kPa, while that of oxygen is 113 kPa − 101 kPa = 12 kPa. The quotient of the partial pressures of the two constituents is approximately equal to their molar ratio: $12/101 \approx \frac{1}{9}$, and so, on Exo-Earth, about 10 % of the atmosphere is oxygen, and the rest is nitrogen.

b) On the basis of the foregoing, we can see that at a pressure of 124 kPa (and a constant temperature of 77.4 K) both gases will be liquid. The two components will begin liquefaction together if their molar ratio is exactly equal to the ratio of their saturated vapour pressures, which is $22.6/101 \approx \frac{2}{9}$. This means that, if the oxygen : nitrogen ratio were about 2 : 9, then, during isothermal compression at 77.4 K, both gases would begin to liquefy at the same time.

> *Note.* In reality, on Earth, the make-up by volume of dry air is 78.09 % nitrogen, 20.95 % oxygen, 0.93 % argon, 0.039 % carbon dioxide and small amounts of other gases. So our assumption that the atmosphere contains only oxygen and nitrogen, and that their ratio is 1 : 4, is only approximately true, and this further increases the uncertainty in our numerical results.

S115 At normal atmospheric pressure, the given temperatures 100 °C and 34.6 °C are the boiling points of water and ether, respectively. The molar mass of water is 18 g mol^{-1} and the molar mass of ether is 74 g mol^{-1}. Although the behaviour of real vapours does not follow the ideal gas laws exactly near a substance's boiling point, it is certainly reasonable to suppose that the density of steam is lower, and that of ether vapour is higher, than the density of air.

For brevity, we will (unscientifically) say that steam is 'lighter' and ether vapour is 'heavier' than air.

a) Consider first the case in which water is used in both flasks. In the straight-necked one, steam will fill the neck of the flask sooner or later, but, with the 'heavier' air above it, the situation is unstable; the steam rises, making space subsequently occupied by air. This convection circulation increases the rate of boiling, so all the water will boil away in a comparatively short time.

In the retort, the steam also occupies the neck of the flask, but the 'heavier' air cannot flow downwards into the retort. After a time, only water and steam are present inside the retort, and the pressure of the steam becomes slightly larger than atmospheric. This extra pressure is the only mechanism responsible for the escape of any steam from the retort. In addition, the pure steam just above the remaining water impedes its vaporisation. So, in summary, for case *a*), the water runs out more quickly in the flask with a straight neck.

b) It is obvious that in this case, in which there is boiling ether in both flasks, the situation is just the opposite. The ether is trapped in the straight neck, and cannot rise up into air of lower density. At the top of the neck, the ether molecules can leave only because of the extra pressure of ether vapour above the boiling liquid.

From the retort, the 'heavy' ether 'flows down and out of the flask', and the resulting convection circulation speeds up the boiling of the remaining ether. So, in case *b*), the retort is the first to be exhausted of ether.

S116   In the narrow test tubes, no convective flow sets in, and the dominant process is diffusion. According to the law of diffusion (*Fick's law*), the speed of material flow is determined by the concentration gradient. Experimentally, these two quantities are, to a good approximation, proportional to each other:

$$\frac{\Delta m}{\Delta t} = -DA\frac{\Delta \varrho}{\Delta z},$$

where Δm denotes the mass of the vapour moving through the cross-sectional area A during a time interval Δt, ϱ is the vapour density at height z, and D is the diffusion coefficient (also called diffusivity). We assume that the diffusion of eau-de-Cologne can be described by an 'average diffusivity' of water and alcohol in air.

Initially, there is no vapour in the test tube, as shown in Fig. 1*a*). For this very reason, strong evaporation begins, and, transitionally, some kind of density distribution is formed in the test tube, as illustrated in Fig. 1*b*). After a very short time, compared to the total evaporation time, a stationary density distribution is established.

This stationary distribution of vapour density can only be one that varies linearly along the tube, as shown in Fig. 1*c*). That this is so can be deduced from the fact that the rate at which vapour enters any arbitrary part of the tube from below is equal

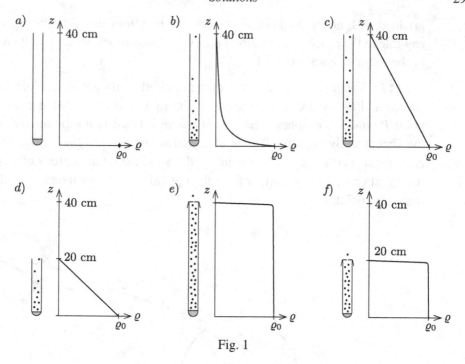

Fig. 1

to the rate at which it leaves its upper boundary; this implies that the concentration gradient is the same everywhere within the tube. At the very bottom of the test tube, just above the liquid, the vapour must be *saturated*, its density depending only on the temperature. At the top of the test tube, where the vapour is in contact with the free atmosphere, its density is practically zero – if it were not (and the density changed sharply there), the vapour would move rapidly and a stationary state would not be possible.

At both the bottom and top of each of the two test tubes, the vapour densities are equal, and so the density gradient – and, consequently, the speed of the material flow – in the longer test tube is one-half of that in the shorter one; this is illustrated in a comparison of Fig. 1c) and 1d). Taking into account that the amount of eau-de-Cologne in the longer tube is twice that in the shorter one, we conclude that total evaporation from the 40 cm tube takes *four times* longer than from the 20 cm one.

What happens when both test tubes are covered, and identical but very small holes are made, one in each cover? If the concentration gradient in the test tube is much lower than that around the hole, the amount of vapour that diffuses through the small cross-section opening will be controlled by the rate at which vapour moves through the wider tube. With the covers in place, the vapour density hardly varies along the tubes (*see* Fig. 1e) and f)) and almost everywhere it is equal to the density of saturated vapour. Only in the immediate vicinity of the hole does it drop to zero, the value of the outside density. Thus there is virtually no concentration gradient and, under such circumstances, the length of the test tube does not play any role in the process. The ratio of the evaporation times depends only on the ratio of the initial masses, i.e. 2 : 1.

S117   *Solution 1*. Take as the optical axis the straight line joining the point source and the centre of the sphere. If the light source is placed at an arbitrary point P_1 inside the sphere, the virtual image formed is not point-like. Although the (backwardly continued) rays of a narrow beam travelling close to the axis are almost focused at a single point on the axis, the continuations of any slightly divergent group of rays further from the optical axis do not intersect each other at that same point.

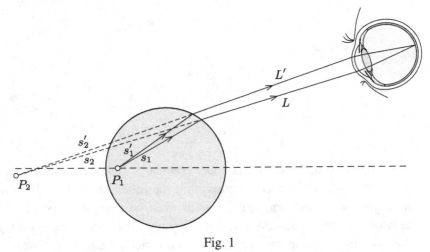

Fig. 1

Consider two light rays moving close to each other, as shown in Fig. 1. The paths covered by the rays are s_1 and s'_1 from the light source to the surface of the sphere, where they are refracted, followed by the optical paths L and L', terminating at the (distant) retina of the eye of the observer. According to Fermat's principle, the light moves along the particular path (among nearby trajectories) that requires the minimal transit time; it is also the path with the shortest optical path length. If the light can traverse two *nearby* trajectories in the same time, then their optical path lengths are equal:

$$ns_1 + L = ns'_1 + L'. \tag{1}$$

The eye observes the light source, actually placed at point P_1, to be at the intersection P_2 of the backward continuations of the two particular rays. This intersection is generally not on the optical axis, and its position depends on how the two light rays are selected (i.e. on the position of the observing eye). If we imagine a light source to be placed at the position P_2 of the virtual image, then we can again apply Fermat's principle to the two rays shown in Fig. 1. The equality of the optical paths from the point P_2 to the retina of our eye gives

$$s_2 + L = s_2' + L'. \qquad (2)$$

Note that s_2 and s_2' are simply physical distances, and do not involve n, even though in Fig. 1 their trajectories do cross the sphere.

Take the difference between equations (1) and (2):

$$ns_1 - s_2 = ns_1' - s_2'. \qquad (3)$$

This relationship characterises the position of the intersection of two *neighbouring* light rays. In general, similar equalities are *not* obeyed for light rays moving at large angles relative to each other. However, there is one special case in which it is – when both sides of equation (3) are equal to zero, i.e. for each ray that is traced, $s_2/s_1 = n$.

This condition can be met if the surface of the glass sphere is an appropriate *Apollonian sphere*, i.e. it is the three-dimensional version of an Apollonian circle. The latter is the locus of the set of points in a plane whose distances from two fixed points (known as foci) have a specified non-unit ratio; this locus is a circle, and an Apollonian sphere is obtained by rotating such a circle around an axis passing through the foci.[57] Working backwards, for a given sphere there are positions P_1 and P_2 that are such that *any* ray from P_1, after refraction at the sphere's surface, appears to the observer to have come from point P_2 on the optical axis.

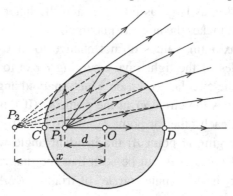

Fig. 2

As we have already shown, for perfect image formation, any arbitrary point on the sphere has to be n times further from the image point than it is from the light source. In particular, this has to be true for the points C and D shown in Fig. 2:

---

[57] *Apollonius of Perga* (*c.* 262 BC–*c.* 190 BC) was a Greek geometer and astronomer noted for his writings on conic sections. It was Apollonius who gave the ellipse, the parabola and the hyperbola the names by which we know them.

$$\frac{x-r}{r-d} = n, \qquad \frac{x+r}{d+r} = n,$$

where x is the distance of the image point from the centre O of the sphere. From these two equations, the two unknown distances can be determined:

$$d = \frac{r}{n} \qquad \text{and} \qquad x = nr.$$

Solution 2. Consider the plane section of the sphere shown in Fig. 3 (this is circle k_1), and denote the positions of the point source of light and its perfect image by S and I, respectively. Consider also the light ray that leaves the sphere tangentially, denoting its exit point by A (the point of tangency). To give perfect image formation, the line of this light ray must appear to come from point I. The refracted ray is perpendicular to the radius OA, and so the angle of refraction is $\beta = 90°$. According to Snell's law

$$\frac{\sin \beta}{\sin \alpha} = n, \tag{4}$$

from which it follows that the angle of incidence α of this particular light ray satisfies $\sin \alpha = 1/n$.

Equation (4) is obeyed, not only by the tangential ray, but also by all other observed light rays emerging from the sphere – their tracks are secants of circle k_1 followed by straight lines that seem to emanate from the point I. Since n is fixed, among all possible light rays, $\sin \alpha$ is maximal for the one for which the value of $\sin \beta$ is as large as possible, and the latter cannot exceed 1. So the ray emerging at A makes the angle α maximal.

Near a maximum, values do not change to first order, which means that the incident angles of the light rays emerging *close* to point A are all the same. It follows that the circular arc k_2, on which marked angle α is subtended by section SO, just touches the circle k_1 at the point A. (If this were not so, circles k_1 and k_2 would cut each other at point A, and so shift by a little the exit point of the light ray emerging at point A; the incident angle would then necessarily either increase or decrease.) It can be seen that the diameter of circle k_2 is the radius AO, and so k_2 is the 'Thales circle' of triangle OSA, and angle $\angle OSA$ is a right angle.

The side SO (opposite angle α) of the right-angled triangle OSA is the required distance d, and so, since

$$\sin \alpha = \frac{d}{r},$$

we have $d = r/n$. Using the similarity of triangles OSA and OAI, the distance x between the image point and the centre of the sphere can be found: $x = nr$.

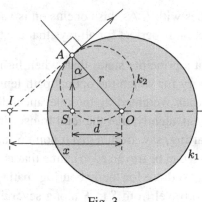

Fig. 3

What still remains to be proved is that the backward continuations of *all* refracted light rays, arising at point S and travelling to the right, meet at point I. We denote by A' the point of refraction of the light ray that is emitted from S at an angle φ to the axis (*see* Fig. 4), and calculate the distance x' from O at which the backward continuation of this light ray cuts the optical axis (at angle ψ say, as shown in Fig. 4).

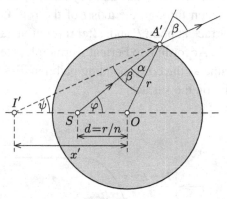

Fig. 4

The sine rule for triangle OSA', combined with the previous result $d = r/n$, gives

$$\frac{\sin \alpha}{\sin \varphi} = \frac{d}{r} = \frac{1}{n}.$$

From this and Snell's law (4), we get $\sin \beta = \sin \varphi$, and, as the angles are acute angles, it follows that $\beta = \varphi$. From the exterior angle property of triangles SOA' and $I'OA'$, we must have $\varphi + \alpha = \psi + \beta$, and so the relationship $\alpha = \psi$ is also valid. From this and the sine rule for the triangle $OI'A'$, it also follows that

$$x' = r\frac{\sin \beta}{\sin \psi} = r\frac{\sin \beta}{\sin \alpha} = nr,$$

and so I' coincides with I. As this conclusion is valid for all the angles φ ($\leq 90°$), the image formation is perfect for all rays that contribute to it.

S118 Fermat's principle states that, when light travels between two points, it follows a trajectory for which the optical path length (the sum of the geometrical path segments, each weighted by the relevant index of refraction) is the shortest available – or, equivalently, that the total transit time is minimal. More explicitly, a ray of light starting, say, from a given point S, and finishing at another point F, follows a path that can be traversed in a time that is *minimal*, when the actual transit time is compared to those for 'neighbouring' paths.

If the light can travel from S to F along several different paths (the task in the problem is to find the conditions for just such a situation), then the optical paths, corresponding to different geometrical paths, must be identical. If any of the optical paths were smaller, or larger, than that of a neighbouring 'trajectory', then the criteria in Fermat's principle would not be met.

We investigate first the parallel light beam, considering it to have come from a very distant source S that lies on the axis of the rod (but is not shown in Fig. 1). It is clear that the situation has axial symmetry, so it is sufficient to investigate a plane section taken through the axis t of the rod. Consider an arbitrary light ray, meeting the surface at point P and, after refraction, taking a straight-line path in the glass to point F. According to Fermat's principle, the optical path length of this ray must be the same as the corresponding path length for any other arbitrarily chosen refraction point on the surface.

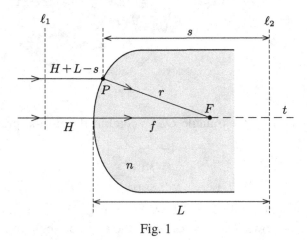

Fig. 1

We now need to express the equality of these optical paths quantitatively. Draw two auxiliary lines perpendicular to the optical axis: one of them (ℓ_1) is placed at a distance H from the peak of the glass rod on the air side, and the other (ℓ_2) is located, on the other side of F, at a distance L from the peak of the glass rod. (The magnitudes of H and L will be fixed later.)

Denote the distance between point P and focus F by r, and the distance between point P and the auxiliary line ℓ_2 by s, as shown in Fig. 1. The optical paths, from the very distant source S (not seen in Fig. 1) to the auxiliary line ℓ_1, are all equal and can be ignored. The criterion for Fermat's principle is that for an arbitrary incident ray

$$H + nf = (H + L - s) + nr.$$

Clearly the (positive) value chosen for H is irrelevant, and this equation can be simplified even further by choosing the arbitrary L to be equal to nf. With this choice, the Fermat criterion takes the form

$$s = nr \qquad \text{or} \qquad \frac{1}{n} = \frac{r}{s},$$

i.e. the ratio of the distance r from P to F must be a fixed fraction ($e < 1$) of the distance s of P from the line ℓ_2. It may seem that the choice of the value of L is 'special' (and devious?), but, if a different value L' were selected, we could replace s in what follows by $s' = s - L' + L$ and arrive at the same conclusion.

It is well known from elementary geometry that an ellipse can be defined as the set of points for which the ratio of the distance of each point on the curve (in the present context, r) from a given point (the focus or focal point, F) to the distance (s) from that same point on the curve to a given line (the directrix, ℓ_2) is a constant ($e < 1$), called the eccentricity of the ellipse.

In our case, the eccentricity of the ellipse is $1/n$, and in the usual notation, in which $2c$ is the distance between the two foci of the ellipse,

$$\frac{1}{n} = \frac{c}{a},$$

and the distance between the straight line ℓ_2 and the centre of the ellipse is

$$L - a = \frac{a^2}{c}.$$

From these expressions, we get the major and minor semi-axes of the ellipse:

$$a = \frac{n}{n+1}f \qquad \text{and} \qquad b = f\sqrt{\frac{n-1}{n+1}},$$

respectively. The required surface is an ellipsoid of revolution (a prolate-elongated spheroid), obtained by rotating the ellipse around the rod's axis. The focus of the glass rod, in an optical sense, is the same point as one of the foci of the ellipsoid (in the geometrical sense). It can be seen (*see* Fig. 2) that the diameter of the light beam brought to a focus cannot be larger than

$$2b = 2f\sqrt{\frac{n-1}{n+1}}.$$

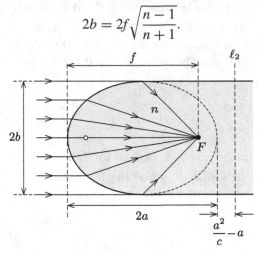

Fig. 2

Note. This problem can also be solved by defining a Cartesian coordinate system whose origin is located at the point at which the rod's axis meets the surface; its positive y-axis is directed axially into the rod, and its x-axis can be in any orthogonal direction. The Fermat criterion then reads

$$y + n(x^2 + (f - y)^2)^{1/2} = nf \qquad \text{for all } x,$$

which, after some careful algebra, can be arranged as

$$\left(\frac{n+1}{nf}\right)^2 \left(y - \frac{nf}{n+1}\right)^2 + \frac{n+1}{f^2(n-1)}\, x^2 = 1.$$

In the notation used in the main solution, this takes the form

$$\frac{(y - a)^2}{a^2} + \frac{x^2}{b^2} = 1,$$

which is the standard equation for an ellipse, except that the major axis is now in the y direction and the centre of the ellipse has been moved along it by a distance a.

Using similar arguments, we can find the conditions needed to focus a light beam propagating inside the glass rod onto a point outside it. Following the notation of Fig. 3, the criterion in Fermat's principle is

$$(H - L)\,n + f = (H - s)\,n + r,$$

and if L is chosen to satisfy

$$L = \frac{1}{n}f,$$

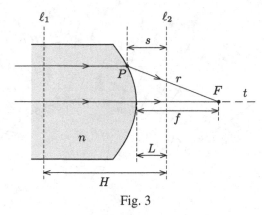

Fig. 3

we get (for arbitrary H)

$$s = \frac{1}{n} r \quad \text{or} \quad n = \frac{r}{s}.$$

Again, from elementary geometry, a hyperbola can be defined as the locus of points for which the ratio of the distances to one focus and to a line (called the directrix) is a constant greater than 1; that constant is the eccentricity of the hyperbola. In our case the eccentricity is n, and one of its foci is the focus of the glass rod in an optical sense. This means that the solution to this part of the problem is (part of) the surface of a hyperboloid of revolution. Now, the diameter of the beam that can be brought to a focus is limited only by that of the rod.

S119 *Solution 1.* According to *Fermat's principle*, light rays very close to each other traverse their trajectories in equal times. We apply this principle to two light rays, one near the surface of the planet of radius R, and the other at an altitude h ($h \ll R$). The first travels along a path of length $2\pi R$ with a speed of $c/n(0) = c/n_0$; the second travels along a path of length $2\pi(R + h)$ with speed $c/n(h)$. Then

$$\frac{2\pi R}{c} n_0 = \frac{2\pi(R + h)}{c} n(h) = \frac{2\pi R n_0}{c} \frac{1 + h/R}{1 + \varepsilon h}.$$

This condition is satisfied for all positive values of h ($\ll R$), provided $R = 1/\varepsilon$.

Solution 2. According to *Huygens' principle*, every point of the wavefront becomes a source of a spherical wavelet, and the new wavefront is formed by the envelope of these wavelets moving with the speed of propagation of the wave. This is the so-called 'phase velocity', which is $c/n(h)$, where c is the speed of light in a vacuum, and $n(h)$ is the refractive index of the medium at the current position.

Consider the cross-section of a wavefront 'orbiting' the planet, which, at some particular time, includes the points marked as A and B in Fig. 1; the wavefront itself is perpendicular to the plane of the figure. A short time interval Δt later, the wavefront has moved to include the points A' and B'. If the radial line $A'B'$ is

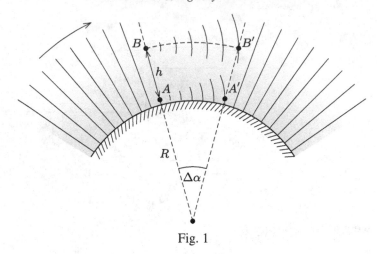

Fig. 1

simply the radial line AB rotated through an angle $\Delta\alpha$, then the light rays, which are perpendicular to it, will move 'horizontally' forwards – as is stated to be the case. This puts conditions on the phase velocities, namely

$$(R + h)\Delta\alpha = BB' = \frac{c}{n(h)} \Delta t,$$

$$R\Delta\alpha = AA' = \frac{c}{n_0} \Delta t.$$

Dividing the first equation by the second:

$$1 + \frac{h}{R} = \frac{n_0}{n(h)} = 1 + \varepsilon h,$$

from which it follows that $R = 1/\varepsilon$, in line with Solution 1.

> *Note.* The two solutions above are not independent of each other, as Fermat's principle (expressed in the framework of geometrical optics) is a consequence of Huygens' principle in wave optics.

Solution 3. The phenomenon of the light 'running around in circles' can also be interpreted as the result of a series of total internal reflections (*see* Fig. 2). These start to occur when the sine of the angle of incidence at a boundary is equal to the ratio of the refractive indices of the materials on either side of the boundary.

We approximate the continuously changing refractive index of the atmosphere by layers of small thickness $h \ll R$, and consider the refractive index to be constant within a layer. Combining the condition for total internal reflection with the geometry indicated in Fig. 2, we have for the lowest layer that

$$\frac{1}{n_0} \frac{n_0}{1 + \varepsilon h} = \frac{n(h)}{n_0} = \sin\alpha = \frac{R}{R + h}.$$

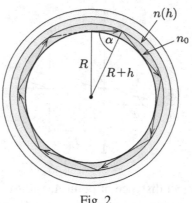

Fig. 2

This condition is satisfied if $R = 1/\varepsilon$. In these circumstances, the light ray, starting horizontally, is totally internally reflected at the top of the layer. It arrives back at its starting point following a series of total internal reflections that take it around the planet. (Of course, for this to happen, the absorption of light needs to be negligibly small, something that does not happen in real circumstances – but then, neither does the existence of planets with such carefully matched atmospheres and radii!)

S120 Assume that the sphere is transparent (so that light can reach the lens from any point of the sphere's surface), but that it is made from material whose refractive index does not differ from that of the air. Refraction of the light rays then occurs only at the two surfaces of the lens. In practice, this situation can be closely reproduced using a sphere of thin, relatively large-meshed wire netting.

Because of the rotational symmetry of the sphere, it is sufficient to determine the image of one of its plane sections – more specifically, the image of one of the great circles that contain the optical axis. The whole image will be the surface of revolution generated by rotating this curve around the optical axis.

Consider an arbitrary point A on the circle of radius r drawn around one of the foci F_1 of the lens, and construct its image point B as determined by two specified rays, as shown in the figure.[58] It is convenient to specify point B by coordinates (x, y) in a Cartesian system that has its origin at the other focus F_2.

Using the similarity of triangles $OA'A$ and $OB'B$, we get

$$\frac{y}{x+f} = \frac{S}{u},$$

and from the similarity of triangles $F_2B'B$ and F_2OQ, it follows that

$$\frac{y}{x} = \frac{S}{f},$$

---

[58] This method is called ray tracing.

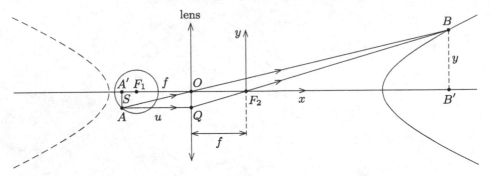

where u is the object distance of point A, and the size of the 'object' is denoted by S (*see* figure). From these ratios S and u can be expressed with the help of x and y:

$$S = f\frac{y}{x}, \qquad \text{leading to} \qquad u = f + \frac{f^2}{x}.$$

The distance between points A and F_1 is given by

$$(u - f)^2 + S^2 = r^2,$$

from which, after substituting for u and S and some algebraic manipulation, we get

$$\left(\frac{rx}{f^2}\right)^2 - \left(\frac{y}{f}\right)^2 = 1.$$

This equation describes a hyperbola with a 'real' (transverse) semi-axis of f^2/r parallel to the optical axis, and a 'virtual' (conjugate) semi-axis of f perpendicular to it.

The image of the whole spherical surface is a surface of revolution created by the rotation of both arms (branches) of the hyperbola around the optical axis; the result is called a *hyperboloid of two sheets*. The image of one half of the sphere (where $u > f$) is a *real* image – its plane section is shown as a continuous line on the right-hand side of the figure. The image of the other half of the sphere, closer to the lens, is a *virtual* image, corresponding to the left-hand side 'sheet' of the hyperboloid – its plane section is shown as a dashed line in the figure.

No image is formed of the points forming the great circle of the sphere that lies in the focal plane of the lens. The images of nearby points are very far away on one of the hyperboloid sheets (and, in the limiting case, at infinity). If the sphere is opaque, then, of course, only the virtual image of the half of it that lies nearer to the lens is formed.

> *Note.* In the solution, we assumed that the lens is thin, and that the image formation is distortionless. For the latter, it is (also) necessary to justify the use of the so-called *paraxial approximation*, which is a small-angle approximation. A paraxial ray is a ray that makes a small angle with the optical axis of the system,

and lies close to the axis throughout the system. In our case, this condition is met if the focal length of the lens is much greater than both the radius of the sphere and the diameter of the lens. The figure in the solution is grossly out of scale; if the real sizes were proportional to those shown, then only a small section of the calculated image surface would accord with reality.

For similar reasons, it does not make much sense to investigate arrangements in which the sphere penetrates the lens, or envelops it ($r \geq f$). Though the investigation of the whole range of the parameters is mathematically interesting, and the case $r > f$ can be achieved physically by caging the lens in wire netting, the analysis of the image formation is largely pointless, because of the difficulties encountered when trying to make experimental observations.

S121   Consider, for example, an astronomical or Keplerian telescope, in which the image is formed by two convex lenses.[59] The image of a very distant object, formed by the objective lens, lies almost exactly in the focal plane of that lens. If we use an ocular (eyepiece) lens to look at this image, its focal plane is made to almost coincide with the position of the image; this is because our eye is naturally accommodated to 'infinity', i.e. to view objects as if they were very distant. Let us denote the focal lengths of the objective and ocular lenses by f_1 and f_2, respectively.

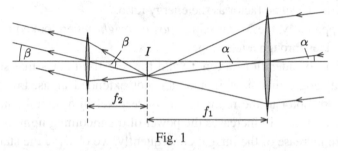

Fig. 1

It can be seen from Fig. 1, which is strongly distorted, that the angular magnification of the telescope (the factor by which the angular diameter of the Moon's circular face appears to be magnified) is

$$M_{\text{ang}} = \frac{\beta}{\alpha} \approx \frac{\tan \beta}{\tan \alpha} \approx \frac{(I/f_2)}{(I/f_1)} = \frac{f_1}{f_2}.$$

Next consider how much greater is the light energy that enters our eye via the telescope compared to what it would be if we looked directly at the Moon. Assume that the diameter d_2 of our pupil is the same in both cases, and that it is smaller than the diameter of the telescope's ocular lens. It can be seen from Fig. 2 (which shows only almost-parallel light rays, and omits those needed for image construction) that the amount of light entering the eye, when using a telescope, is a factor of

---

[59] Other telescopes can be analysed in similar ways.

$$\left(\frac{d_1}{d_2}\right)^2 = \left(\frac{f_1}{f_2}\right)^2 = M_{\text{ang}}^2$$

times larger than it would be without the telescope.

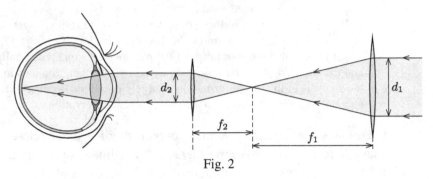

Fig. 2

Finally, we compare the sizes of the images formed on the retina in the two cases. The linear size of the image is, to a good approximation, directly proportional to the visual angle (angular diameter) of the object, so the ratio of the areas of the two different images is equal to the square of the angular magnification, that is, M_{ang}^2 – exactly the same factor as the energy ratio.

Apparently, the Moon seems *just as bright* when looked at with the naked eye as it does through a telescope!

Our considerations are not valid for celestial bodies that subtend such a small visual angle that their images are not extended areas, but cover only a single photoreceptor in the retina (or a single pixel in a digital camera). In such cases the telescope still increases the power of the incoming light, but it is not diluted by a size increase of the image; consequently, we observe the star to be *brighter* than it appears to the naked eye.

> *Notes.* 1. The human eye is a sophisticated optical system containing several refractive interfaces, which cannot be described by an equivalent single thin lens. The light rays entering the eye are refracted at two distinct places: first at the cornea, and then at the eye lens. The refractive indices of the two media filling the rest of the eye (aqueous fluid and vitreous humour) are significantly different from that of air, with values of about 1.3. It follows that most of the light refraction occurs at the air–cornea interface. The thick eye lens (with its different refractive index) only 'modifies' the directions of the light rays passing through it to achieve a sharp image.
>
> Nevertheless, image construction by the human eye can be approximated quite well by a single spherical 'substitute' refractive medium, in which the light rays, going through its centre C, strike the retina without any change of direction (*see* Fig. 3). Accordingly, it follows that, for both naked eye and telescopic observations, the sizes of the images formed on the retina are directly proportional to the visual angles α and β.

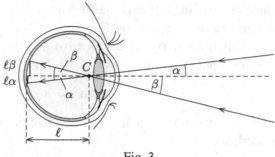

Fig. 3

2. If the diameter of the ocular lens were smaller than the distended pupil (this does not happen in practice!), then the Moon would seem fainter through the telescope than with the naked eye. The same effect is brought about by the automatic (reflex) reduction in pupil size that occurs when we look at a strong light source.

S122 The two halves of the lens form two point-like, real images of the source, which, acting as two coherent light sources, then create an interference pattern on the screen. So we first investigate what kind of interference pattern is formed by two coherent monochromatic point sources, a distance d apart, on a screen placed a distance h ($\gg d$) from them.

Maximal constructive interference occurs at those points for which the *difference* in path lengths to the two sources is an integral multiple of the wavelength λ. The mathematical implications of this in three dimensions are complex, but a detailed analysis shows that constructive interference takes place on the surfaces of a series of two-sheet hyperboloids, whose common axis of symmetry is the line passing through the two light sources.

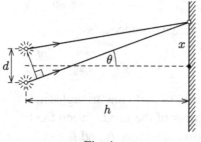

Fig. 1

On the screen, whose plane is parallel to this axis, a section of the hyperboloids can be observed as very slightly curved hyperbola-shaped lines. The hyperboloids, and hence the hyperbolic fringes, do not intersect each other, and the spacing of the latter is smallest along their common symmetry axis. As we are interested only in the maximum number of observable fringes, we need only consider rays that lie in the plane that contains that axis; this is the one that is perpendicular to the screen and contains the two sources (*see* Fig. 1).

The condition for maximal constructive interference for rays travelling at an angle θ to the optical axis, as shown in Fig. 1, is

$$d \sin \theta = n\lambda,$$

where n is an integer. For small values of angle θ, the approximation $\sin \theta \approx x/h$ can be used, and so

$$\frac{xd}{h} = n\lambda, \qquad \text{that is} \qquad x = n\frac{\lambda h}{d}.$$

Thus, the distance between neighbouring interference fringes is $\Delta = \lambda h / d$.

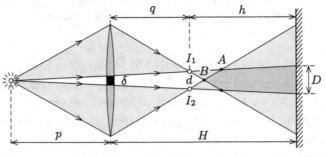

Fig. 2

Now consider the image formed by one of the converging half-lenses with focal length f. We apply the thin lens formula for the point-like source at object distance p:

$$\frac{1}{p} + \frac{1}{q} = \frac{1}{f},$$

from which the image distance q is

$$q = \frac{pf}{p-f}.$$

Using the similarity of particular triangles in (the not-to-scale) Fig. 2, it can be seen that, if the thickness of the gap between the two half-lenses is δ, then the distance d between the formed images I_1 and I_2 can be written as

$$\frac{d}{\delta} = \frac{p+q}{p},$$

from which

$$d = \frac{p+q}{p}\delta = \frac{p\delta}{p-f}.$$

The images I_1 and I_2, formed by the lens, produce interference fringes on the screen with separation (as we have shown earlier) $\Delta = \lambda h/d$, where h is the distance between the images and the screen:

$$h = H - q = \frac{H(p-f) - pf}{p-f}.$$

So the distance between the neighbouring interference fringes is

$$\Delta = \frac{\lambda(H(p-f) - pf)}{p\delta}.$$

However, interference can be observed only on that part of the screen where the light beams from the two image points overlap. Again, using the similarity of suitable triangles, we can get the width D of the overlapping zone:

$$D = \delta\frac{H+p}{p}.$$

The number of interference fringes observable on the screen is therefore

$$N \approx \frac{D}{\Delta} = \frac{\delta^2}{\lambda}\frac{H+p}{H(p-f) - pf}.$$

Substituting the given data into the formula, we get the result $N \approx 46.7$, and so about 47 interference fringes are visible on the screen.

> *Note.* With the given numerical data, the screen is located further from the lens than the point marked A in Fig. 2. If this were not so, the size D of the overlapping zone would need to be calculated in a different way, and, in addition, the diameter of the lens would be required. If the screen is located closer to the lens than point B, there can be no interference at all.

S123 First, let us suppose that alternate slits (the second, fourth and so on) are covered. We then have an ordinary grating with slit separation $4d$, and the position on the screen x_n of the nth-order maximum can be found using the standard equations

$$4d \sin \theta_n = n\lambda \qquad \text{and} \qquad \sin \theta_n \approx x_n/L$$

to give

$$x_n = n\frac{\lambda L}{4d}. \tag{1}$$

The diffraction pattern would be the same if the other series of slits (the first, third and so on) were covered.

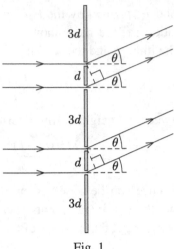

Fig. 1

To tackle the original problem, we need to investigate the net amplitudes of the light in the directions determined by equation (1), taking into account the relative displacement of the two gratings by a distance d. This displacement will add (or subtract) an additional $d \sin \theta$ to the path length of the light from one of the gratings, causing a phase difference between the light from the two sources of $2\pi d \sin \theta / \lambda$ for light diffracted through a general angle θ. For the nth-order spectrum at angle θ_n, this phase difference is

$$\frac{2\pi(n\lambda/4)}{\lambda} = \frac{n\pi}{2}.$$

Thus, there are four different cases to consider, depending on the form of the integer n:

- If $n = 4k$, then the phase difference is a multiple of 2π and there is perfect constructive interference between the light from the two sources; consequently, the net amplitude is double that produced by a single series of slits, and the intensity is four times larger.
- If $n = 4k+1$, the phase difference is $2\pi k + \pi/2$; clearly, the $2\pi k$ can be discarded. The amplitude of the sum of two waves with the same amplitude E_0 and a phase difference of $\pi/2$ between them (at any particular position) is

$$E_0 \sin(\omega t) + E_0 \sin(\omega t + \pi/2) = E_0 \sin(\omega t) + E_0 \cos(\omega t)$$

$$= \sqrt{2}E_0\left(\frac{1}{\sqrt{2}} \sin(\omega t) + \frac{1}{\sqrt{2}} \cos(\omega t)\right) = \sqrt{2}E_0 \sin(\omega t + \pi/4).$$

So, the amplitude is $\sqrt{2}$ times larger than that from a single series of slits, and the intensity is therefore doubled.

- If $n = 4k + 2$, then the phase difference between the light from the two series of slits is π, and there is perfect destructive interference. No light travels in these directions

- If $n = 4k + 3$, the phase difference is $3\pi/2$, so the light amplitude (as in the case of a $\pi/2$ difference) is $\sqrt{2}$ times larger than that due to a single series of slits, and the intensity is correspondingly doubled.

In summary, the intensity distribution of the diffraction pattern can be seen in Fig. 2; the distance between the high-intensity maxima is $\lambda L/d$.

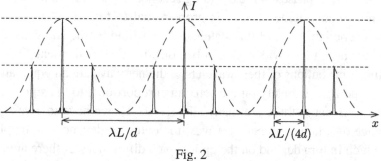

Fig. 2

Notes. 1. This 'unusual' optical diffraction grating can be considered as a grating with a spacing of $4d$, but with complex slits, each slit being a (*Young's*) *double slit* with a slit separation of d. The intensity distribution of a normal optical grating (of very narrow slits) with slit spacing $4d$ has sharp peaks of identical height, separated from each other by $\lambda L/(4d)$. It can be proved that the diffraction pattern of one double slit with slit separation d is a cosine function – shown in Fig. 2 as a dashed line. The intensity distribution of the unusual grating can be found if the two intensity functions (mentioned above) are multiplied by each other.

2. It is particularly striking that the four times longer repeat period of the optical grating produces a four times denser diffraction pattern. This kind of inverse proportionality holds generally for diffraction phenomena: the features with the largest period determine the finest details of the diffraction pattern, and, vice versa, the smallest periodic signals produce the large-scale structure of the pattern. In the current problem, along with periodicities of d and $4d$, there is a third size scale, one that has so far been neglected – the width of the slits. If this were also taken into account, then it would result in a long-period modification of the diffraction pattern; the intensity distribution shown in Fig. 2 would have to be multiplied by the intensity pattern of a single slit. The latter would be almost constant over a significant range of diffraction angles centred on $0°$, but would drop off at larger angles.

3. The intensity peaks are not infinitely 'sharp', but have a finite width, which is related to the width of the illuminating laser beam. For instance, if the illuminating laser beam is thin (and the distance to the screen is large), then the light passes through only a few slits, and this causes a broadening of the intensity peaks. This statement can be illustrated by reference to a normal optical grating (with simple

periodic slits). If N slits of such a grating are illuminated, then, along the direction of an interference maximum, the amplitude is the sum of the contributions from all N slits. The net amplitude is proportional to N, and the *height* of the intensity peak is proportional to N^2. The transported power of light passing through the grating is proportional to N, and the integrated area under an intensity peak must follow suit (since energy is conserved). This is possible only if the *width* of an intensity peak is proportional to $1/N$.

S124 The monochromatic light falls perpendicularly onto the grating, and so the (initial) phases of the outgoing secondary waves from the slits are all the same, and the amplitude of each is proportional to the width of the relevant (narrow) slit. The latter part of this statement is justified by the Huygens–Fresnel principle, since an n times wider slit can be considered as n identical slits side by side, and the contributions of their wavelets to the net wave are all equal and in phase.

The light intensity on the screen is proportional to the square of the net amplitude. Because there are many slits ($N \gg 1$), we have to investigate the interference of a large number of waves; the outcome depends on the phase differences, which in turn depend on the optical path differences. If there are phase differences between the waves coming from adjacent slits, then, in general, the contributions from the many waves cancel each other out. The only exception is the case in which the waves from alternate slits reach the screen in phase. Then, the waves originating from both the even-numbered and the odd-numbered slits (separately) show constructive interference, and their net amplitudes are

$$E_{\text{even}} = K\frac{N}{2}b \quad \text{and} \quad E_{\text{odd}} = K\frac{N}{2}a.$$

The constant K depends on both the intensity of the grating illumination and the distance of the screen, but its precise value is not important in the following analysis.

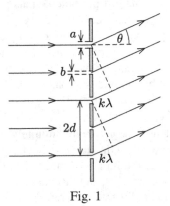

Fig. 1

What are the directions for which the above condition is fulfilled? It can be seen from Fig. 1 that, if, for waves travelling at an angle θ to the normal, the equality

$$2d \sin \theta = k\lambda \qquad (k = 0, \pm 1, \pm 2, \ldots)$$

holds, then, separately, all the even-numbered waves and all the odd-numbered waves are in phase. Consequently, for each set, the amplitudes simply add up arithmetically, producing totally constructive interference.

The task has now been reduced to combining only two waves, the net amplitude E_{even} of the even-numbered slits and the net amplitude E_{odd} of the odd-numbered ones. What is the phase difference between these two waves? As the path difference between every second slit is $k\lambda$, that between adjacent ones is $k\lambda/2$ (as can be seen in Fig. 1). For even k, this is a whole number of wavelengths, and so the two waves are in phase and are to be *added*; for odd k, they are out of phase and their net amplitude is their *difference*. Accordingly, the intensity of the light at angle θ_k (i.e. the intensity of the kth-order maximum) is

$$I_k \sim \begin{cases} (a + b)^2, & \text{if } k \text{ is even,} \\ (a - b)^2, & \text{if } k \text{ is odd.} \end{cases}$$

The kth-order intensity peak is formed at a distance

$$x_k = L \tan \theta_k \approx L \sin \theta_k = k\frac{L\lambda}{2d}$$

from the central maximum on a screen placed a distance L behind the grating. Here, we have used the fact that, because $d \gg \lambda$, the diffraction angles are small.

On the screen, we can see a line of equidistant, relatively sharp light bands, but their intensities are *not equal*: stronger and weaker bands follow each other, alternately (*see* Fig. 2).

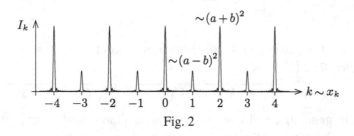

Fig. 2

There are two cases to consider: *a*) if $a \approx b$, then the intensity distribution shown in Fig. 3*a*) will be observed; while *b*) if $a \gg b$, then Fig. 3*b*) shows the intensity distribution on the screen.

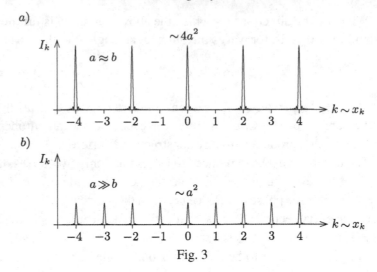

Fig. 3

S125   The location of the holes on the sheet can be specified by the two-dimensional position vector $r = (x, y)$ (*see* Fig. 1), where x and y are integral multiples of the lattice constant d.

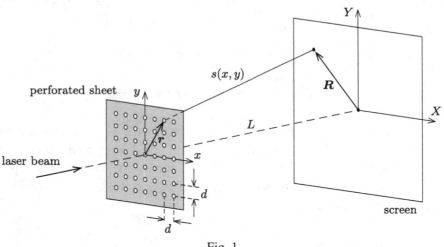

Fig. 1

The distance of a particular hole from a point on the screen given by the position vector $R = (X, Y)$ is

$$s(x, y) = \sqrt{L^2 + (X - x)^2 + (Y - y)^2} \approx \sqrt{L^2 + X^2 + Y^2 - 2(xX + yY)},$$

as, in general, x and y are much smaller than X and Y (except for $X = Y = 0$). Since $L \gg d$, this distance can be further approximated:

$$s(x, y) \approx \sqrt{L^2 + X^2 + Y^2} \sqrt{1 - 2\frac{xX + yY}{L^2 + X^2 + Y^2}}$$

$$\approx \sqrt{L^2 + X^2 + Y^2} - \frac{xX + yY}{\sqrt{L^2 + X^2 + Y^2}}.$$

The path difference between light arriving from a hole with coordinates x and y and an arbitrarily chosen reference wave, say, that from the hole $x = y = 0$, is

$$\Delta s = -\frac{xX + yY}{\sqrt{L^2 + X^2 + Y^2}} \approx -\frac{xX + yY}{L}. \tag{1}$$

In the final step, we have assumed that L is much larger than the size of the screen, and hence much larger than any feasible value of X or Y.

If the path difference (1) is an integral multiple of the wavelength for all permitted values of x and y, then all of the waves arrive at (X, Y) in phase, and there is an interference maximum there. The condition for this is

$$\boldsymbol{r} \cdot \boldsymbol{R} = xX + yY = \text{integer} \times (\lambda L). \tag{2}$$

Taking account of the possible values of x ($= n_x d$) and y ($= n_y d$), condition (2) will always be satisfied if

$$X = n\frac{\lambda L}{d} \qquad \text{and} \qquad Y = m\frac{\lambda L}{d},$$

where n and m are arbitrary whole numbers. It follows that the diffraction pattern is also a square grid – one with 'lattice constant' $\lambda L/d$, as shown in the upper part of Fig. 2.

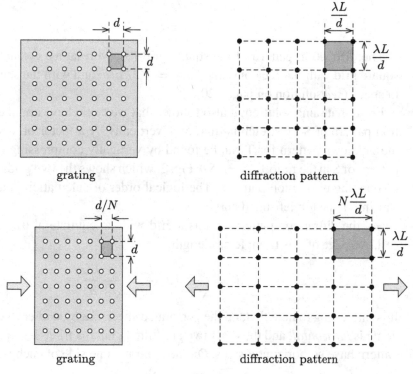

Fig. 2

If the square grid is *compressed* along the x-axis by a factor of N, then condition (2) still holds, provided the X coordinates of the maxima are *increased* by a factor of N. So, the diffraction pattern of light passing through a rectangular grid is also a 'rectangular grid'. However, the scales in any particular direction – of the holes on the sheet and the intensity maxima on the screen – are inversely related: the smaller the hole separation on the sheet, the further apart are the locations of the bright spots on the screen (*see* lower part of Fig. 2).

S126   In the triangular grid (T), consider two adjacent 'unit cells' that share a common horizontal side. If this formation is compressed vertically by a factor of $\sqrt{3}$, then we get a square of side $a = d/\sqrt{2}$ (*see* Fig. 1).

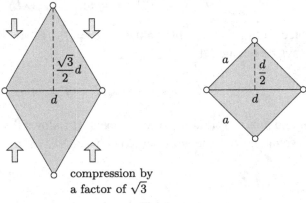

compression by
a factor of $\sqrt{3}$

Fig. 1

The diffraction pattern of the square grid (S), with lattice constant a, is also a square grid with 'lattice constant' $\lambda L/a = \sqrt{2}\lambda L/d$, and with the same orientation in space (*see* solution on page 320).

From that same solution, it also follows that we could have obtained the diffraction pattern of S by stretching that of T vertically by a factor of $\sqrt{3}$. Conversely, the unknown pattern for T can be found by vertically compressing the diffraction pattern of S in the ratio $1 : \sqrt{3}$. (*See* Fig. 2, which shows the two grids and their corresponding diffraction patterns. The logical order of calculation is anticlockwise, starting in the top left-hand corner.)

So, the diffraction pattern of T is a grid with an equilateral triangle as its unit cell. An edge of this triangle has length

$$ d^* = \frac{1}{\sqrt{3}}\sqrt{2}\frac{\lambda L}{a} = \frac{2}{\sqrt{3}}\frac{\lambda L}{d}. $$

It is interesting to note that, on the perforated sheet, one side of each equilateral triangle is *horizontal*, and the other two are 'tilted', but the triangles in the diffraction pattern have no horizontal sides. On the contrary, one side of each is *vertical*.

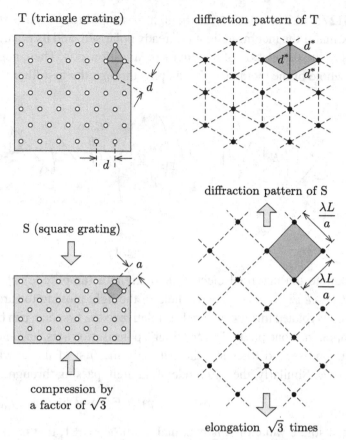

Fig. 2

Note. In the above solution, we have used the inverse relationship between the grid constant of the hole pattern in the sheet and that of the diffraction pattern (in the same direction) on the screen. In the solution on page 320, this was only shown to hold for a rectangular grid, compressed or elongated in an axial direction, whereas in the current problem the compression is in a direction that makes non-zero angles with both axes (45° with each, in this case).

However, the result is still valid, as can be shown by recognising that the diffraction pattern on the screen is essentially a spatial representation ($\boldsymbol{R}$) of the Fourier transform (a function of wavenumber ($\boldsymbol{k}$)) of the hole pattern. The compression is a particular example of a so-called *affine* transformation – one that, in general, involves a reversible linear mapping followed by a translation, and also preserves straight lines as straight lines. Under such a transformation, a Fourier transform keeps its original form, but its arguments are those obtained by applying the inverse mapping to the original arguments – more explicitly, the values of wavenumber components k_i that are needed to keep the kernel of the transformation (a function of $\boldsymbol{k} \cdot \boldsymbol{r}$) unchanged. For a simple compression by a factor N, the inverse mapping is that describing an elongation, in the same direction, and by the same factor. This behaviour of Fourier transforms is directly reflected in that of diffraction patterns.

S127 *a*) The intensity of the light passing through the first polarising filter is maximal if the incoming light is already polarised, and its polarisation plane is parallel to the orientation of the filter's polarisation axis. Then, neglecting absorption, the whole of the incident light can pass through the first filter.

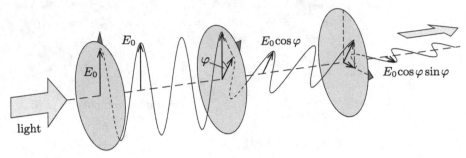

Fig. 1

Denote the maximal electric field strength (amplitude) of the incoming light by E_0 (*see* Fig. 1). The electric field of the light moving towards the second filter, which is rotated through an angle φ relative to the first one, can be resolved into two components: one parallel to that filter's polarisation axis, and another perpendicular to it. The second filter lets through only the first of these, which has amplitude $E_0 \cos \varphi$. Similarly, the amplitude of the light passing through the third filter is

$$E = E_0 \cos \varphi \cos(90° - \varphi) = E_0 \cos \varphi \sin \varphi = \tfrac{1}{2} E_0 \sin 2\varphi.$$

The absolute value of E is maximal if $\sin 2\varphi = \pm 1$, that is, $\varphi = \pm 45°$. Then the amplitude of the transmitted light is one-half of the incoming amplitude, and so the light intensity (proportional to the square of the amplitude) decreases to one-*quarter* of its original value.

b) As in part *a*), maximal light passes through the first polarising filter if the polarisation plane of the incoming light is parallel to the filter's polarisation axis.

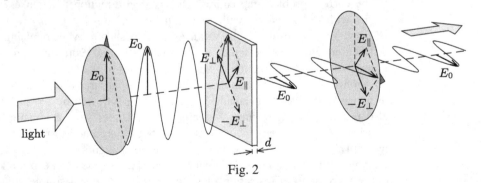

Fig. 2

The amplitude of the light moving towards the birefringent plate can be thought of as being resolved into two components: one, $E_\parallel$, is polarised parallel to the direction of e, and the second, $E_\perp$, is polarised perpendicular to it (*see* Fig. 2). Because of the two different refractive indices, the two components entering the plate leave it with different phases – how different depends upon the thickness of the plate. At the exit face of the plate, the superposition of the two components, with their different polarisations, determines the polarisation of the ongoing light; it can be linear, circular, or even elliptical, depending upon the size of the phase difference.

If the phase difference between the two components passing through the bi-refringent plate is π, or an odd multiple of π, then, on leaving the plate, the two components are superimposed with opposite phases, which is equivalent to one of the components (say, $E_\perp$) changing sign. The condition for this is that the two optical path lengths differ by an odd number of half-wavelengths:

$$|n_1 d - n_2 d| = (2k + 1)\frac{\lambda}{2}, \qquad \text{that is} \qquad d = (k + \tfrac{1}{2})\frac{\lambda}{|n_1 - n_2|}, \qquad (1)$$

where k is a whole number. In this situation, the polarisation remains linear, the light's amplitude does not change, but the plane of its polarisation *is* rotated.

If the plate's birefringence axis e makes an angle of 45° with the polarisation directions of both filters, then the polarisation plane of the light is rotated by 90° (*see* Fig. 2) as it passes through the plate. As a consequence, the third polarising filter does not reduce the light amplitude. Thus, if the thickness of the plate is as given by formula (1), and the orientation of the plate is $\varphi = \pm 45°$, then, provided absorption is negligible, 100 % of appropriately polarised light can pass through the system.

> *Notes.* 1. If the incoming light is unpolarised (i.e. a mixture of the two uncorre-lated light waves, polarised perpendicularly to each other, and changing rapidly but uniformly in time), then the first polarising filter alone reduces its intensity by one-half, however that filter is oriented. Then, in case *a*), the intensity of the transmitted light is at most one-eighth of the incident intensity, and in case *b*), at most one-half.
>
> 2. In optical experiments requiring the rotation of the polarisation plane of linearly polarised light, birefringent plates with a thickness calculated according to formula (1) are often used. Because of the optical path difference for the two components with perpendicular polarisations, they are known as *half-wave plates*. There are also quarter-wave and three-quarter-wave plates; with the help of these, linearly polarised light can be made circularly polarised, and vice versa (*see* the problem on page 41).

S128 First we need some technical information. 3D movies are shot using two motion-picture cameras with slightly different 'viewpoints'. In the cinema (movie theatre), the footage from the two cameras is projected onto the screen using two separate projectors: one image is for viewing by the left eye, and the other is for the right eye. By making use of the properties of polarised light, it can be arranged that each of the viewer's eyes receives only light from the appropriate image.

In older 3D cinemas, each projector has a linear polarising filter placed in front of it, with the two polarisation axes mutually perpendicular. The diffuse light reflected back from the screen, which is a special textile covered by metal grains, mainly aluminium, retains its plane of polarisation. In the spectacle frames worn by the audience, there are also two linearly polarised filters with perpendicular orientations. They are aligned with the polarisation directions of the projected images, which are usually at angles of 45° to the horizontal, one clockwise, the other anticlockwise. With this arrangement, both the right and left eyes see only the images intended for them (*see* left-hand side of Fig. 1, where the orientations are vertical and horizontal).

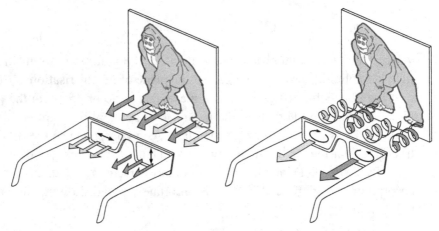

Fig. 1

The disadvantage of using linearly polarised light appears when viewers' heads are tilted, because then some unintended light passes through each filter, and ghost images are formed. Recently, the use of circularly polarised light has been introduced into more up-to-date 3D cinemas, with right-handed and left-handed circularly polarised light projected onto the silver screen by the two projectors. Removal of the non-required part of the reflected light is provided by the films fitted in the 3D specs (*see* right-hand side of Fig. 1).[60] But these films – as we will see – are not simple linear polarising filters, because such filters would pass some of the light of each (circular) polarisation, and the viewer's eyes would not each be presented with a separate image.

Now, we can discuss the analysis of Nick's 'experiments'!

---

[60] In the figure, 'right-handed polarised light' is presented to the right eye, and 'left-handed light' to the left one. But, in practice, it could equally well be the other way round.

a) The old 3D spectacles, found by Nick when tidying up his room, were undoubtedly fitted with linear polarising filters. The unpolarised light rays starting from Nick's closed (say, left) eye, first pass through the filter and are transformed into linearly polarised rays. The direction of their polarisation is not changed by the reflection from the mirror (*see* left-hand side of Fig. 2), and so these rays cannot pass through the right-eye polarising filter (which is oriented perpendicularly to the left-eye filter). Nick's closed eye does not appear in the mirror, and he sees only the darkened 'lens' in its place. However, (one-half of) the light originating from his open (right) eye, and subsequently reflected from the mirror, can pass both ways through the right-eye polarising filter, and Nick can see the reflection of his open eye.

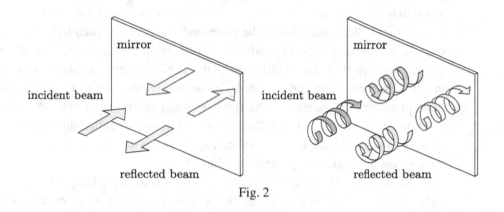

Fig. 2

b) The fact that the new 3D spectacles produce polarisation effects totally contrary to those shown by the old pair indicates that, in the later 'experiment', reflection from the mirror modifies not only the direction of propagation, but also some other relevant property of the light. As the polarisation of linearly polarised light is not changed by reflection from a mirror, we must be dealing with circularly polarised light.

Circularly polarised light can be produced by a linear polarised filter and a so-called quarter-wave plate, aligned with an angle of 45° between their respective axes, as shown in Fig. 3. A quarter-wave plate is a parallel-sided flat plate made of birefringent material, which has a different refractive index for light polarised parallel to its orientation direction than it has for light polarised perpendicular to it (as described in Note 2 on page 325). This gives rise to an optical path difference between two components with different polarisations; by an appropriate choice of plate thickness, this difference is made to be just one-quarter of the relevant wavelength.

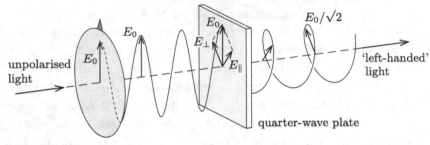

Fig. 3

The incoming unpolarised light, passing through an oriented (say, vertically) polarising filter, becomes linearly polarised. On then passing through the quarter-wave plate, a phase difference of $\pi/2$ arises between the electric field components $E_\parallel$, parallel to the orientation of the plate, and $E_\perp$, perpendicular to it – the result is circularly polarised light. Depending on which polarisation component of the light propagates faster in the birefringent plate, the emerging circular polarisation is either 'left-handed' or 'right-handed'. In the case shown in Fig. 3, the quarter-wave plate produces 'left-handed' light. If we wanted to form 'right-handed' light using the same birefringent material, then we would need a so-called three-quarter-wave plate; this has a different thickness (chosen to make the phase difference between the two electric field components equal to $3\pi/2$).

In Nick's new 3D spectacles, there are films that can selectively transmit or block right-handed or left-handed circularly polarised light. These circular 'analysers' work very similarly to the arrangement shown in Fig. 3, but the circularly polarised light now passes through the birefringent plate *before* reaching the linear filter. If the incoming light striking the birefringent quarter-wave plate shown in Fig. 3 is 'left-handed' (for which the phase difference between $E_\parallel$ and $E_\perp$ is $-\pi/2$), then the reverse phase change process takes place, and light linearly polarised in a vertical plane is formed (*see* Fig. 4). This light is let through by the vertically directed polarising filter.

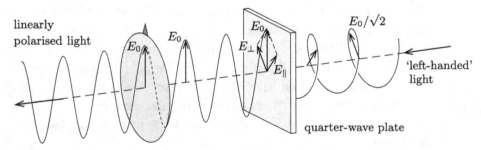

Fig. 4

But if the light striking the quarter-wave plate is 'right-handed' (i.e. the phase difference between $E_{\parallel}$ and $E_{\perp}$ is $+\pi/2$), then, because of the phase shift of $+\pi/2$ produced by the plate, horizontally polarised light is formed; this is not transmitted by the polarising filter (*see* Fig. 5). So, the spectacle 'lens' with this arrangement lets through 'left-handed', but not 'right-handed', light. If the quarter-wave plate is replaced by a three-quarter-wave plate, then the 'lens' works in the reverse way: it lets through 'right-handed' light, but not 'left-handed'. So, a 3D movie, using circularly polarised light, works on the principle illustrated in the right-hand half of Fig. 1.

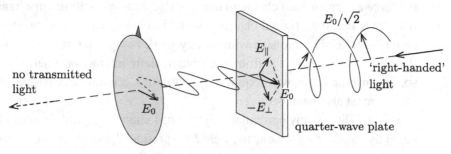

Fig. 5

Finally, we are able to explain the outcome of Nick's second 'experiment'. The light rays starting from his closed (say, left) eye are formed into (say, left-handed) circularly polarised light going through the left-hand side 'lens' of the glasses (since they pass through the polarising filter first, and then through the birefringent plate). But the 'left-handed' light is reflected back from the mirror as 'right-handed' (in the same way as the mirror reflection of the left-handed glove is a right-handed glove, *see* right-hand side of Fig. 2), and so it can pass through the right-hand side 'lens' of the spectacles without impediment. This is why Nick can see his closed eye in the mirror.

Conversely, he cannot see his open (right) eye, since the light starting from it is formed into 'right-handed' light when it first passes through the film of the spectacles, and is then reflected back as 'left-handed' light. As such, it cannot pass back through the right-hand 'lens' of the glasses, and Nick cannot see his own open right eye – even though he can see the closed left one!

> *Notes.* 1. It follows from the operating principle of birefringent quarter-wave plates that they are able to transform linear polarised light into circular polarised light (and vice versa) only for a particular wavelength (i.e. a given colour). However, by applying thin coatings to it, a plate can be made to work 'reasonably effectively' across almost the whole range of the visible spectrum.
>
> 2. Some 3D cinemas (movie theatres) and home televisions solve the problem of separating the two images arriving at our eyes, not with polarising filter spectacles, but with glasses containing active (computer-controlled) liquid-crystal

'lenses'. On the screen of such 3D television sets, the images for the two eyes are flashed alternately at a frequency in the range 100–200 Hz. The glasses can be synchronised with the flashing images using radio waves, infrared signals or cables. These signals are used to switch the voltage applied to the liquid-crystal 'lenses', making them alternately dark and light, and so provide the two separated stereoscopic images, which together give a 3D visual impression.

S129 Increasing all the distances between the charges in the same proportion, say by a factor of λ, changes the forces acting between them by a factor of $1/\lambda^2$. Since, in the original arrangement, every charge was in stationary equilibrium, the net force acting on each charge in the new ('scaled-up') setting must again be zero. It follows that the original charge distribution could be 'blown up' to arbitrarily (even 'infinitely') large size without any work being done on, or by, the system. But, for charges very far from each other, their interaction energy is zero, and so, because no work was required, the system's original electrostatic interaction energy must also have been *zero*.

The stability of any particular (say, the ith) charge's equilibrium can be investigated by considering the electric field $E(r)$ produced, in the vicinity of the charge's equilibrium point r_i, by all the other charges. The force acting on this particular charge is

$$F(r) = Q_i \sum_{k \neq i} E_k(r) = Q_i E(r).$$

The equilibrium will be stable if the force field $F(r)$ is such as to return the charge Q_i to r_i, if it were to be displaced from there in any direction. If this is the case, then the total electric flux of $E(r)$ over a small closed surface around the (vacated) point r_i cannot be zero. The flux is either all into, or all out of, the surface, and the integrated flux has to be a positive or negative number, depending upon the sign of Q_i. However, such a conclusion stands in clear contradiction to Gauss's law: the integrated flux of the electrostatic field over a closed surface without any charge inside it is always zero. Accordingly, the equilibrium of the charge system cannot be stable, and must therefore be *unstable*.

> *Notes.* 1. The general considerations discussed above can be illustrated by a simple example (*see* figure).

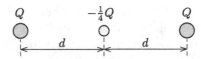

If three point-like bodies, carrying charges Q, $-Q/4$ and Q, are arranged along a straight line, with the two outside ones a distance d from the middle one (*see* figure), then each of the three experiences a zero net force. The electrostatic interaction energy of the system is

$$W_{\text{el}} = k_e \frac{(-\frac{1}{4}Q)Q}{d} + k_e \frac{(-\frac{1}{4}Q)Q}{d} + k_e \frac{Q^2}{2d} = 0,$$

in accord with our general result.[61] Further, if the middle (negative) charge is moved to the right by a small distance x ($x \ll d$), then the net force acting on it is

$$F(x) = k_e \frac{Q^2}{4} \left[\frac{1}{(d-x)^2} - \frac{1}{(d+x)^2} \right] \approx k_e \frac{Q^2}{d^3} x > 0.$$

This force acts to the right, and so tends to move the middle charge even further away from its equilibrium position, demonstrating that the equilibrium *is* unstable, as our general result says it must be.

2. The conclusion reached, about the instability of any equilibrium arrangement of point charges, can be considered as a special case of the so-called *Earnshaw's theorem*.[62] This theorem states that, for arbitrarily complex electrostatic, magnetic and gravitational force fields (or for any combination of them), neither point-like nor extended bodies can be maintained in a stable stationary equilibrium configuration if their motion can be described by Newtonian mechanics. In the late 1800s, this theorem greatly intrigued physicists, because, at that time, they did not know of any other force fields or of any non-Newtonian mechanics, and yet there was no doubting the stability of everyday materials. The inexplicable contradiction was resolved by the introduction of quantum theory, in which the stability of atoms is explained by the non-Newtonian behaviour of electrons.

In the predecessor of this book, one of the problems[63] presents an example in which a superconducting ring is able to sustain stable 'levitation' in stationary magnetic and gravitational fields; at first sight, this might appear to violate Earnshaw's theorem. However, the current flowing in the ring changes with time during the ring's small oscillations around its equilibrium position, and so there is no violation of the theorem, which deals only with stationary (time-independent) fields.

S130 In the absence of external forces, the centre of mass (CM) of the three-pearl system remains at rest. It is therefore convenient to choose it as the origin of a vector coordinate system in which at any given time the position vectors of the pearls are r_i, and their distances from each other are d_i ($i = 1, 2, 3$) (*see* figure).

The net force acting on pearl 1 is the vector sum of the electrostatic forces exerted on it by the other two pearls:

$$F_1 = m_1 a_1 = k_e \frac{Q_1 Q_2}{d_3^3} (r_1 - r_2) + k_e \frac{Q_1 Q_3}{d_2^3} (r_1 - r_3). \tag{1}$$

Because the vector origin has been chosen to be at the position of the CM:

$$m_1 r_1 + m_2 r_2 + m_3 r_3 = \mathbf{0}.$$

---

[61] Although it is of no consequence in the current problem, more explicitly, $k_e = 1/(4\pi\varepsilon_0)$.

[62] Samuel Earnshaw (1805–1888) was an English clergyman and mathematician, noted for his contributions to theoretical physics.

[63] See 'Problem 182' in P. Gnädig, G. Honyek & K. F. Riley, *200 Puzzling Physics Problems* (Cambridge University Press, 2001)

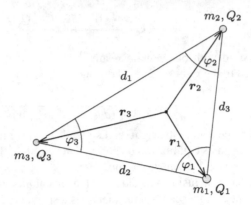

Substituting for r_3 from this into equation (1) gives the equation of motion for pearl 1 as

$$m_1 a_1 = k_e Q_1 \left(\frac{Q_2}{d_3^3} + \frac{Q_3}{d_2^3} \left[\frac{m_1}{m_3} + 1 \right] \right) r_1 + k_e Q_1 \left(\frac{Q_3}{d_2^3} \frac{m_2}{m_3} - \frac{Q_2}{d_3^3} \right) r_2. \qquad (2)$$

Now, the condition for a straight-line trajectory for pearl 1 is that the direction of its acceleration should be the same as that of its instantaneous position vector. Consequently, the factor multiplying r_2 on the right-hand side of equation (2) must be zero. This happens if $Q_2 d_2^3 / m_2 = Q_3 d_3^3 / m_3$. Similar conditions can be written for the other two pearls. So we get the requirement that, at any instant,

$$\frac{Q_1}{m_1} d_1^3 = \frac{Q_2}{m_2} d_2^3 = \frac{Q_3}{m_3} d_3^3 = \lambda, \qquad (3)$$

where λ has the same value for all three bodies. The common value λ is not a constant of the motion, but changes with time, as the distances between the charges increase.

If the equalities (3) are satisfied, then the accelerations of the pearls can, after a bit of careful algebra, be shown to be given by the following formula:

$$a_i = k_e \frac{Q_1 Q_2 Q_3 (m_1 + m_2 + m_3)}{\lambda m_1 m_2 m_3} r_i \qquad \text{for } i = 1, 2, 3.$$

Accordingly, the ratio of the accelerations of the pearls is the same as the ratio of the lengths of their position vectors, and the proportionality factor is the same for all three bodies at any given time; consequently, the ratio of the distances from the origin of the three pearls remains *constant in time*.

These ratios can be found from condition (3) and the given proportion for the mass-to-charge ratios of the three pearls:

$$d_1 : d_2 : d_3 = \sqrt[3]{\frac{m_1}{Q_1}} : \sqrt[3]{\frac{m_2}{Q_2}} : \sqrt[3]{\frac{m_3}{Q_3}} = 1 : \frac{1}{\sqrt[3]{2}} : \frac{1}{\sqrt[3]{3}}.$$

With such distance and charge-to-mass ratios, the three pearls move in such a way that at any moment the triangle they form is similar to their initial triangle, and (using the cosine law) the angles of that triangle are

$$\varphi_1 = \arccos\left(\frac{d_2^2 + d_3^2 - d_1^2}{2d_2 d_3}\right) = 84.2°, \qquad \varphi_2 = 52.2°, \qquad \varphi_3 = 43.6°.$$

Note. Because of the similar forms of Coulomb's law and Newton's law of universal gravitational, the above solution can also be applied to three point masses moving in each other's gravitational fields. Their trajectories can be straight lines if the connection between their distances and their 'mass-to-gravitational-charge' ratios are in accord with equalities (3). As the 'mass-to-gravitational-charge' ratio is actually the inertial-mass-to-gravitational-mass ratio, which is the same for every body (for the sake of simplicity and practical common sense, this ratio is unity), rectilinear motion of the bodies, in the gravitational field of each other, can occur only if $d_1 = d_2 = d_3$, that is, the bodies are at the vertices of an equilateral triangle. Of course, the gravitational force is always attractive, and so the distance between the bodies decreases after they are released from rest.

S131   The trajectory of the electrons can be divided into three significant segments: the first is from the electron gun to the capacitor, the second segment is inside the capacitor, and the last one is from the capacitor to the screen. The electrons, emerging from the gun, move along an essentially rectilinear trajectory with constant speed until they reach the capacitor. Their trajectory is only 'essentially' rectilinear, because, even here, the 'upstream' fringing field of the capacitor causes the electrons to deviate in the direction of the positively charged plate, and at the same time increases their speed a little.

When the electrons enter the capacitor (with its approximately homogeneous electric field inside), they start accelerating in a direction perpendicular to the plates. The force component parallel to the plates is zero, and so the parallel component of the electrons' velocity remains constant, and the trajectory of the electron beam is a parabola. Because of the work done by the electric field, the speed of the electrons as they *just* leave the capacitor is greater than it was when they entered.

Along the trajectory segment between the capacitor and the screen, only the weak fringing electric field of the capacitor acts on the beam, and so one might think that the electrons hit the screen with a speed that is obviously larger than that with which they left the gun.

But let us recognise that the fringing field between the capacitor and the screen acts on the electron beam for a significantly longer time than does the strong electric field between the plates. Also, since the electrons are moving slantwise relative to the axis of the cathode ray tube, it may possibly be that the fringing field of the capacitor *decelerates* them. This could happen because, since the beam is no longer centrally positioned between the plates, it is in a region in which the component of the fringing electric field, parallel to the velocity of the electrons, is not zero. If this component (defined, conventionally, for a notional positive 'test charge') and the velocity vector point in the same direction, the fringing field will decelerate the *negative* electrons!

A solution to the overall problem that avoids these conflicting notions can be found using energy considerations. Initially, the electrons emerge from an electron gun placed on the symmetry plane between the parallel plates. Because of the symmetry, the potentials produced by the two plates have the same magnitudes but opposite signs, and so the electric potential energy is zero where the beam is. When they impact the screen, the electrons are very far from the capacitor, and so their potential energy in its field is again (approximately) zero. It therefore follows from the conservation of energy that when the electrons hit the screen they have the *same speed* as they had when they emerged from the electron gun.

S132   We are going to prove that the electric field strength is zero at the so-called *incentre*, the centre of the triangle's inscribed circle (which has radius r in the figure).

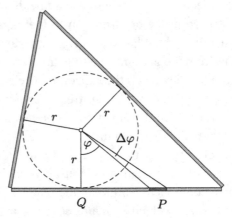

Let us consider a small length of rod at position P on one of the sides of the triangle; let it subtend an angle $\Delta\varphi$ at the incentre (*see* figure). Its distance from the incentre is $r/\cos\varphi$. Its small length Δx can be found by noting that P is a distance $x = r\tan\varphi$ along the rod from the fixed point Q and so $\Delta x = (r\Delta\varphi)/(\cos^2\varphi)$. Consequently the charge it carries is

$$\Delta q = \frac{\lambda r \Delta\varphi}{\cos^2\varphi},$$

where λ is the linear charge density on the rods. The magnitude of the elementary contribution of this small piece to the electric field at the incentre is

$$\Delta E = \frac{1}{4\pi\varepsilon_0} \frac{\Delta q \cos^2\varphi}{r^2} = \frac{1}{4\pi\varepsilon_0} \frac{\lambda r \Delta\varphi}{r^2}.$$

It can be seen from this result that the same electric field (in both magnitude and direction) would be produced by an arc of the inscribed circle that subtends $\Delta\varphi$ at the circle's centre and carries the same linear charge density λ as the rod.

Summing up the contributions of the small arc pieces corresponding to all three sides of the triangle, we will, because of the circular symmetry, obtain zero net field. It follows that the electric field strength produced by the charged sides of the triangle is also zero at the incentre.

S133 *Solution 1.* Since each rod can be regarded as infinitely long, it produces an electric field that is perpendicular to the rod, and has a magnitude that varies inversely with distance from the rod (as can be proved by applying Gauss's law). As the charge on the rods is all of the same sign, it is clear that the net electric field strength cannot be zero anywhere outside the plane containing them, and the point charge could be in equilibrium only if it is placed inside the triangle.

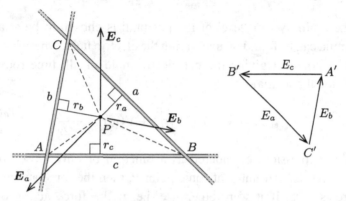

Assume that, at some point P inside the triangle, the net electric field strength is zero. Denote the sides and vertices of the triangle, and the distances of P from its sides, in the way shown in the figure. The magnitudes of the electric fields of the rods are

$$|E_a| = \frac{\lambda}{2\pi\varepsilon_0}\frac{1}{r_a}, \qquad |E_b| = \frac{\lambda}{2\pi\varepsilon_0}\frac{1}{r_b}, \qquad |E_c| = \frac{\lambda}{2\pi\varepsilon_0}\frac{1}{r_c},$$

where λ is the common uniform linear charge density. Accordingly,

$$E_a : E_b : E_c = \frac{1}{r_a} : \frac{1}{r_b} : \frac{1}{r_c}. \tag{1}$$

If the net electric field is zero at P, i.e. $E_a + E_b + E_c = 0$, then a graphical addition of the three corresponding vectors produces a closed triangle. If this vector triangle is now rotated anticlockwise by $90°$, we obtain the triangle $A'B'C'$ shown on the right-hand side of the figure. The sides of this triangle are parallel to the sides of the original triangle ABC, and so the two triangles are similar and, consequently, their side ratios are identical:

$$E_a : E_b : E_c = a : b : c. \tag{2}$$

Combining this result with (1), we obtain

$$a r_a = b r_b = c r_c.$$

According to this, the three straight-line segments that connect point P with the vertices of triangle ABC divide that triangle into three parts with identical areas. As the triangle's centroid (and only its centroid) has this property, the equilibrium position of the point charge must be the *centroid* of the triangle.

Solution 2. We use the figure and notations of Solution 1. The electric potential at a distance r from a single long rod, carrying a uniform linear charge density λ, is

$$V(r) = -\frac{\lambda}{2\pi\varepsilon_0} \ln\left(\frac{r}{r_0}\right),$$

if the arbitrary zero level of the potential is chosen to be at a distance r_0. (This formula can be found by integrating the electric field, which is proportional to $1/r$.)

The potential giving the net electric field of all three rods is the sum of the individual potentials:

$$V_{net}(r) = V(r_a) + V(r_b) + V(r_c) = -\frac{\lambda}{2\pi\varepsilon_0} \ln\left(\frac{r_a r_b r_c}{r_0^3}\right). \tag{3}$$

If this expression, considered as a function of only those points that lie in the plane, has an extremum at some point P, then the force acting on a charge placed there is zero. If it were otherwise, i.e. if the force acting on the point charge were not zero, then the charge could be moved an arbitrarily small distance in the direction opposite to that of the force, only by doing positive work on it. But this would contravene the fact that, near the position of an extremum, to first order, the potential does not change.

We are going to prove that the potential in formula (3) is maximal at the centroid of the triangle. Since the potential $V(r)$ depends on the position of point P only through the product $S = r_a r_b r_c$, and the logarithm is a monotonic function, the extremum of the potential can be found by maximising S. To do this, we employ the 'trick' of multiplying S by an expression with a known value, and which is independent of the position of P:

$$S\frac{abc}{8} = \frac{a r_a}{2}\frac{b r_b}{2}\frac{c r_c}{2} \equiv T_a T_b T_c,$$

where T_a, T_b and T_c denote the areas of triangles BPC, CPA and APB, respectively. Now, the sum of these areas is a fixed value (the area of triangle ABC), and so – because of the general inequality that governs the arithmetic and geometric means of any set of positive numbers – their product is maximal when all three areas are equal.

Accordingly, the charge can be in equilibrium only at a point for which it is true that the line segments connecting the point to the vertices of the triangle divide the latter into three parts with identical areas. The only point to do this is the *centroid* of the triangle.

> *Note.* The equilibrium of the point charge placed at the centroid of the triangle – if it were allowed to move only in the plane of the rods – would be *stable* if it had the same sign as the charge on the rods, but would be *unstable* if they carried unlike charges. But if motion perpendicular to the plane of the rods is a possibility, it is clear that the equilibrium is *always unstable* – in line with Earnshaw's theorem (*see* Note 2 on page 331).

S134 The electric field vector of one of the rods (say, the horizontal one in the problem's figure) is perpendicular to the rod, and has a magnitude, as a function of distance r, that is given by Gauss's law as

$$E(r) = \frac{\lambda}{2\pi\varepsilon_0} \frac{1}{r}.$$

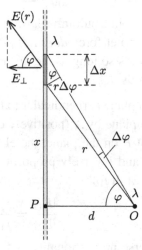

We re-sketch the original arrangement in such a way that one of the rods is perpendicular to the plane of the drawing, and appears as the point O in the figure. The other rod lies in the plane of the drawing and the nearest point on it to O is denoted by P.

The amount of charge on a small piece with length Δx, at a distance x from the point P and subtending an angle $\Delta\varphi$ at point O, is $\Delta Q = \lambda \Delta x$. The force exerted on the charge by the electric field of the other rod has a component perpendicular to the rod given by

$$\Delta F = E_\perp(r)\Delta Q = \frac{\lambda^2}{2\pi\varepsilon_0} \cdot \frac{1}{r}\cos\varphi \cdot \Delta x.$$

(The force components parallel to the rod cancel each other out when the net force is calculated.)

The length of the small piece (marked by a darker segment in the figure) can be expressed in terms of angle $\Delta\varphi$:

$$\Delta x = \frac{r\Delta\varphi}{\cos\varphi}.$$

So, the perpendicular force component acting on the charge carried by the particular small segment is

$$\Delta F = \frac{\lambda^2}{2\pi\varepsilon_0}\Delta\varphi.$$

The total net force acting on the insulating rod can be calculated as the sum of the elementary forces ΔF that act on its constituent small pieces. As the angle subtended at O by the very long rod is π, the magnitude of the net force of repulsion is

$$F = \sum \Delta F = \frac{\lambda^2}{2\pi\varepsilon_0} \sum \Delta\varphi = \frac{\lambda^2}{2\varepsilon_0} = 2\pi k_e \lambda^2,$$

where k_e is the constant in Coulomb's law.

It is curious that the net force *does not depend* on the distance d between the rods – a result that holds so long as the distance d is much smaller than the 'very long' lengths of the rods.

S135   Consider a plane perpendicular to the rods, and denote the points where the rods intersect the plane by A (positively charged) and B (negatively charged). For an arbitrary point P on the plane, the electric field E_i produced there by one of the rods is radial and inversely proportional to the distance r_i of P from the particular rod ($i = 1, 2$). It follows that

$$\frac{|E_1|}{|E_2|} = \frac{r_2}{r_1}.$$

We now use this inverse proportionality and the equality of the angles made with parallel lines by another line that crosses them, to deduce the geometry of the field lines.

Although representing a mixture of physical lengths and electric field vectors (*see* Fig. 1.), the three triangles formed by the following pairs of vectors (and their closures) are similar triangles:

$$\overrightarrow{AP} \text{ and } \overrightarrow{AB}, \qquad E_2 \text{ and } E, \qquad (E_1 - E) \text{ and } -E.$$

In particular, the angle between E and E_2 is equal to angle $\angle PAB$.

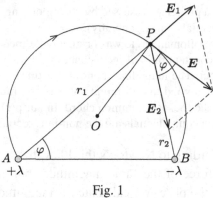

Fig. 1

Now, for any triangle *KLM*, by drawing the radii of its circumscribed circle to each of the three vertices, and so forming three isosceles triangles, it is straightforward to show that the tangent to that circle at any vertex, say *K*, makes an angle with side *KL* that is equal to angle $\angle KML$. In the present context, this means that the line of action of the net electric field vector at point *P* is tangent to the circumscribed circle (with centre *O*) of triangle *ABP*.

It follows that the net electric field vector at an arbitrary point of the plane is tangential to an arc of the circle that circumscribes the triangle whose vertices are *A*, *B* and that particular point. Conversely, the curve to which any particular field line is always a tangent is an *arc of a circle*. The electric field lines can be seen in Fig. 2, which also shows, as dashed lines, two sample equipotentials – they too are circles, but complete ones.

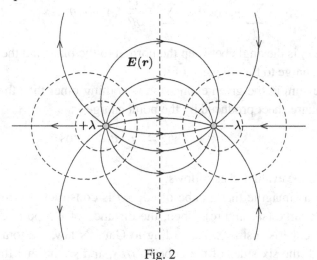

Fig. 2

Note. The two sets of circles, one of field lines, the other of equipotentials, shown in Fig. 2, together form what is known as a set of *Apollonian circles*. Every member of one set intersects every member of the other – and does so at right angles (a property possessed by corresponding sets of electrostatic field lines and equipotentials, more generally).

An Apollonian circle was originally defined as the locus of a point that maintains a constant ratio of its distances from two fixed points; the latter are known as its foci. The individual (non-intersecting) circles that make up one of the sets each correspond to a different value of the ratio. The other set all have the line joining the foci as a common chord. In our problem, the foci are the points A and B and the equipotentials are the non-intersecting set of circles.

S136   According to Newton's third law, the insulating plate acts on the point charge with a force of the same magnitude (but opposite direction) as the point charge does on the plate. We calculate the magnitude of this latter force.

Divide the plate (notionally) into small pieces, and denote the area of the ith piece by ΔA_i. Because of the uniform charge distribution, the charge on this small piece is

$$\Delta Q_i = \frac{Q}{d^2} \Delta A_i,$$

and so the electric force acting on it is $F_i = E_i \Delta Q_i$, where E_i is the magnitude of the electric field produced by the point charge q at the position of the small piece.

The force acting on the insulating plate, as a whole, can be calculated as the vector sum of the forces acting on the individual pieces of the plate. Because of the axial symmetry, the net force is perpendicular to the plate, and so it is sufficient to sum the perpendicular components of the forces:

$$F = \sum_i F_i \cos \theta_i = \sum_i E_i \frac{Q}{d^2} \Delta A_i \cos \theta_i = \frac{Q}{d^2} \sum_i E_i \Delta A_i \cos \theta_i,$$

where θ_i is the angle between the normal to the plate and the line that connects the point charge to the ith piece of it.

The sum in the given expression is nothing other than the electric flux through the square sheet produced by the point charge q:

$$\Psi_\square = \sum_i E_i \Delta A_i \cos \theta_i,$$

and can be evaluated as follows.

Let us imagine that a cube of edge d is constructed symmetrically around the point charge (*see* figure). Then, the distance of the point charge from each side of the cube is just $d/2$. According to Gauss's law, the total electric flux passing through the six sides of the cube is q/ε_0, and so the flux through a single side is one-sixth of this:

$$\Psi_\square = \frac{q}{6\varepsilon_0}.$$

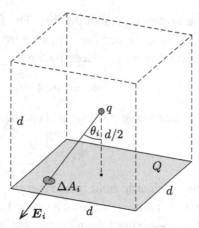

Using this and our previous observations, we calculate the magnitude of the force acting on the point charge due to the presence of the charged insulating plate as

$$F = \frac{Qq}{6\varepsilon_0 d^2}.$$

S137 Because of the symmetry, the electrostatic force acting on any particular face of the cube is perpendicular to that face. This is why the magnitude of the force acting on any one face can be found by summing the perpendicular components of the forces that act on the elementary pieces of that face:[64]

$$F = \sum_i \sigma E_i \Delta A_i \cos \theta_i = \frac{Q}{d^2} \sum_i E_i \Delta A_i \cos \theta_i. \qquad (1)$$

Here ΔA_i is the surface area of the ith elementary piece, $\sigma = Q/d^2$ is the surface charge density, E_i is the magnitude of the electric field strength at the position of the ith piece, and θ_i is the angle between the electric field vector there and the normal to the square face. In the above expression, the summation is only over the electric flux Ψ_5 that is produced by five faces of the cube, and passes through the sixth face. To find the flux, consider the next figure.

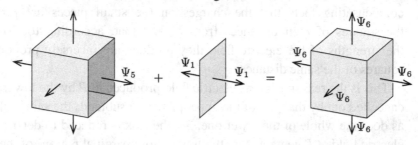

---

[64] In a similar way to that employed in the solution on page 340.

In the figure the arrows represent the flux. The flux, Ψ_6, that comes out through one of the faces of the cube consists of two parts: the flux Ψ_1 of the face's own electric field, and the flux Ψ_5 produced by the other five faces:

$$\Psi_6 = \Psi_5 + \Psi_1.$$

The flux Ψ_1 of the face's own electric field can be calculated from Gauss's law:

$$\Psi_1 = \frac{1}{2}\frac{Q}{\varepsilon_0},$$

since the numbers of electric field lines emerging from the two sides of the face are equal. The flux Ψ_6 coming out from any one face of the cube is, because of the symmetry, equal to one-sixth of the total flux emerging from the cube:

$$\Psi_6 = \frac{1}{6}\frac{6Q}{\varepsilon_0}.$$

Accordingly, the electric flux through the sixth face produced by the other five faces is

$$\Psi_5 = \Psi_6 - \Psi_1 = \frac{Q}{\varepsilon_0} - \frac{Q}{2\varepsilon_0} = \frac{Q}{2\varepsilon_0}.$$

Inserting this into expression (1), we get

$$F = \frac{Q^2}{2\varepsilon_0 d^2}$$

for the magnitude of the electrostatic force acting on each face of the cube.

S138   Small pieces of the upper (positively charged) and lower (negatively charged) plates that subtend the same solid angle $\Delta\Omega$ at the point P – they just cover each other when viewed from P – create electric fields at P that have the same magnitude, but opposite directions. This conclusion follows from the compensating facts that the charges on the small pieces are proportional to the squares of their distances from P, and that, according to Coulomb's law, the strengths of the electric field they produce are inversely proportional to the squares of the same distances.

This is the reason why the electric field produced at P by the lower plate is fully cancelled out by that part of the upper plate that subtends the same solid angle at P as does the whole of the lower one. So, the task is reduced to determining the net electric field at P generated by the 'remaining marginal region' of the upper plate, marked off by a dashed line in Fig. 1. In this figure, the plates have dimensions $a \times b$, with P above the midpoint of one of the edges of length a.

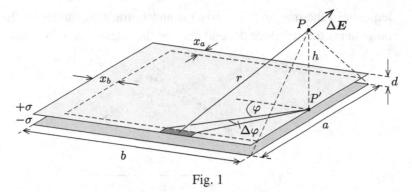

Fig. 1

Because P is not centrally placed above the plate, the width of the margin is not the same everywhere. From the ratio property of similar triangles, its width along the most distant edge of the upper insulating plate is

$$x_b = \frac{bd}{h},$$

whereas along the other two edges it is

$$x_a = \frac{ad}{2h}.$$

Consider now an element of the margin that both subtends an angle $\Delta\varphi$ at the point P' shown in Fig. 1 and is located along one of the edges of length b. The distance of this small piece from the point P is (using the notation of the figure)

$$r = \sqrt{\left(\frac{a}{2\sin\varphi}\right)^2 + h^2} \approx \frac{a}{2\sin\varphi},$$

where the fact that $h \ll a$ has been used. The length of the small piece is

$$\Delta s = \frac{a}{2\sin^2\varphi}\Delta\varphi,$$

and the charge it carries is

$$\Delta Q = \sigma x_a \Delta s = \sigma \left(\frac{a}{2\sin\varphi}\right)^2 \frac{d}{h}\Delta\varphi.$$

The magnitude of the electric field strength produced at point P by the margin element is

$$|\Delta E| = \frac{1}{4\pi\varepsilon_0}\frac{\Delta Q}{r^2} = \frac{\sigma}{4\pi\varepsilon_0}\frac{d}{h}\Delta\varphi,$$

and, because $h \ll a$, its direction is roughly horizontal (although this cannot be seen in our figure, because its scale is heavily distorted). This result is independent of the edge lengths, a and b, and also of the angle φ. The same is true for all other elements of the margin – for given d and h, the magnitudes of the electric field components they produce depend only on the angles $\Delta\varphi$ they subtend at P'.

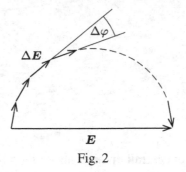

Fig. 2

Next, notionally divide the whole margin into elements that subtend identical angles $\Delta\varphi \ll \pi$ at the point P'. The electric field vector components produced by these small elements can be (vectorially) added to each other (*see* Fig. 2), yielding one half of a regular polygon. Refining the subdivision ($\Delta\varphi \to 0$), the semipolygon approximates a semicircle with 'a curved perimeter' of length

$$\frac{\sigma}{4\pi\varepsilon_0}\frac{d}{h}\sum \Delta\varphi = \frac{\sigma}{4\pi\varepsilon_0}\frac{d}{h}\pi.$$

The length of its 'diameter' represents the magnitude of the electric field strength, and is

$$|\boldsymbol{E}| = \frac{\sigma}{2\pi\varepsilon_0}\frac{d}{h}.$$

The direction of the net electric field is approximately horizontal, and perpendicular to the edges of the plates below the point P. Fig. 3 shows a side view of the electric field lines in the vicinity of the plate edges.

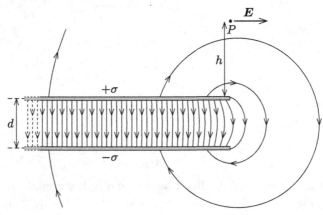

Fig. 3

S139 The electric field of the charged holed spherical shell can be found if it is imagined that the missing piece of the shell (approximately a small disc) has been replaced into the hole and carries the same uniform surface charge density as the rest of the shell (*see* Fig. 1).

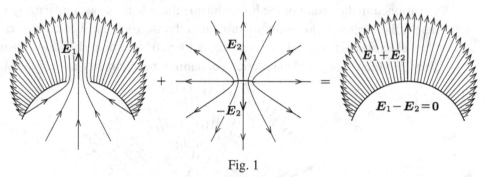

Fig. 1

Denote by E_1 the electric field at the centre of the aperture in the holed shell. If we consider the electric field of the missing small disc alone, then, very close to its centre, the magnitude $|E_2|$ of the electric field is the same on both sides of the disc. If (in thought) the missing piece were replaced into the hole, then we would get a uniformly charged insulating spherical shell, for which the electric field is zero everywhere inside. So, from the superposition principle it follows that

$$E_1 - E_2 = 0. \tag{1}$$

Outside the spherical shell, the electric field is the same as if the charge Q were all condensed into a point charge at the centre of the sphere, and so the electric field at a distance R from the centre (near the surface of the sphere, but outside it) is

$$|E_1 + E_2| = \frac{1}{4\pi\varepsilon_0}\frac{Q}{R^2}. \tag{2}$$

From equations (1) and (2), we find that, in the case of the holed shell, at the centre of the aperture the magnitude of the electric field is

$$|E_1| = \frac{1}{8\pi\varepsilon_0}\frac{Q}{R^2},$$

and that (because of the symmetry) its direction is radial. Note that this is just one-half of the electric field due to the whole charged sphere, and as $E_1 = E_2$, the contributions to the net electric field of the small disc and the holed shell are equal.

The sketching of the electric field lines of the holed shell still remains to be done. From the superposition argument, it follows that the electric flux passing through the hole – i.e. the number of electric field lines that cross the hole – is just one-half of the flux through a piece of the shell (with the same surface area as the hole) that is situated far away from the hole.

But where do the electric field lines, crossing the hole, start from? They start on the inner surface of the charged insulating spherical shell! Inside the shell, far away from the hole, the electric field is similar to what it would be if a point charge of the same magnitude, but opposite sign, to the charge on the missing piece were positioned in the centre of the hole. Outside the sphere, as we move away from the shell, the electric field gradually turns into the isotropic field of a point charge Q. The electric field lines of the charged, but holed, spherical shell are illustrated in Fig. 2, which shows a plane section containing the symmetry axis of the shell.

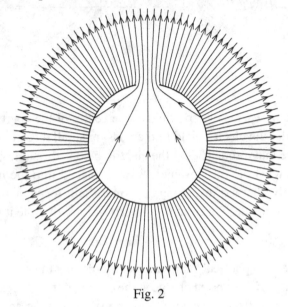

Fig. 2

S140   *a*) We investigate first the special case in which $q = Q$. In this case, the electric field becomes that due to a single uniformly charged insulating spherical shell carrying a charge $2Q$. Anywhere outside the shell, the electric field can be calculated as if the charge $2Q$ were all concentrated at the centre of the sphere, and so the electric field just outside the shell at a distance from the centre marginally greater than R is

$$E = \frac{1}{4\pi\varepsilon_0} \frac{2Q}{R^2},$$

and inside the sphere it is zero. The uniform charge density on the surface of the insulating shell is

$$\sigma = \frac{2Q}{4\pi R^2} = \frac{Q}{2\pi R^2} = \varepsilon_0 E.$$

The electric field E exerts a force of $\Delta F = \frac{1}{2}\Delta Q E$ on any small piece of the spherical shell with surface area ΔA, and charge $\Delta Q = \sigma \Delta A$. The factor of $\frac{1}{2}$

comes from the fact that the magnitude of the electric field is E only on the outer side of the surface element, whereas it is zero inside, leading to an average value of $E/2$.

> *Note.* The factor of $\frac{1}{2}$ can also be justified, perhaps more rigorously, by using the argument given on page 345 in connection with a holed sphere. Close to a small piece of the spherical shell, the electric field receives significant contributions from both the small surface element considered, and all the other parts of the spherical shell. It was shown that the contribution ($E/2$) of the small piece to the electric field is the same as that due to the rest of the sphere (also $E/2$). This conclusion followed from the fact that the electric field is zero inside the surface of the sphere, and E outside it. The force acting on the charge ΔQ situated in an electric field with strength $E/2$ is, as stated, $\Delta F = \frac{1}{2}\Delta QE$.

Thus the force acting on the charge carried by unit area of the shell is

$$p = \frac{\Delta F}{\Delta A} = \frac{1}{2}\sigma E = \frac{1}{2}\varepsilon_0 E^2.$$

Since this force is the same for all parts of the shell's surface, i.e. is isotropic, the situation is just the same as if there were a gas at gauge pressure p inside the sphere.

The net force acting on each of the hemispherical shells is the same (in magnitude) as the force that acts on a (theoretical) circular plate that closes it. This has to be the case because, if it were not, we could create *perpetuum mobile* by filling a closed hemispherical shell with gas at a non-zero gauge pressure. So, the force exerted by one of the two identical hemispherical shells, each carrying uniformly distributed charge Q, on the other is

$$F_{Q\rightarrow Q} = p \cdot \pi R^2 = \frac{1}{2}\sigma E \cdot \pi R^2 = \frac{1}{8\pi\varepsilon_0}\frac{Q^2}{R^2}.$$

Let us now return to our original question, in which the two hemispherical shells carry different charges, namely Q and q. The electrostatic force acting on a body is directly proportional to the charge it carries, and so

$$F^{(a)} = F_{Q\rightarrow q} = F_{q\rightarrow Q} = \frac{q}{Q}F_{Q\rightarrow Q} = \frac{1}{8\pi\varepsilon_0}\frac{Qq}{R^2}.$$

b) Solution 1. Complement the arrangement of hemispherical shells of different radii with its 'mirror image' (*see* Fig. 1)! Denote the net force exerted by the two left-hand hemispherical shells on the two right-hand shells by F_0. This force consists of four components:

$$F_0 = F_{Q\rightarrow Q} + F_{q\rightarrow q} + F_{Q\rightarrow q} + F_{q\rightarrow Q}.$$

The last two terms in the sum are equal to each other, and both of them are just the required force $F^{(b)}$. So, if F_0 could be calculated, then with the help of $F_{Q\rightarrow Q}$ and $F_{q\rightarrow q}$ (found in part *a*)) the 'wanted' force $F^{(b)}$ could be determined.

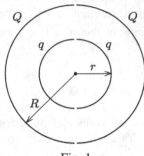

Fig. 1

For the calculation of F_0, consider the electric field produced by two concentric spherical shells, an inner shell of radius r and charge $2q$, and an outer one of radius R and charge $2Q$. In essence, this is also the electric field produced by the four hemispherical shells.

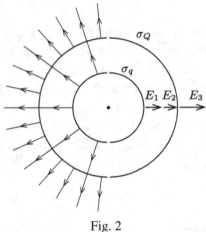

Fig. 2

Inside the smaller shell, the electric field is zero. The charge density on the surface of this shell is (*see* Fig. 2)

$$\sigma_q = \frac{q}{2\pi r^2},$$

and the electric field just outside the surface of the sphere is given by

$$E_1 = \frac{1}{4\pi\varepsilon_0} \frac{2q}{r^2}.$$

The electric field of this smaller sphere at the surface of the larger sphere – strictly speaking, just inside its surface layer of charge – is E_2, where

$$E_2 = \frac{1}{4\pi\varepsilon_0} \frac{2q}{R^2}.$$

The charge density on the surface of the larger sphere is

$$\sigma_Q = \frac{Q}{2\pi R^2}.$$

Outside the larger sphere, both spheres are responsible for the electric field. Consequently, the magnitude of the electric field here (close to the surface) has the form

$$E_3 = \frac{1}{4\pi\varepsilon_0} \frac{2q + 2Q}{R^2}.$$

With the help of the above expressions – and following the argument applied in part a) – the force F_0 can be calculated as follows:

$$F_0 = \tfrac{1}{2}\sigma_q E_1 \pi r^2 + \tfrac{1}{2}\sigma_Q (E_2 + E_3)\pi R^2.$$

Note that, in line with our discussion in part a), we can assume that the average electric field inside the surface layer of charge $2q$ on the smaller sphere is $E_1/2$, and that inside the charge layer on the larger sphere is $(E_2 + E_3)/2$.

Substituting the expressions for σ_q, σ_Q, E_1, E_2 and E_3 into the formula for F_0, we get the force between the two double hemispherical shells:

$$F_0 = \frac{1}{8\pi\varepsilon_0} \frac{q^2}{r^2} + \frac{1}{8\pi\varepsilon_0} \frac{Q(2q + Q)}{R^2}.$$

Using the results from part a), we also have that

$$F_{Q\to Q} = \frac{1}{8\pi\varepsilon_0} \frac{Q^2}{R^2} \qquad \text{and} \qquad F_{q\to q} = \frac{1}{8\pi\varepsilon_0} \frac{q^2}{r^2}.$$

Finally, from these we obtain the force in question:

$$F^{(b)} = F_{Q\to q} = F_{q\to Q} = \frac{1}{8\pi\varepsilon_0} \frac{Qq}{R^2}.$$

The result is surprising, since the force $F^{(b)}$ is *independent* of r. Accordingly, the force between the charged hemispherical shells is the *same* in both cases – $F^{(b)} = F^{(a)}$!

Solution 2. Let us notionally enclose the arrangement, seen in Fig. 3a) (in which, for the sake of simplicity, only the force acting on the smaller hemispherical shell is indicated), with a spherical shell that carries a uniformly distributed charge $-2Q$, and has a radius that is 'just a hair' larger than R, as shown in Fig. 3b). When this is done, the force between the hemispherical shells does not change, as the electric field inside a uniformly charged spherical shell is zero. This arrangement is equivalent to that shown in Fig. 3c), from which we get the situation shown in Fig. 3d) by changing the sign of $-Q$.

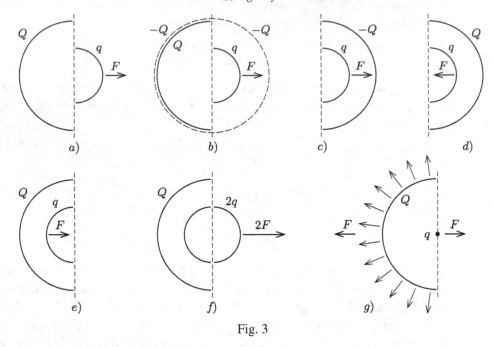

Fig. 3

Now, reflection of the whole system in the imaginary boundary plane of the hemispherical shells gives the arrangement in Fig. 3e), and shows that the (uniformly charged) hemispherical shell carrying charge Q exerts the same force on a concentric smaller one carrying charge q when the latter is 'inside' it, as it does when it is 'outside', i.e. the force is the same in both magnitude and direction in Figs. 3a) and 3e). This is perhaps somewhat surprising, as the charges are clearly closer together in Fig. 3e) than they are in Fig. 3a), and this will lead to larger forces. However, in Fig. 3a) all the forces, though weaker, are more nearly directed along the axis of symmetry and so reinforce, rather than cancel, each other, as the components perpendicular to that axis do in Fig. 3e). It seems that these two effects exactly compensate for each other.

By adding the effect that the charge Q carried by the larger hemisphere has on the 'outside' small hemisphere (Fig. 3a)) to its effect on the 'inner' small hemisphere (Fig. 3e)), we obtain Fig. 3f), which shows that the larger shell would exert a force of $2F$ on a complete sphere carrying charge $2q$. On a sphere with half that charge, i.e. q, the force would be halved, that is, F.

According to Newton's third law, the magnitude of the force, acting on the larger hemispherical shell with charge Q, and exerted by the sphere with charge q, is just the same as the required $F^{(b)}$. But, outside its surface, the electric field of the sphere (with radius less than R) can be replaced by that of a point charge at its centre (this is where the dependence on radius r disappears). This point charge exerts the force $F^{(b)}$ on the larger hemispherical shell as shown in Fig. 3g). The final result can be found by applying Coulomb's law and using the analogy of a gas under positive gauge pressure:

$$F^{(b)} = \frac{1}{4\pi\varepsilon_0} \frac{q}{R^2} \frac{Q}{2\pi R^2} R^2\pi = \frac{1}{8\pi\varepsilon_0} \frac{Qq}{R^2}.$$

Note. Using similar 'reflection methods', it can be shown that the magnitude of the force between the hemispherical shells is given by the above expression, even if the symmetrical axes of the two shells make an arbitrary angle α with each other; the magnitude of the force depends only on the product of the charges, and the radius of the larger shell. The direction of the force is parallel to the axis of symmetry of the larger hemispherical shell, but its line of action does not, in general, go through the common centre of the shells (*see* Figs. 4a) and 4b)). This is a strange result, because if r and R are almost equal, but one of them is 'just a hair' larger than the other, then the *direction* of the force changes precipitously, depending on which radius is that little bit larger (*compare* Figs. 4b) and 4c)).

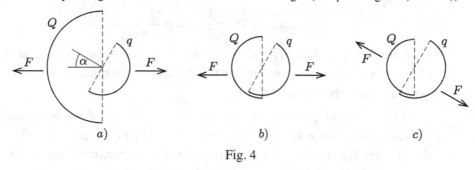

Fig. 4

S141 At the centre of a cube carrying a homogeneous charge density, the potential V_{centre} can depend only on Coulomb's constant $k_e = 1/(4\pi\varepsilon_0)$, the charge Q on the cube and the length a of one of its edges. If the units of the quantities involved, i.e.

$$[V_{\text{centre}}] = \frac{\text{N m}}{\text{C}}, \qquad [k_e] = \frac{\text{N m}^2}{\text{C}^2}, \qquad [Q] = \text{C}, \qquad [a] = \text{m},$$

are taken into account, the functional dependence can only be of the form

$$V_{\text{centre}}(a, Q) = ck_e\frac{Q}{a}, \tag{1}$$

where c is a dimensionless constant (characterising the geometry of the cube).

Imagine a larger cube with edge length $2a$, built up from eight identical smaller cubes with edge length a, and each carrying a charge Q uniformly distributed throughout it (*see* figure). The potential at the centre of this larger cube is

$$V_{\text{centre}}(2a, 8Q) = ck_e\frac{8Q}{2a} = 4ck_e\frac{Q}{a}.$$

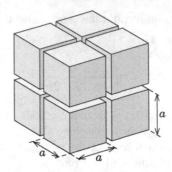

The centre of the large cube is one of the vertices of each of the smaller cubes. Now, further imagine that *seven* of the eight small cubes are removed to a great distance; the potential at the position of the centre of the large cube decreases to one-eighth of its original value. So the potential at the vertex of the remaining small cube is

$$V_{\text{vertex}}(a, Q) = \frac{1}{8} \times 4ck_e \frac{Q}{a} = \frac{1}{2}ck_e \frac{Q}{a}. \tag{2}$$

Comparing expressions (1) and (2), it can be seen that the potential at the centre of the cube carrying a homogeneous volume charge distribution is exactly *twice as large* as that at its vertex.

S142 A small piece of the surface of a charged conductor experiences an outward normal force that is directly proportional to the square[65] of the local charge density (*see* Fig. 1). This, in turn, depends on the geometrical shape of the surface.

In our case, the charged conductor is the mercury (as charge moves onto it when the high voltage is switched on), and its shape is shown in Fig. 2. Inside the part of the capillary tube that is below the outer mercury level, the electric field is practically zero, because this region is inside the conductor, and the Faraday cage effect comes into play. We could also come to the same conclusion by arguing that, if there were a significant electric field inside the recess, then there would be a potential difference between the top and the bottom of the cavity, but – because of the metallic character of mercury – this cannot happen.

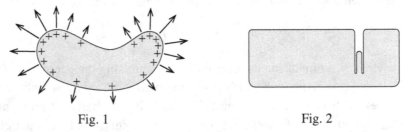

Fig. 1   Fig. 2

---

[65] Both the local electric field strength, and the charge on which that field acts, are proportional to the local charge density.

So, in practical terms, there is no electric field in the capillary tube that could exert an upward force on the mercury. But, on the other parts of the mercury surface, there is such a force, acting vertically upwards. As the total volume of the liquid mercury is given, the mercury level in the capillary tube must *sink*.

> *Note.* The phenomenon of thermal conduction can be described with equations formally similar to the laws of electrostatics. In the analogy, the temperature corresponds to the electric potential, and the analogue of the electric field is a vector proportional to the heat flux (current density).
>
> The given electrostatic problem corresponds to the following thermal question: 'What is the temperature distribution in the vicinity of a good thermal conductor (with the shape shown in Fig. 2), if the body of the conductor is slightly warmer than its surroundings, which have only moderate thermal conductivity?' It is obvious that heat is being given out by the body, but inside the cavity significant heat currents cannot be formed, as the temperature of the cavity wall is practically uniform.

S143 In the derivation of the approximate capacitance formula for a parallel-plate capacitor, it is assumed that the electric field strength is constant inside the capacitor, that it is zero outside (the fringing fields are neglected) and that the electric field lines are perpendicular to the plates. These are good approximations when the width of the plates is much greater than their separation d.

Within this approximation, the magnitude of the uniform electric field E can be found if Gauss's law is applied to a rectangular cuboid that surrounds one of the plates of the capacitor, and has two of its faces parallel to the plate. In accord with the approximations and Gauss's law,

$$EA = \frac{Q}{\varepsilon_0},$$

where A is the inner surface area of the plate and Q is the charge on it. The potential difference (voltage) between the plates is $\Delta V = Ed$. In this approximation, the capacitance is

$$C_{\text{appr}} = \frac{Q}{\Delta V} = \varepsilon_0 \frac{A}{d}.$$

We now assume that Q is given, and estimate (without the approximations) the real voltage between the identical plates. We note that their charge distributions cannot be uniform, because the repulsive forces between the charges mean that their densities are larger near the edges of the plates, and smaller in the middle, than their average value. For rectangular or circular plates, the electric field along the symmetry axis that passes through both plate centres must be normal to the plates, and its magnitude must be smaller than $Q/(\varepsilon_0 A)$ (because of the less-than-average surface charge density there).

As we move along the symmetry axis from the plates to the centre of the capacitor, the density of the electric field lines cannot increase (in fact, it decreases), and so the magnitude of the electric field does not reach the value of $Q/(\varepsilon_0 A)$ at any point on the symmetry axis. It follows that the potential difference (voltage) between the plates – which can be calculated along any arbitrary path, in particular along the symmetry axis – must be less than $Qd/(\varepsilon_0 A)$. Since Q is given, we immediately conclude that the real capacitance is *greater* than $\varepsilon_0 A/d$.

The electric field inside the capacitor may or may not be homogeneous to a good approximation, but the fringing field outside the capacitor inevitably contributes to the (fixed) Gauss integral, and reduces the amount of flux attributable to the field between the plates. This is why, in reality, both E and ΔV are smaller than predicted by the approximate calculation. An alternative view is to consider the fringing field as due to a small capacitor connected in parallel with an ideal capacitor; when two capacitors are connected in parallel, their capacitances are added and the equivalent capacitance is always greater than either.

> *Note.* The result of a more accurate calculation of the capacitance depends on the shape of the plates. In the case of circular plates ($A = \pi R^2$), the approximate result can be corrected by a factor that depends on the ratio d/R. As examples, if $d/R = 0.2$, then the factor is 1.286; for $d/R = 0.01$, it is only 1.023.
>
> For rectangles, the calculation is much more sophisticated. Quite a good approximation is obtained if we work with an enlarged plate area given by enlarging each of its real dimensions by $3d/8$.[66]

S144   For exercises in physics textbooks, the charges on the plates of a capacitor nearly always have equal magnitudes but opposite signs. But here, the signs of Q_1 and Q_2 are unknown, and their magnitudes can be different or equal. If the charges on the plates had values $+Q$ and $-Q$, then the potential difference would have the well-known value of

$$\Delta V = \frac{Q}{C}.$$

And if both plates carried charges of $+Q$, then because of symmetry, the potential difference between the plates would be zero.

We now apply the principle of superposition and incorporate both of these observations. Notionally, we add a charge of $-(Q_1 + Q_2)/2$ to each plate. The potential difference between the plates cannot change because of this, but the charges on the plates change from Q_1 to $+(Q_1 - Q_2)/2$, and from Q_2 to $-(Q_1 - Q_2)/2$. We then have a 'traditionally' charged capacitor with a potential difference across it of

---

[66] See R. P. Feynman, R. B. Leighton & M. Sands, *The Feynman Lectures on Physics, including Feynman's Tips on Physics: The Definitive and Extended Edition* (Addison-Wesley, 2006).

$$\Delta V = \frac{|Q_1 - Q_2|}{2C}.$$

So this must also be the potential difference across the parallel-plate capacitor when it was charged asymmetrically.

> *Note.* From the solution found, the charge distribution on the plates can be determined. Because of charge induction, on the inner surfaces of the plates, charges of $+(Q_1 - Q_2)/2$ and $-(Q_1 - Q_2)/2$ have replaced Q_1 and Q_2, respectively. But, on the outer surfaces, the amounts of charge are identical, both being $(Q_1 + Q_2)/2$.

S145 *Solution 1.* Consider first a scenario in which the two metal plates, connected by a wire, are *grounded*, and charges can move freely onto and off the plates. This is *not* the situation in the problem, but it is nevertheless quite instructive. Because of their electrical connection, the two plates behave as if they were a single, very large (essentially infinite), grounded metal plate. The point charge Q produces the same electric field above the (combined) plate as would be produced by a system consisting of itself and a 'mirror point charge' $-Q$, situated on the same normal to the plate, and a distance d below the plate (*see* Fig. 1).

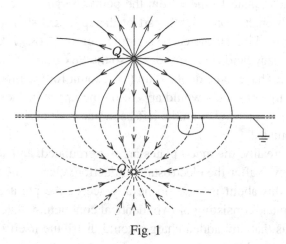

Fig. 1

The charge distribution formed on the plate is determined by the (non-uniform) electric field strength on the 'upper' surface of the metal plate. According to Gauss's law, the amount of charge on a given surface area of the plate is proportional to the number of electric field lines passing through that area. As all of the electric field lines must terminate somewhere on the 'infinitely large' plate, the total charge on the whole (grounded) metal plate must be $-Q$.

Just as the electric field can be found by superposing the fields of the real charge and the image charge, so can the number of electric field lines that pass through a given surface area of the plate be determined. But the two contributions are equal to each other, and so the total electric flux arriving at any given area of the plate is *twice* the flux due to the real point charge Q.

We next investigate how, if the charges induced on the plate were not present, the terminations of the Q/ε_0 electric field lines originating (with spherical symmetry) from the point charge Q would be distributed between the two plates. To determine this, imagine that Q is surrounded by a spherical shell of radius d, as seen in Fig. 2, with half of the field lines, $Q/(2\varepsilon_0)$ in number, passing through the lower hemispherical shell.

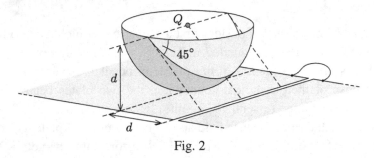

Fig. 2

The metal plate furthest from the point charge intercepts those field lines that emerge through a one-eighth part of the spherical shell (a 'melon shell', shaded light grey in Fig. 2); the corresponding flux is $Q/(8\varepsilon_0)$. On the other plate, three times as many field lines arrive, and so the flux passing through it is $3Q/(8\varepsilon_0)$. As these fluxes have to be doubled up (to account for the image charge), we conclude that a charge of $-Q/4$ would appear on the upper surface of the right-hand plate, and a charge of $-3Q/4$ on the left-hand one – all of this calculated as if the plates were grounded.

But, in reality, the metal plates are *not* grounded, and so their total charge must be zero, even after the electrostatic induction; clearly the total charge to be added to bring this about is $+Q$. In the absence of the point charge – i.e. for a large isolated plane consisting of two identical conducting plates, electrically joined – it is obvious that any added charge would distribute itself equally between the two plates. Our problem is the superposition of the above two scenarios. So, if we add $+Q/2$ to the charges calculated above for each of the two plates, then we get the solution to the original problem: the total charge on the left-hand plate is $-Q/4$, and that on the right-hand plate is $+Q/4$.

> *Note.* If charge Q were divided into two equal portions, and one of the halves were moved parallel to the thin gap between the plates, then, although the charges would rearrange themselves, the total charge on each individual plate would *not* change. Repeating this process many times, charge Q can even be 'smeared' uniformly along a thin, straight, insulating rod, and the problem essentially transformed into a two-dimensional one, as illustrated in Fig. 3. The solution to

this modified question remains the same as that for the original one, i.e. if the plates are grounded, the net charge ratio on the plates is 3 : 1 (in line with the ratio of opening angles of the dark and light grey segments in Fig. 3.)

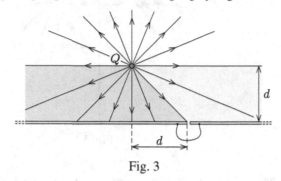

Fig. 3

Solution 2. Denote the amount of charge on the left-hand metal plate by $-q$. Since the combined plate is not earthed, the electrons making up this charge will have been provided by the right-hand plate, which must therefore now carry a charge $+q$. Next consider placing a charge $-Q$ at the position shown in the centre of Fig. 4; this generates charge $+q$ on the right-hand plate and $-q$ on the left-hand one. The two arrangements can now be superimposed, with the result shown on the right of Fig. 4.

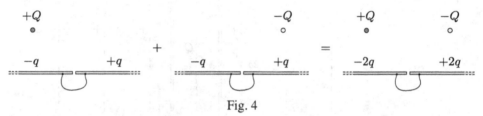

Fig. 4

The electric field of this superimposed arrangement can be determined by using the method of image charges (*see* Fig. 5), as in Solution 1. The ('vertical') plane delineated by the dashed line in the figure is an equipotential – as would be any real metal plate inserted there. This hypothetical plate would be an equipotential surface at zero potential – just like the real 'horizontal' plate. From symmetry – up and down on the left-hand side, as compared to left and right above the plate – a further charge of $-2q$ would appear on the 'left-hand surface' of the upper part of this hypothetical plate. As all of the field lines that start on charge $+Q$ must terminate on one of the two semi-planes, we have that $-2q - 2q = -Q$, that is, $q = Q/4$.

As it must, this approach yields the same conclusions as did Solution 1. Charge $-Q/4$ appears on the plate closer to the charge Q; charge $+Q/4$ appears on the other plate, as the latter supplies the electrons needed to produce the extra charge on the former.

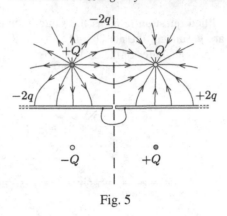

Fig. 5

S146 The potentials of the capacitor plates are both zero, and, for this reason, the method of image charges can be applied. The first few of the infinite number of image charges of $+Q$ and $-Q$ are indicated in the figure by open circles. (For the moment, the black circles also shown in the figure are to be ignored, as they are not true image charges.)

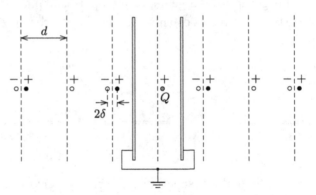

A little careful calculation shows that the open circles carrying a positive sign are symmetrically placed about the pearl, and so their effects cancel each other out in pairs. Further, there can be no change in the force acting on the pearl if this symmetric (about the pearl) set of positive charges $+Q$ is replaced by another symmetric set of positive charges $+Q$ – not least because neither set has any net effect on the pearl. The positions of the replacement charges are marked in the figure by the black circles. (The positively marked open circles are now to be ignored.)

The net effect of these observations and charge replacements is to reduce the image charges to a series of electric dipoles, each dipole being 'formed from a negative open circle and a nearby positive black circle'. We note that all the dipoles are aligned in the same direction. As the fields produced by dipoles decay with distance much more rapidly than those due to isolated charges, the summation over the infinite series involved should converge much more rapidly.

Finally, we have to explicitly sum the effects on the pearl of electric dipoles, each with dipole moment $p = Q \cdot 2\delta$, which are (to a good approximation) located at distances $\pm d, \pm 3d, \pm 5d, \pm 7d, \ldots$ from the pearl. The magnitude of the force[67] acting on the pearl due to a single dipole at a distance r is

$$F_{\text{dipole}}(r) = \frac{1}{2\pi\varepsilon_0} \frac{pQ}{r^3},$$

and so the effect of all the dipoles is a force

$$F = \frac{2}{2\pi\varepsilon_0} \frac{2Q^2\delta}{d^3} \left(1 + \frac{1}{3^3} + \frac{1}{5^3} + \cdots\right) = \frac{2}{\pi\varepsilon_0} \frac{Q^2\delta}{d^3} \sum_{n=0}^{\infty} \frac{1}{(2n+1)^3},$$

and this force is directed towards the closer plate. When the first few terms of this sum are calculated, it soon becomes clear that the series converges very rapidly, and we get the following result for the force acting on the pearl:

$$F \approx 2.1 \frac{Q^2\delta}{\pi\varepsilon_0 d^3}.$$

Note. The sum above can be expressed in terms of the Euler–Riemann zeta function, which can be defined as the sum

$$\zeta(s) = \sum_{n=1}^{\infty} \frac{1}{n^s}$$

for all values of s that make the series convergent (for real numbers, this means $s > 1$). It can be proved that our result (involving only the odd integers) can be written in the form

$$F = \frac{7\zeta(3)}{4} \frac{Q^2\delta}{\pi\varepsilon_0 d^3} = 2.1036 \frac{Q^2\delta}{\pi\varepsilon_0 d^3}.$$

It is interesting to note that calculating only the sum of the first few terms gives quite an accurate result.

S147 The electric field between the plates, generated by the charged rod and the grounded plates, is just the same as that produced by the charged rod and a set of imaginary (linear) image charges, with appropriate positions and signs; none of them may be located in the space between the plates. In case *a*) there is in fact only a single plate, and the required zero potential at the plate can be provided using only a single 'image rod' (*see* Fig. 1).

---

[67] See, for example, 'Problem 183' in P. Gnädig, G. Honyek & K. F. Riley, *200 Puzzling Physics Problems* (Cambridge University Press, 2001).

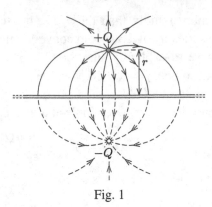

Fig. 1

In case *b*), the potential on the plates can be made zero if the charged rod is reflected in either metal plate, and the 'image rods' are then also reflected in the plates. The process is repeated, theoretically for ever, but if, at some stage, all the new images coincide in position with previous ones, the procedure is halted, and we have an appropriate set of image rods. The charge always changes sign at each reflection.

As noted earlier, this method can be applied only if, after the multi-reflections, *none* of the image charges lie in the segment occupied by real charges (in our case, between the metal plates, where the charged rod is located). This condition is fulfilled if $2n\theta = 360°$, where n is a positive integer (*see* Fig. 2).

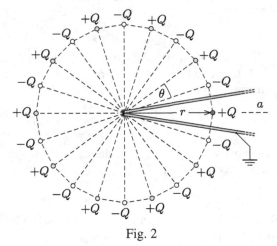

Fig. 2

The electric field produced by a rod of length L, and carrying uniformly distributed charge Q, can be calculated as a function of the distance $x \ll L$ from the axis of the rod, using Gauss's law. The electric flux passing through the lateral area of a right-circular cylinder with base radius x and length L, and having its axis coincident with the rod, is $E \cdot 2\pi x L$. This must be equal to Q/ε_0, and so the magnitude of the field is

$$E = \frac{Q}{2\pi\varepsilon_0 Lx}.$$

In case *a*), the electric field of the image rod, at the position of the real rod, can be found by setting $x = 2r$, and so the magnitude of the force acting on the charged rod is

$$F = \frac{Q^2}{4\pi\varepsilon_0 Lr},$$

and its direction is towards the image charge, i.e. towards the metal plate. (In reality, of course, this force on the rod is due to the electrostatically induced charges on the earthed metal plate.)

In case *b*) (when viewed parallel to the rod), the real rod and the image rods define a regular 20-sided polygon. The force on the original (real) rod is the net force exerted by all 19 image rods, which acts – because of the symmetry of the arrangement – perpendicularly to the rod, and in the direction of the bisecting line (axis) between the plates. So, it is sufficient to calculate the (*a*-directed, *see* Fig. 2) components of the forces exerted by the image charges that point along that axis, and then, finally, to sum them.

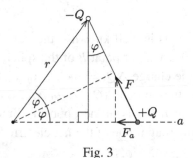

Fig. 3

Consider the image rod shown in Fig. 3, which subtends an angle 2φ at the intersection line (axis) of the plates, measured from the azimuthal direction (marked *a*) of the original (real) rod. The distance between this image rod and the original one is $x = 2r\sin\varphi$, and so the magnitude of the force acting between them is

$$F(\varphi) = \frac{Q^2}{2\pi\varepsilon_0 Lx} = \frac{Q^2}{4\pi\varepsilon_0 Lr\sin\varphi}.$$

Its direction is attractive or repulsive, depending on the sign of the image charge. The component of this force parallel to the bisector *a* of the metal plates is

$$F_a = F(\varphi)\sin\varphi = \pm\frac{Q^2}{4\pi\varepsilon_0 Lr}.$$

The magnitude of this component *does not depend* on the angle φ. Only its direction changes, according to the sign of the image charge (for unlike charges, it is directed towards the intersection line, whereas for like charges, it points in the opposite direction).

As 10 of the 19 image rods have charges $-Q$, and only nine of them have $+Q$, the net force is the same as it would be if only a single image rod were present. So, the magnitude and direction of the force acting on the original rod in case *b*) is equal to that calculated in case *a*)!

> *Note.* It can be seen that the same result is obtained for any arbitrary angle θ satisfying $\theta = 180°/n$ (where n is a positive integer). It can be proved, using more sophisticated mathematics, that the electrostatic force acting on the charged rod is totally independent of the angle between the metal plates, i.e. even if n is not a whole number, and the method of image charges cannot be applied, the result still holds.

S148   Consider first a metal sphere of radius R and total charge Q, which (because of symmetry) is distributed uniformly on its surface, i.e. the surface charge density is

$$\sigma_0 = \frac{Q}{4\pi R^2} = \text{constant.}$$

Inside the sphere – as is well known – the net electric field is zero. This can be proved at any arbitrary inner point P of the sphere, by calculating the electric field contributions of the charge $\Delta Q_1 = \sigma_0 \Delta A_1$ on a (very small) surface piece of area ΔA_1, and that of the charge $\Delta Q_2 = \sigma_0 \Delta A_2$ on an area ΔA_2 that subtends the same solid angle at P but is in the opposite direction (*see* Fig. 1). According to Coulomb's law, the magnitude of the net electric field is

$$|E_1 + E_2| = k_e \frac{\Delta Q_1}{r_1^2} - k_e \frac{\Delta Q_2}{r_2^2} = k_e \sigma_0 \left(\frac{\Delta A_1}{r_1^2} - \frac{\Delta A_2}{r_2^2} \right) = 0.$$

In the final step we used the fact that, from geometrical considerations,

$$\frac{\Delta A_1}{\Delta A_2} = \frac{r_1^2}{r_2^2}.$$

> *Note.* To prove the above equality, it is necessary to take into consideration that the area of the small surface piece, around an arbitrary point A on the surface of the sphere, subtending a given (small) solid angle at point P, is directly proportional – in accord with the definition of solid angle – to the square of the length of line segment PA, if the surface is *perpendicular* to the direction of $\overrightarrow{PA}$. Even though in general the tangent plane of the spherical surface is not perpendicular to $\overrightarrow{PA}$, and makes a different angle with it, this angle is the same at that made with the opposite surface piece, and so accordingly the 'distortion factor' cancels out in the quotient.

The field effects of the charges on the small surface pieces opposite each other cancel out in pairs, so the net electric field of all the charges on the sphere is *zero* at any point inside the sphere.

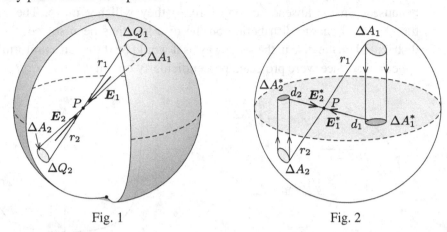

Fig. 1   Fig. 2

Now let us move on to the case of the electrically charged disc. We show that its charge distribution can be found from the uniform surface density of the charged metal sphere. Let us (notionally) 'glue into position' the charges on the surface of the sphere, and then project the points of the spherical surface perpendicularly onto a plane that contains the centre of the sphere and the point P investigated earlier (*see* Fig. 2).

As the charges cannot move, the charge ΔQ_1, originally on the surface area ΔA_1, when projected onto the plane, occupies a somewhat smaller area ΔA_1^*, and so the surface charge density (with an unchanged amount of charge) increases. In addition, its distance from the point P also changes, from the original value of r_1 to the smaller one of d_1. The corresponding thing happens with the charge ΔQ_2 originally located on the opposite side of the sphere.

What electric field is produced at P by the two charged pieces (projected onto the plane, and still glued to the disc)? The net field can be calculated again from Coulomb's law:

$$|E_1^* + E_2^*| = k_e \frac{\Delta Q_1}{d_1^2} - k_e \frac{\Delta Q_2}{d_2^2} = k_e \sigma_0 \left(\frac{\Delta A_1}{d_1^2} - \frac{\Delta A_2}{d_2^2} \right) = 0.$$

In the last step, the geometrical results

$$\frac{\Delta A_1}{\Delta A_2} = \frac{r_1^2}{r_2^2} \qquad \text{and} \qquad \frac{r_1}{r_2} = \frac{d_1}{d_2}$$

were used. The same zero-field result holds for all the other charge pairs, and so also for the whole charge distribution; the electric field is everywhere *zero* on the plane of the disc.

We have obtained the interesting result that the electric field produced by all the other charges does not produce any force on the charge fixed at P, for *any* P. Consequently, we can even abolish the notional fixing (gluing) of the charges, because – in the absence of any force – they will not move. The question of the electrical charge distribution on the disc has thus been solved: the charge is distributed on the disc in the same way as it would be if that on a uniformly charged spherical surface were projected perpendicularly onto it.

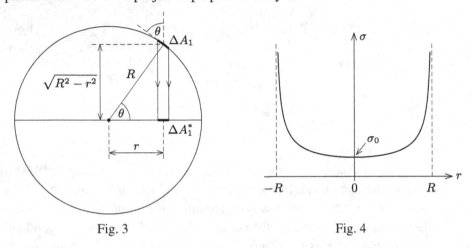

Fig. 3   Fig. 4

The surface charge density can be found with the help of a 'side view' sketch of a sphere of radius R, in which the particular surface pieces appear as line segments (*see* Fig. 3). The charges, originally situated on the area ΔA, are, after projection, at a distance r from the centre of the disc of radius R, and occupy an area

$$\Delta A^* = \Delta A \sin\theta = \frac{\sqrt{R^2 - r^2}}{R}\Delta A$$

at their new position. So the charge density on the disc at a distance r from its centre is

$$\sigma(r) = \frac{\Delta Q}{\Delta A^*} = \sigma_0 \frac{\Delta A}{\Delta A^*} = \frac{Q}{4\pi R^2}\frac{R}{\sqrt{R^2 - r^2}}$$

(*see* Fig. 4). As expected, the charge density increases away from the centre of the disc, and at its rim becomes 'infinitely large'. But this infinity should not be taken seriously, because, when r approaches the radius R of the disc, the disc's thickness (neglected until now), as well as edge effects, become important, and the above calculation method fails.

Attention! The derived formula represents the charge density on one of the sides (say, the upper side) of the metal disc, as it was derived from the 'projection' of the charge $Q/2$ located on the upper hemisphere. The other (lower) side of the disc also carries charge $Q/2$, with the same distribution as on the upper side.

Note. An interesting question might present itself. 'What happens if we project the charge on a uniformly charged sphere onto one of its diameters, i.e. onto a straight-line segment?' Superficially, the force pairs cancel each other out as before, and so we might conclude that, on a very thin metal filament (e.g. on a knitting needle), the linear charge density is *uniform* (since the area $2\pi Rh$ of an azimuthal strip is proportional to the height h of the segment of a sphere, of radius R, that it covers). But this certainly cannot be true, as, for instance, at the trisection point of the needle, the force produced by the charges on the two-thirds section of it would be larger than that due to those on the remaining one-third of the needle's length.

Where is the mistake? That is left to the reader!

S149 Let us denote the radii of discs A and B by R, the distance between them by d ($d \ll R$), and the radius of the third disc C, between A and B, by $R^* = \lambda R$. Here λ is a dimensionless quantity. We will also assume that q is positive ($q > 0$).

We can find the force acting on one of the discs (for example, on the left-hand disc A) if we can calculate the charge distribution on the surface of that particular disc. When the surface charge density is σ at any given point, then very close to that point the outside electric field is σ/ε_0, while inside the metal disc it is zero. So the average field at the surface (where the surface charges reside) is $\sigma/2\varepsilon_0$, and the outward force per unit area (the 'negative or inner pressure') is $p = \sigma^2/2\varepsilon_0$. The total force can be found by integrating these stresses across the whole surface of the disc.

The crux of the solution is to consider the system of three discs (as viewed from outside) as a single metal disc of radius R^* carrying a charge of $2q$. This is a good approximation because the distances between the discs are very small compared to their radii. Of course, there is electrostatic induction, which causes charge separation on all three discs, and so, for example, the central part of the middle disc is negatively charged, whereas the ring around this centre is positively charged. In the case of the outer (and smaller) discs, some of the positive charges on their inner surfaces are face-to-face with the middle disc, and so the charges on their outer surfaces are smaller than q. From a distant point of view, the system looks like a single disc that is positively charged, and the two parallel-plate capacitors inside the system are not apparent.

How do we find the surface charge distribution on a thin metal disc of radius R^* carrying a charge $2q$? This was exactly the task that was solved in the solution on page 362. Using the result given there, the surface charge density on one side of our 'single disc' is

$$\sigma(r) = \frac{q}{2\pi R^*} \frac{1}{\sqrt{R^{*2} - r^2}},$$

where r is the distance from the centre of the disc.

Now, considering one half of the whole arrangement (say, the left-hand side), let us denote the charge on the outer surface of disc A by q_1. This means that on the outer ring of the middle disc C the charge is $q - q_1 \equiv q_2$. On the inner surface of disc A the charge is also $+q_2$, and so the central part (a circular plate with radius R) of disc C will carry a charge $-q_2$. This part of C and the whole of A together act as a simple parallel-plate capacitor with a homogeneous field and hence with a homogeneous surface charge density (which we denote by σ^*). It is obvious that this surface charge density is

$$\sigma^* = \frac{q_2}{\pi R^2}.$$

Let us now calculate the charges involved:

$$q_1 = \int_0^R \sigma(r)2\pi r\,dr = \frac{q}{R^*}\int_0^R \frac{r\,dr}{\sqrt{R^{*2} - r^2}} = q\left[1 - \sqrt{1 - \frac{R^2}{R^{*2}}}\,\right]$$

$$= q\left[1 - \sqrt{1 - \frac{1}{\lambda^2}}\,\right],$$

$$q_2 = q - q_1 = q\sqrt{1 - \frac{1}{\lambda^2}},$$

where we have used the notation $R^* = \lambda R$.

The force F acting on disc A can be calculated using the surface charge densities σ and σ^*, where

$$\sigma^* = \frac{q_2}{\pi R^2} = \frac{q\sqrt{1 - \lambda^{-2}}}{\pi R^2},$$

and the positive direction points outwards,

$$F = \frac{1}{2\varepsilon_0}\int_0^R (\sigma^2 - \sigma^{*2})2\pi r\,dr = \frac{q^2}{4\pi\varepsilon_0 R^2}\left[\frac{1}{2\lambda^2}\ln\frac{\lambda^2}{\lambda^2 - 1} - 2\left(1 - \frac{1}{\lambda^2}\right)\right].$$

This force is zero if the term in the large square brackets vanishes. Using graphical or numerical methods, we find that the force is zero if

$$\lambda = \frac{R^*}{R} = 1.1584 \approx 1.16.$$

This means that the net force acting on disc A (or B) is zero if the middle disc C is about 16 % larger in diameter than discs A and B.

> *Notes.* 1. If the middle disc C is very large (so λ approaches infinity) then the net force exerted on A and B becomes
>
> $$F = -\frac{q^2}{2\pi\varepsilon_0 R^2} \qquad \text{as} \qquad \lambda \to \infty.$$

This result is reassuring because it is the well-known formula for the attractive force between the plates of a charged circular parallel-plate capacitor.

2. If λ approaches 1, the second (attractive) term of the force formula approaches zero. This means that there are no induced electric charges inside the discs (as might be expected in this case). But the first (repulsive) term of the force formula becomes divergent and the force approaches infinity, showing that fringing effects cannot be neglected in this case. If we want to find a good estimate for the magnitude of the force in these circumstances, we have to introduce a cut-off for the integration range at $R - d$, where d is the distance between the plates:

$$F \approx \frac{q^2}{8\pi\varepsilon_0 R^2} \ln\frac{R}{2d} \qquad \text{if} \qquad \lambda \to 1.$$

3. One can see clearly from the formula for F that, if the middle disc C is small, then the net forces acting on discs A and B are repulsive and also that, if the middle disc C is very large, then the net forces are attractive.

S150   *a*) The charge $+Q$ on the (left-hand) solid spherical cap induces electrostatic charge separation in the neutral (right-hand) spherical dome. The insulated cut between the two parts of the sphere is actually a parallel-plate capacitor with negligible plate separation. If we denote the charges on this 'capacitor' by q and $-q$, the overall charge distribution on the sphere is as shown in Fig. 1.

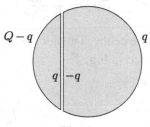

Fig. 1

Of course, because both parts of the sphere are made from metal, all the charges shown are surface charges. It should be noted that the total charge on the outer surface of the sphere is $+Q$, independent of the value of q.

Now consider the energy of the system. Because of the negligible separation between the left cap and the right dome, the capacitance of the 'capacitor' is very large, and the energy stored in it can be neglected. The same is true of the net interaction energy between the charge $+Q$ on the outer surface of the sphere and the charges $\pm q$ on either side of the 'gap'. Since we are seeking the equilibrium charge state by minimising the total electrostatic energy of the system, the question comes down to finding the minimal energy of a conducting sphere when the total amount of charge on its surface is $+Q$. The answer is simple: the minimal energy corresponds to a *uniform* surface charge density – just the same as for a truly solid conducting sphere.

This means that the charge on the cap's surface area of πR^2 (one-quarter of the total surface area) is $Q/4$, and that on the dome's outer surface area is $3Q/4$. Since the total charge on the (neutral) larger part of the sphere must be zero, we have $3Q/4 - q = 0$, and so $q = 3Q/4$, in accord with Fig. 1 and the conclusion reached in the previous sentence.

We can also reach the same conclusion in a different way. In electrostatics, solid metals are always equipotentials, and inside them the electric field is always zero. As the gap between the two parts of the sphere is very small, and the electric field within the gap is finite, the potential difference between the two parts is zero, and the whole sphere is an equipotential! The condition $V = $ constant on a sphere carrying a given amount of charge Q can be fulfilled in only one way, and that is with a (spherically symmetric) *uniform* surface charge density σ. That density is

$$\sigma = \frac{Q}{4\pi R^2} = \text{constant.}$$

The uniform surface charge distribution results in a zero electric field everywhere inside the sphere (except for the volume within the 'parallel-plate capacitor'). Outside, it closely approximates the well-known Coulomb field with magnitude

$$E(r) = \frac{1}{4\pi\varepsilon_0}\frac{Q}{r^2} \qquad (r > R)$$

at a distance r from the centre. Outside the sphere, the electric field lines are very close to those illustrated in Fig. 2.

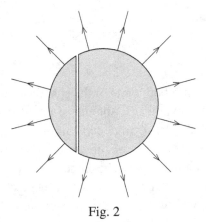

Fig. 2

b) The electrostatic interaction force between the cap and the dome has two components: the attraction between the two 'plates of the capacitor', and the repulsion between the like (positive) charges on the two parts of the sphere's outer surface.

Denote the height of the cap by h (in the actual problem $h = R/2$, since the curved surface area of the cap $2\pi Rh$ must be equal to πR^2). The attractive force component between the 'plates of the capacitor' is

$$F_{attr} = \frac{1}{2}qE = \frac{q^2}{2\varepsilon_0 A},$$

where q is the charge on, and $A = \pi r^2$ is the area of, the capacitor plate; here r is the radius of the spherical cap's base and is given by $r = \sqrt{2Rh - h^2}$.

The amount of charge on the 'capacitor' is

$$q = \sigma(4\pi R^2 - 2\pi Rh) = \frac{Q}{4\pi R^2}(4\pi R^2 - 2\pi Rh) = Q\left(1 - \frac{h}{2R}\right),$$

and so the attractive force component is

$$F_{attr} = \frac{Q^2(2R - h)}{8\pi\varepsilon_0 R^2 h}.$$

The surface charge density σ on the spherical cap is

$$\sigma = \frac{Q}{4\pi R^2},$$

and the amount of charge on an area ΔA is $\Delta Q = \sigma\Delta A$. The electric field strength, which is E outside the sphere, and zero inside, has an average value of $E/2$, and so exerts a force on this charge of magnitude $E\Delta Q/2$, directed along the outward normal to ΔA.

Since the force acting on surface area ΔA is proportional to ΔA, the quantity

$$p = \frac{E\Delta Q}{2\Delta A} = \frac{\sigma E}{2} = \frac{Q}{4\pi R^2}\frac{Q}{8\pi\varepsilon_0 R^2} = \frac{Q^2}{32\pi^2\varepsilon_0 R^4}$$

has the characteristics of a pressure. So far as the cap is concerned, it is as if it were being pushed horizontally to the left by a gas at pressure p. If a closed vessel with thin walls containing gas at pressure p were formed from the surface of the cap and a flat circular closing plate of radius r, then the net pressure forces on the cap and on the plate must balance – or the vessel would move of its own accord! It follows that the magnitude of the net force acting on the cap is

$$F = p\pi r^2.$$

Note. This same result can be found by summing the horizontal components of the forces acting on the various surface pieces. The net vertical component must be zero because of the azimuthal symmetry:

$$F = \sum p\Delta A\cos\theta = p\sum\Delta A\cos\theta,$$

where θ is the angle the tangent plane at a surface piece makes with the vertical. As $\Delta A\cos\theta$ is the area of the vertical projection of the surface piece, the sum of these vertical projections is simply the projected area of the whole spherical cap, i.e. it is equal to the area πr^2 of the cap's base, which has radius r.

Accordingly, the repulsive force component is

$$F_{rep} = p \, \pi r^2 = \frac{Q^2}{32\pi^2 \varepsilon_0 R^4} \pi(2Rh - h^2) = \frac{Q^2(2R - h)h}{32\pi\varepsilon_0 R^4}.$$

Since physically $0 < h < 2R$, it seems that the attractive force is always greater than the repulsive one. In the limiting case ($h = 0$, implying that r is also zero), the attractive force would appear to be infinitely large, and the repulsive force zero, but this situation is unphysical because it is impossible to concentrate a finite amount of charge into an infinitesimally small volume. In the other limiting case ($h = 2R$), both forces are zero because we would have only a uniformly charged 'whole' sphere. So, in fact, the net force is always attractive and equal to

$$F_{net} = F_{attr} - F_{rep} = \frac{Q^2(2R - h)(4R^2 - h^2)}{32\pi\varepsilon_0 R^4 h}.$$

In the actual problem, $h = R/2$, and so the net attractive force is

$$F_{net}^{(h=R/2)} = \frac{45Q^2}{128\pi\varepsilon_0 R^2}.$$

S151 When the ring is very far from the metal sphere, the electric potential at its centre is

$$V_0 = \frac{1}{4\pi\varepsilon_0} \frac{Q}{R},$$

where Q is the total charge on the ring.

If the ring is placed above the metal sphere in the way described in the problem, then the value of the potential at its centre must become zero, since the centre of the ring is just at the top of the grounded metal sphere. The charges on the ring cause electrostatic induction in the metal sphere, and an unknown charge distribution forms on its surface. These surface charges on the sphere, together with those on the ring, produce zero potential at the centre of the ring. If we knew this non-uniform surface charge density, we could calculate the total charge on the sphere – but finding it requires sophisticated analysis. Fortunately, we can answer the question posed without having to find the distribution!

We search for a point inside the sphere at which the contributions to the potential from the charges on both the ring and the sphere's surface can be calculated easily, i.e. without knowledge of the precise surface charge distribution. The centre of the sphere is just such a point, because all parts of the spherical surface are the same distance r from it.

If the total charge induced on the sphere, is $-q$ (the negative sign recognises that the charge on the sphere must have the opposite sign to that on the ring), then the potential due to the surface charges is

$$V_{\text{sphere}} = -\frac{1}{4\pi\varepsilon_0}\frac{q}{r}$$

at the centre of the sphere.

The centre of the sphere is at a distance $\sqrt{r^2 + R^2}$ from every point of the ring, and so the contribution of the charged ring to the potential at the centre of the sphere is

$$V_{\text{ring}} = \frac{1}{4\pi\varepsilon_0}\frac{Q}{\sqrt{r^2 + R^2}}.$$

As each point of the grounded metal sphere (including its centre) has zero potential, it follows that

$$V_{\text{sphere}} + V_{\text{ring}} = 0,$$

and, on substituting for the two terms, the induced charge on the sphere is obtained as

$$-q = -4\pi\varepsilon_0 V_0\frac{rR}{\sqrt{r^2 + R^2}}.$$

S152 According to the principle of superposition, if charges are in equilibrium in an arrangement, and the sizes of all the charges are increased proportionally (say λ times), then the equilibrium is maintained in the new arrangement, with the values of the electric fields and potentials λ times larger than in the original arrangement. Moreover, it is also the case that superimposing two equilibrium arrangements leads to a third one in which the net charge arrangement will again be in equilibrium, and in which the field strengths and the potentials at any particular point are the vector and scalar sums (respectively) of the original values.

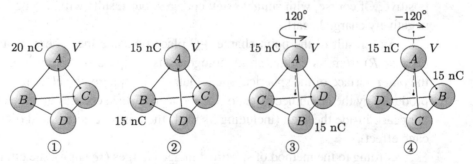

Let us call the first charge arrangement (when only one sphere is charged) simply 1, and the second (when spheres A and B are charged) 2 (*see* figure). Rotate arrangement 2 by 120° around the line joining the centre of A to the centre of the tetrahedron, and denote the resulting configuration by 3. Similarly produce arrangement 4 by rotating 2 by $-120°$.

If we are to achieve our aim of making the potential at A equal to V using three equally charged spheres, say those in the plane originally occupied by A, B and D, then clearly we do not need any contribution from 4. Equally clearly, the contributions from 2 and 3 must be equal. Moreover, the total charge on A must be made the same as the (equal) charges on the other two spheres in the plane. Finally, we must ensure that the superimposed potentials at A add up to the original value of V.

With these considerations in mind, superimpose λ_1 times state 1, and λ_2 times states 2 and 3. The required conditions on the charge on A, and the potential of A, are satisfied if

$$20\lambda_1 + 15\lambda_2 + 15\lambda_2 = 15\lambda_2$$

and

$$\lambda_1 V + \lambda_2 V + \lambda_2 V = V.$$

The solution to these simultaneous equations is

$$\lambda_1 = -\frac{3}{5} \quad \text{and} \quad \lambda_2 = \frac{4}{5},$$

and, accordingly, each of the three superimposed charged spheres carries a charge of $15 \times (4/5) = 12$ nC.

Similarly, with an appropriate superposition of arrangement 1, and states 2, 3 and 4, it can be arranged that all four spheres have the same charge, and that the potential of sphere A is V. In this case, the charge on the spheres turns out to be 10 nC, but the detailed calculation of this is left as a task for the reader.

S153 For the sake of definiteness, let us suppose that the charge on the pearl is positive. Of course, with suitable sign changes, our results will also be valid for a negatively charged pearl.

a) As a result of the point charge $+Q$ placed outside the sphere, at a distance d $(d > R)$ from its centre, an inhomogeneous charge distribution is induced on the *outer* surface of the spherical shell. Outside the sphere, the electric field is that produced by the combined effect of the induced surface charges and the charge on the pearl. Inside the shell (including its wall) there is no electric field (the Faraday cage effect).

According to the method of spherical image charges (*see* hint), the effects of the surface charges on the shell can be replaced by those due to a single point charge $-q$, provided its magnitude and position are chosen appropriately. One condition that must be satisfied is that the substitute charge must *not* be placed in the same region of space as the real charges present.

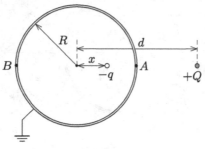

Fig. 1

We use the notation of Fig. 1, and let the distance between the image charge $-q$, which *must* be positioned *inside* the shell, and the centre of the spherical shell be x. As the sphere is grounded, the value of the potential at any point on its surface, and in particular at points A and B, is zero. So, we must have

$$k_e \frac{Q}{d-R} - k_e \frac{q}{R-x} = 0 \quad \text{and} \quad k_e \frac{Q}{d+R} - k_e \frac{q}{R+x} = 0.$$

From these equations, we get the following expressions for the magnitude of the image charge, and the distance x:

$$-q = -\frac{R}{d}Q \quad \text{and} \quad x = \frac{R^2}{d}.$$

As the force acting on the pearl does not depend upon whether the electric field, at the position of the pearl, is produced by the surface charges on the spherical shell, or by the imaginary image charge, that force is given by

$$F_a = -k_e \frac{Qq}{(d-x)^2} = -k_e Q^2 \frac{Rd}{(d^2-R^2)^2}.$$

The force is always attractive, and it approaches infinity if $d \to R$. The electric field lines outside the sphere are illustrated in Fig. 2.

b) If the spherical shell is uncharged and not grounded, then its potential is not zero, but it is still an equipotential surface. What needs to be done is to modify our answer to part *a*) in such a way that the sphere remains an equipotential surface, but the net charge it encloses becomes zero. The obvious way to do this is to add a second image charge, of magnitude $+q$, at the centre of the sphere (an allowed region for an image charge). In this way both conditions – being an equipotential and being uncharged – are satisfied.

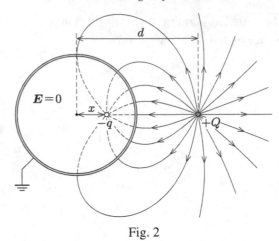

Fig. 2

The electric field inside the sphere is still zero, but outside, the field is the same as if it were produced by three point charges: the pearl carrying charge $+Q$, an image charge $-q$ positioned as in part *a*), and an image charge $+q$ at the sphere's centre. The force acting on the pearl can be found by adding to the force calculated in part *a*) the repulsive force due to the notional charge $+q$ at the centre of the sphere:

$$F_b = F_a + k_e \frac{Qq}{d^2} = -k_e Q^2 \frac{R}{d^3} \left[\frac{d^4}{(d^2 - R^2)^2} - 1 \right] = -k_e Q^2 \left(\frac{R}{d} \right)^3 \frac{2d^2 - R^2}{(d^2 - R^2)^2}.$$

This expression is always negative, since $d > R$, so this force is always attractive, as it was in the case of the earthed sphere in part *a*).

c) When the sphere carries (on its surface) a real charge Q', the force on the pearl can be found using an almost identical approach to that used in case *b*) (*see* Fig. 3); we have to imagine a charge of $Q' + q$ at the centre of the sphere! Inside the sphere the electric field is still zero, but outside, the field is the same as if it were produced by three point charges: the pearl carrying charge $+Q$, an image charge $-q$ positioned as in part *a*), and an image charge $Q' + q$ at the sphere's centre.

Now the force acting on the pearl is

$$F_c = F_b + k_e \frac{QQ'}{d^2} = -k_e Q^2 \frac{R}{d^3} \left[\frac{d^4}{(d^2 - R^2)^2} - 1 - \frac{Q'}{Q} \frac{d}{R} \right].$$

It is interesting to note that, even if Q and Q' have the same signs, F_c can still be negative – and the interaction *attractive* – provided d/R is not too large, i.e. if the pearl is sufficiently close to the surface of the sphere.

It still remains to investigate how our results in parts *a*), *b*) and *c*) change if the pearl is not outside, but inside, the spherical shell. With the pearl inside the sphere, the image charge is outside, but its magnitude and position are still described by the same formulae (though now $d < R$):

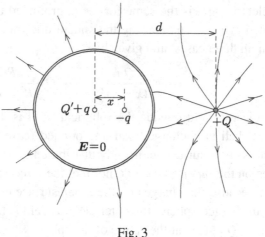

Fig. 3

$$-q' = -\frac{R}{d}Q \quad \text{and} \quad x = \frac{R^2}{d}.$$

If the metal shell is grounded, then, inside it, the electric field is just the same as the combined field produced by the pearl and the image charge of $-q'$ (which is now *outside* the shell). The electric field lines, originating from the point charge $+Q$ carried by the pearl, terminate on charge $-Q$ situated on the *inner* surface of the spherical shell; this is the only way that the electric field can be zero inside the shell wall. Outside the sphere there is no electric field, the charges on the inner surface of the shell totally shield the field due to the pearl (if this were not the case, the potential of the spherical shell could not be zero). The electric field of the grounded shell (case a)) is illustrated on the left-hand side of Fig. 4.

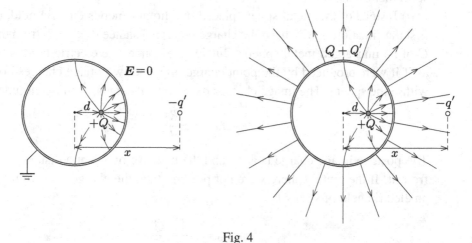

Fig. 4

The charge distribution formed on the inner surface of the spherical shell is independent of whether or not the spherical shell is grounded, and so inside the shell the electric field is the same for the ungrounded uncharged case as it is for the grounded one – even charging the sphere does not alter this. So the force is identical in all three cases, and given by

$$F = -k_e \frac{Qq'}{(x-d)^2} = -k_e Q^2 \frac{Rd}{(R^2-d^2)^2}.$$

But the electric field outside the spherical shell is different in all three cases. If the metal shell is uncharged and ungrounded (case b)), then a charge of $+Q$ appears on its *outer* surface, uniformly distributed. The explanation for this is that the charges on the *outer* surface do not feel the effect of the charged pearl inside the sphere, because the charges on the *inner* surface completely shield it. In this case of an uncharged sphere, the outer electric field is as if it were produced by a point charge $+Q$ placed at the centre of the sphere.

When the spherical shell has a net charge of Q' (case c)), then the amount of charge on its *outer* surface is $Q' + Q$, distributed evenly, and the field outside the sphere is identical to the electric field of a point charge $Q' + Q$, as is shown on the right-hand side of Fig. 4.

> *Note.* The electric charge distribution on the inner and outer surfaces of the spherical metal shell can be calculated using Gauss's law. The magnitude of the *real* surface charge density is proportional to the *real* electric field strength ($\sigma = \varepsilon_0 E$) obtained by superimposing the fields of the real and image charges.

S154   In the solution on page 373, among other things, the interaction of an uncharged spherical metal shell and a point charge is described. As, inside metal bodies, there is no electric field, the results from that solution are also valid for solid metal spheres.

a) Instead of the metal sphere placed in a homogeneous electric field, consider first the situation in which a point charge $+Q$ is a distance d ($d > R$) from the centre C of an uncharged metal sphere. Outside the sphere the electric field is the same as if it were produced by the point charge $+Q$, and two image charges positioned within the sphere. The image charges have opposite signs, with magnitudes

$$\pm q = \pm \frac{R}{d} Q. \tag{1}$$

The positive image charge is at C, and the negative one is at a distance $x = R^2/d$ from it. If the metal sphere were not present, then the charge $+Q$ would produce an electric field of strength

$$E_0 = \frac{1}{4\pi\varepsilon_0} \frac{Q}{d^2} \tag{2}$$

at the position of the centre of the sphere (*see* Fig. 1).

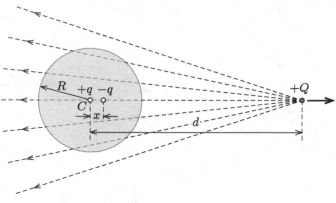

Fig. 1

Let us notionally move the real point charge away from the metal sphere, but in such a way that the electric field it produces at C remains constant, with strength E_0. To bring this about, it is necessary to increase the amount of charge Q with the distance d according to the following prescription (derived from equation (2)):

$$Q(d) = 4\pi\varepsilon_0 E_0 d^2.$$

As the distance d is increased, the structure of the electric field, produced at C by the real point charge, becomes more and more like a homogeneous electric field.

During the distancing of the charge, the absolute value of the two image charges increases in accord with equation (1):

$$q(d) = 4\pi\varepsilon_0 E_0 Rd, \tag{3}$$

while their separation x decreases inversely with d. However, their product qx (the electric dipole moment of the two image charges) remains constant at

$$p = q(d)x = 4\pi\varepsilon_0 E_0 R^3. \tag{4}$$

In the limiting case, in which $d \to \infty$, the electric field outside the sphere can be found by superimposing on a homogeneous field E_0 the field of an electric dipole of moment p placed at C. Of course, inside the sphere the electric field is still zero, showing that the surface charges on the metal sphere produce an electric field of $-E_0$ inside the sphere (*see* Fig. 2).

b) We have to find a surface charge distribution that produces the electric field shown in the first sketch of Fig. 2: that of a dipole field outside the sphere, and a homogeneous field of constant strength E_0 inside.

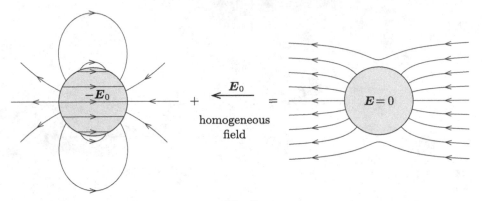

Fig. 2

Let us go back to the case in which the uncharged metal sphere is placed, not in a homogeneous field, but in the vicinity of a point charge $+Q$. Up until now, the external effect of the sphere's surface charges has been described using two point-like image charges with magnitudes $+q$ and $-q$, which are imagined to be inside the sphere (*see* Fig. 1). This method also works if the image charges are imagined, not as point-like objects, but as two extended spheres, each carrying a homogeneous volume charge distribution, and total charges $+q$ and $-q$. Moreover, there is no additional problem if the two image spheres overlap.

So, for the two point image charges, we substitute two spheres of radius R with their centres a distance x apart and carrying total charges of $+q$ and $-q$, uniformly distributed throughout their volumes. The charge density of each is

$$\varrho = \frac{q}{(4\pi/3)R^3} = \frac{3\varepsilon_0 E_0 d}{R^2} = \frac{3\varepsilon_0 E_0}{x}, \tag{5}$$

where we have used equation (3) and the fact that $xd = R^2$.

The field of the 'dilated' charge spheres must be the same in the region outside the spheres as the field produced by the point charges. But what about in the region where the charged spheres overlap? Here the net charge density is zero, but, even so, there is a homogeneous non-zero electric field (as we might have guessed from Solution 1 that appears on page 143). To prove this, consider Fig. 3.

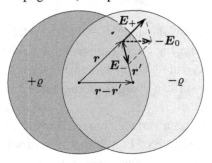

Fig. 3

If only the sphere with charge density $+\varrho$ were present, then the field strength at a point inside it, displaced by r from its centre, can be calculated using Gauss's law:

$$4\pi r^2 E_+ = \frac{1}{\varepsilon_0}\frac{4\pi r^3}{3}\varrho\frac{r}{r},$$

from which

$$E_+ = \frac{\varrho}{3\varepsilon_0}r.$$

Similarly, if only a negatively charged sphere were present, then, inside it, the field strength would be

$$E_- = -\frac{\varrho}{3\varepsilon_0}r',$$

where r' is the position vector from the centre of that sphere to the given point. As both spheres are present, the field in the overlapping region is

$$E_+ + E_- = \frac{\varrho}{3\varepsilon_0}(r - r') = E_0\frac{d}{R^2}(r - r') = -E_0,$$

where (5) and the fact that $|r - r'| = x = R^2/d$ have been used.

So, in the overlapping region, the electric field is in fact homogeneous, and its strength is just E_0.

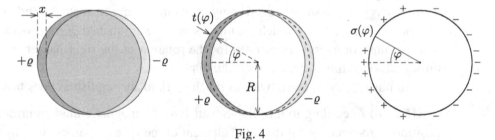

Fig. 4

When the charge $+Q$ is moved away from the sphere – in the way described in part *a*) – the situation approaches that of a metal sphere placed in a homogeneous field. At the same time, the distance between the centres of the 'dilated' charged spheres gets smaller, and their charge densities increase according to equation (5) (and approach infinity), whereas the thickness of their charge layers decreases (and approaches zero). In the limiting case, the volume charge distribution turns into a surface charge distribution. The amount of charge per unit area can be calculated as the product of the volume charge density and the thickness t of the non-overlapping parts of the spheres (*see* Fig. 4).

So, in the limiting case of the metal sphere placed in a homogeneous field, the charge density on a part of the surface whose 'polar' direction with respect to the external electric field is φ is

$$\sigma(\varphi) = \varrho t(\varphi) = \varrho x \cos\varphi = 3\varepsilon_0 E_0 \cos\varphi.$$

Notes. 1. The surface charge density may also be found from the formula $\sigma = \varepsilon_0 E$, where E is the signed magnitude of the electric field just outside the metal sphere.

2. Using similar methods to those employed here, it is also possible to find the electric field and charge distribution of a long, neutral, metal cylinder placed perpendicularly to a homogeneous electric field.

S155   We first note that, in either case, the work done increases the magnetic interaction energy.

Consider the following thought experiment: the right-hand magnet is removed very far from the left one along the straight line connecting the magnets; the work done against the *attraction* between the two magnets is, say, one unit. This is the work needed to perform task *b*).

Next, the distant right-hand magnet is rotated through $180°$; no work is needed for this because now the interaction between the magnets is negligible. Finally, the rotated magnet is brought back to its original place; on symmetry grounds, the work done against the mutual *repulsion* while doing this is again one unit. This completes task *a*), and all told the work that was required was two units.

The magnetostatic field is conservative, i.e. any change in the magnetic interaction energy of a system depends only on its initial and final states, and does not depend on the process by which the final situation was reached. It follows from this that two units of energy are needed for the rotation of the right-hand bar magnet through $180°$, if this is done at its original position.

In summary, task *a*) takes twice as much work to accomplish as does task *b*).

S156   *a*) According to the Biot–Savart law, the magnetic field contribution at position r produced by a differential element of the wire – characterised by vector $\Delta\ell$ in the direction of the conventional current I – is

$$\Delta B = \frac{\mu_0 I}{4\pi} \frac{\Delta\ell \times r}{|r|^3},$$

where r is defined relative to the position of the wire element. The magnetic field of a complete circuit can be found in simple (symmetrical) cases by summing these ΔB elementary field contributions, or, in more general cases, by evaluating a line integral of such terms.

At the point P_1, which lies on the loop's vertical axis of symmetry and at a distance L from its centre, symmetry considerations dictate that the net magnetic field vector is directed along that axis (parallel to the dipole moment $\boldsymbol{m}$ of the current loop). Consequently, it is sufficient to calculate the axial components of the $\Delta\boldsymbol{B}$ contributions, and then sum them (*see* Fig. 1).

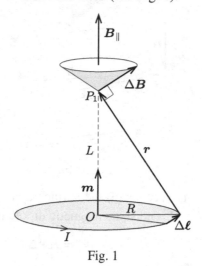

Fig. 1

Using the notation of the figure:

$$B_{\parallel} = \sum \frac{\mu_0 I}{4\pi} \frac{\Delta\ell}{r^2} \frac{R}{r} = \frac{\mu_0}{4\pi} \frac{IR}{r^3} \sum \Delta\ell = \frac{\mu_0}{4\pi} \frac{2}{r^3} \underbrace{(\pi R^2 I)}_{|\boldsymbol{m}|}.$$

The factor R/r in the second expression represents the cosine of the angle between $\Delta\boldsymbol{B}$ and the symmetry axis that arises when the former is resolved in the direction of the latter. The final part of the last term in the parentheses is just the magnitude of the magnetic dipole moment $\boldsymbol{m}$ of the current loop. So the value of the magnetic field strength at point P_1 is

$$B_{\parallel} = \frac{\mu_0}{4\pi} \frac{2m}{\left(R^2 + L^2\right)^{3/2}} \approx \frac{\mu_0}{2\pi} \frac{m}{L^3},$$

and using $|\boldsymbol{m}| = \pi R^2 I$, its magnitude is

$$B_{\parallel} = \frac{\mu_0 I R^2}{2L^3}.$$

b) It is a more difficult task to determine the magnetic field vector $\boldsymbol{B}_{\perp}$ in the plane of the circular current loop.[68] To do this, we replace the circular loop with a circular ring sector of the same surface area (and dipole moment), as shown in Fig. 2, and

---

[68] The $\perp$ subscript indicates that the magnetic field being investigated is in a *direction* perpendicular to the direction of the magnetic dipole moment $\boldsymbol{m}$ – not that the field is necessarily perpendicular to $\boldsymbol{m}$.

calculate its magnetic field at the centre P_2 of the sector. In this substitute wire-frame loop, current I flows along arcs with radii L and $L + \Delta L$ ($\Delta L \ll L$), and along straight-line segments with a small angle ε between them.

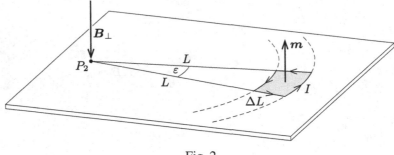

Fig. 2

As the angle ε is very small, the surface area of the wire frame is, to a good approximation, $L\varepsilon \Delta L$, and so its magnetic dipole moment is

$$|m| = IL\varepsilon \Delta L.$$

The magnetic field vector at the centre P_2 of the wire-frame arcs is perpendicular to the plane of the current loop, and its direction is opposite to that of the dipole moment m. Only the currents flowing in the curved arcs produce contributions to it:

$$B_\perp = \frac{\mu_0}{4\pi} \frac{I\varepsilon L}{L^2} - \frac{\mu_0}{4\pi} \frac{I\varepsilon(L+\Delta L)}{(L+\Delta L)^2} \approx \frac{\mu_0}{4\pi} \frac{1}{L^3} \underbrace{(IL\varepsilon \Delta L)}_{|m|}.$$

The last expression (in parentheses) is precisely that for the magnitude of the magnetic dipole moment, which must be the same for the original circular current loop and the substitute ring sector loop. So the value of the magnetic field at P_2 (taking into account its directions) is

$$B_\perp = -\frac{\mu_0}{4\pi} \frac{m}{L^3},$$

which can be expressed in terms of the data of the given circular current loop as

$$B_\perp = \frac{\mu_0 I R^2}{4L^3}.$$

The points P_1 and P_2 are known, respectively, as the Gaussian first and second principal positions; the field at the first position is exactly *twice* that at the second position.

> *Note.* By utilising the results for the magnetic field vectors in the two Gaussian principal positions, the magnetic field due to a current loop (or, more generally, to any magnetic dipole) can be found at an arbitrary place. For example, if we are seeking the magnetic field strength vector B at the point P, at a distance L

from the dipole, and in a direction making an angle ϑ with the dipole axis, then it is convenient to resolve vector $\boldsymbol{m}$ into two components as shown in Fig. 3 with magnitudes $m_1 = m \cos \vartheta$ and $m_2 = m \sin \vartheta$.

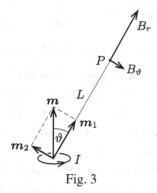

Fig. 3

Point P is then at the first principal position of m_1 and at the second principal position of m_2, and so the 'radial' and 'tangential' components of the magnetic field there are

$$B_r = \frac{\mu_0}{2\pi} m \frac{\cos \vartheta}{L^3} \quad \text{and} \quad B_\vartheta = \frac{\mu_0}{4\pi} m \frac{\sin \vartheta}{L^3}.$$

S157 *Solution 1.* Because of the symmetry of the given arrangement, points C and D are equivalent, and are therefore at the same potential; consequently no current flows between them. The resistances of the circuit elements ACB and ADB are twice that of element AB, and so the currents flowing in the individual edges of the tetrahedron are

$$I_{AC} = I_{CB} = I_{AD} = I_{DB} = \tfrac{1}{4}I, \qquad I_{AB} = \tfrac{1}{2}I.$$

According to the Biot–Savart law, the contribution to the magnetic field of a straight piece of wire of finite length is proportional to the current in the wire, and the direction of the field at some point P is perpendicular to the plane containing both the wire and the point. For example, in this problem, the contribution of edge AC to the magnetic field at the centre O of the tetrahedron can be written as

$$\boldsymbol{B}_{AC} = kI_{AC} \cdot \overrightarrow{BD},$$

where k is a positive number, whose value depends upon, among other things, the size of the tetrahedron, but not upon which edge is under consideration. Here we have used the facts that the edges of a regular tetrahedron are all equal in length and in distance from its centre, and that opposite edges are perpendicular to each other. The vector $\overrightarrow{BD}$ therefore has just the right direction for expressing both the Biot–Savart law and the right-hand rule. The contributions of the other edges can be calculated similarly, and the total magnetic field at point O is given by

$$\boldsymbol{B} = \tfrac{1}{4}kI\left(\overrightarrow{BD} + \overrightarrow{AD} + \overrightarrow{CB} + \overrightarrow{CA} + 2\overrightarrow{DC}\right).$$

We note that the incoming and outgoing long, straight currents make no contribution to the magnetic field at the centre O. The right-hand side of the above equation has zero value because

$$\overrightarrow{AD} + \overrightarrow{DC} = \overrightarrow{AC} = -\overrightarrow{CA} \qquad \text{and} \qquad \overrightarrow{BD} + \overrightarrow{DC} = \overrightarrow{BC} = -\overrightarrow{CB}.$$

We conclude that the magnetic field at the centre of the tetrahedron is zero.

Solution 2. It can be shown that, for *any* regular polyhedron (not only for a tetrahedron, but also for a cube, octahedron, dodecahedron or icosahedron), the following statement is true: the magnetic field at the polyhedron's centre is *zero* if 'centrally directed' incoming and outgoing currents are connected to *any* two vertices of the polyhedron.

To see this, consider a polyhedral arrangement of conducting wires that distributes electricity in a way reminiscent of the way a 'cut diamond' distributes light, with a single input but multiple outputs. Current I is conducted *into* one vertex (say, A) through a long, straight wire directed towards the centre O of the polyhedron, while from all the other (N, say) vertices identical currents (of I/N) are conducted away through similarly directed straight wires. Consider also a further physically identical arrangement, but one in which a current I is conducted *away from* vertex B through a long, straight, also 'centrally directed' wire, while identical currents, adding up to I, are fed into all the other N vertices. In both cases, the sum of incoming and outgoing currents is zero.

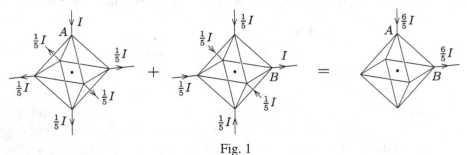

Fig. 1

The superposition of these two arrangements gives the result described pictorially in Fig. 1 for a current I' (equal to $[(N+1)/N]I$, i.e. $\frac{6}{5}I$ in the case illustrated) flowing through the circuit using just two wires. It follows that proving that the magnetic field at O is zero in one of the 'cut-diamond' arrangements implies that the same result holds for the original problem with two conducting wires.

In view of the above result, we examine the fields involved in a 'cut-diamond' arrangement of wires. Suppose that when current I is fed into vertex A (and conducted away equally through all the other vertices), a magnetic field $\boldsymbol{B}$ is generated at the centre O of the polyhedron. This vector $\boldsymbol{B}$ must be directed parallel to AO; there cannot be a net component of it perpendicular to the axis AO because every polyhedron has an *n*-fold rotational symmetry (for some *n*) around that axis. For example, if, in Fig. 1, the octahedron ($n = 4$) were rotated through 90°, an identical arrangement to the original one would be obtained and the magnetic field would necessarily be the same as the original one. This is impossible unless the component of the field perpendicular to the axis is zero.

As a final step, let us imagine that a charged particle is moving near to O in a circular orbit around axis AO. By choosing appropriate values for the angular velocity and rotational direction of the particle, the magnetic field (parallel to the axis) can be made to provide the required centripetal force for uniform circular motion.

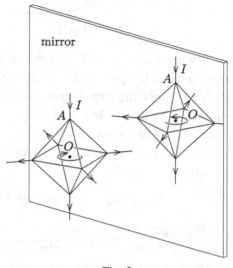

Fig. 2

Let us further imagine that we view the whole arrangement (the distributed currents, the resulting magnetic field and the circulating particle) in a mirror (*see* Fig. 2). The image of the polyhedron is the same as the original, the current distribution is also unaltered, but the direction of the particle's rotation is reversed. To maintain this reversed rotation would require a reversed magnetic field. However, the direction of a finite magnetic field cannot change; the only vector that is equal (-1) times itself is the zero vector. So the magnetic field at the polyhedron's centre O has both its perpendicular and parallel components equal to zero.

In summary, the magnetic field at the centre of a single 'cut-diamond' arrangement of a conducting wire polyhedron can only be zero, and, as shown earlier, linear superposition then means that the same is true for the 'two-wire case'.

S158 *Solution 1.* Let the unit vector e represent the direction of the current in the wire, and R be the position vector of P relative to the junction of the wire and plate (*see* Fig. 1). The magnetic field vector must be determined by these two vectors (and the given current magnitude I) as a single-valued function $B(e, R)$, i.e. the magnetic field is a vector function of two vector variables.

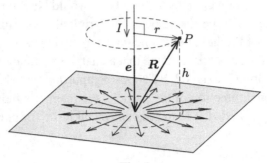

Fig. 1

To find the direction of the magnetic field, the 'right-hand (corkscrew) rule' must play a central role, and this fact greatly constrains the form of the function $B(e, R)$. Using just these two vectors, the only formulation for a magnetic field vector that is in accord with the right-hand rule is their vector product. For this reason, the magnetic field vector must be proportional to $e \times R$, and the proportionality factor must be a scalar function that does not depend upon the (arbitrary) sign convention in the right-hand rule:

$$B(e, R) = (e \times R) f(e \cdot R, R^2, I).$$

The variables in the function f that involve e and R must be scalar quantities constructed from them. They could be scalar products with each other or with themselves (but not e^2, since this equals unity, and is not a variable). For given r and h, these scalars have fixed values:

$$R^2 = r^2 + h^2 \quad \text{and} \quad e \cdot R = -h.$$

Consequently, so does the function f, independent of the azimuthal position of P.

Thus, the magnetic field vector at point P is perpendicular to the plane containing e and R, and it follows that the direction of the magnetic field is tangential to the circle shown in Fig. 1. The magnitude of the magnetic field is given by applying Ampère's law to a circuit defined by that circle:

$$B(r) \times 2\pi r = \mu_0 I \quad \longrightarrow \quad B(r) = \mu_0 \frac{I}{2\pi r}.$$

This means that the magnetic field (above the plate) is the same as the field due to an infinitely long, current-carrying, straight wire.

There is no magnetic field anywhere below the plate because, there, the current passing through any surface 'bounded' by any circle is zero; hence, by Ampère's law, so is the field.

Solution 2. The magnetic field – even though it is conventionally symbolised by the vector notation **B** – is not a directed line segment, i.e. it is not an arrow pointing from an initial point to a terminal point (as does the position vector or an electric field vector), but it can be represented by a directed circular contour and a magnitude (in a similar way to that in which an angular velocity or a torque can be represented). Such quantities are called *axial vectors*. The directed circular contour associated with the magnetic field at a given point is provided by consideration of a charged particle moving (with a suitable speed) in a stable circular orbit in the vicinity of the point. The normal to the plane in which the orbit lies and the direction of circulation of the particle provide the necessary directional characterisation.

For the wire-and-plate problem, consider the situation that would result from (hypothetically) reflecting the whole arrangement in the plane S defined by the straight wire and the given point P. After reflection, the current distribution would remain exactly the same as originally, and so the magnetic field must do the same. We will now use this idea to put limits on the configurations the magnetic field might take.

Consider first a possible 'radial' component of **B**, which would be represented by a circular contour in a plane perpendicular to the shortest radial line joining the wire to the given point P, as shown in Fig. 2a). On reflection of the system in S, the direction of rotation of the particle would be reversed, implying that this component would change sign. But we have already established that it will remain unchanged; this can only be so if the component has zero value.

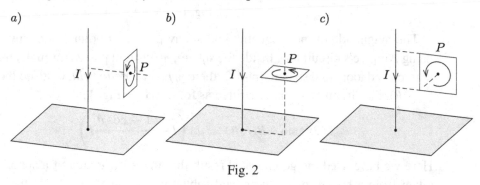

Fig. 2

The same reasoning can be applied to any 'longitudinal' component of the magnetic field, which, as shown in Fig. 2b), can be represented by a circular contour with a plane perpendicular to the straight wire. Under the reflection process, it too would both change sign and remain unchanged; again this implies a zero value.

The same reasoning *cannot* be applied to the third, 'azimuthal' component of the magnetic field. Its representative circular orbit lies in the reflection plane S and remains unaltered by the reflection (*see* Fig. 2c)). Hence it is not possible to deduce anything about the value of this component, which could be zero or non-zero.

The considerations above apply for any arbitrary point P, and so throughout the whole space only an 'azimuthal' magnetic field can be present. The field lines must be concentric circles, with the wire as their common centre. The strength of the field can be found, as in Solution 1, using Ampère's law:

$$B(r) = \mu_0 \frac{I}{2\pi r}$$

above the plate, and zero below it.

S159   We first calculate the magnetic field of an arrangement in which a *single* straight input wire provides a current that is subsequently led away (to infinity) radially and uniformly in all directions, i.e. in a spherically symmetric pattern.[69]

For a single straight current-carrying wire, the magnetic field has cylindrical symmetry, with the wire as its 'axis' (*see* solution on page 385).

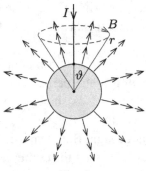

Fig. 1

The magnitude of the magnetic field at any particular point can be investigated using Ampère's circuit law. Inside the sphere, *no* current passes through any imaginary closed loop, and so here ($r < R$) there is *no magnetic field*. Outside the sphere ($r > R$), the circuit law can be written as follows (*see* Fig. 1):

$$2\pi r \sin \vartheta \, B(r, \vartheta) = \mu_0 \left(I - \frac{1 - \cos \vartheta}{2} I \right).$$

Here we have used the geometrical result that the surface area of a spherical cap, taken from a sphere of radius r, and subtending an angle of 2ϑ at the sphere's centre, is $2\pi r^2 (1 - \cos \vartheta)$. Since the surface area of the corresponding whole sphere is $4\pi r^2$, the fraction of the current I that flows out through the cap is $(1 - \cos \vartheta)/2$. The circular base of the cap, which has radius $d = r \sin \vartheta$, is shown by the dashed line in Fig. 1.

---

[69] This current distribution can be closely simulated in practice if a sphere with very good surface conductivity is placed in an 'infinitely' large medium with some electrical conductivity, and a potential difference between the sphere and the medium's boundary is provided.

Rearranging the previous equation gives the magnitude of the magnetic field strength as

$$B(r, \vartheta) = \frac{\mu_0 I}{4\pi} \frac{1 + \cos\vartheta}{r \sin\vartheta} = \frac{\mu_0 I}{4\pi} \frac{\cot(\vartheta/2)}{r}.$$

Near the current-carrying wire ($\vartheta \approx 0$), for which the approximations $\sin\vartheta \approx \vartheta$ and $\cot(\vartheta/2) \approx 2/\vartheta$ are valid, we recover the expression $\mu_0 I/(2\pi d)$ for the magnetic field around a very long, straight wire; diametrically opposite to the incoming current, as $\vartheta \to \pi$, the magnetic field gradually diminishes to zero.

Let us now apply the same calculation to the wire that carries the current away from the globe (and is fed by spherically symmetric incoming currents), and then superimpose the magnetic fields of the two arrangements as shown in Fig. 2. Clearly, the first calculated field has to be rotated by 90°, to obtain the second one, before combining them. In the superposition, the two symmetric current distributions cancel each other, and only the currents in the wires are left – matching the given physical situation.

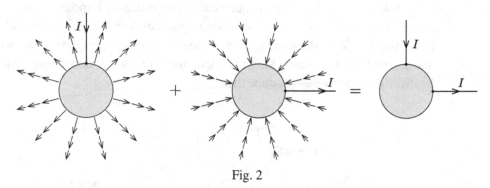

Fig. 2

Inside the sphere, the magnetic field remains *zero* everywhere. At the specific point P given in the question (marginally outside the globe), the two contributions reinforce (rather than oppose) each other, and the relevant angle ϑ is $\pi/4$ for each:

$$B_P = 2B(R, \pi/4) = \frac{\mu_0 I}{2\pi R} \cot(\pi/8) = \frac{\mu_0 I}{2\pi R}(\sqrt{2} + 1).$$

Note. On the surface of the sphere, the currents flow from the input junction to the output one along stationary current-streamlines. It can be shown that these streamlines (running on the curved surface) are plane curves, namely parts of *circles*, one of which is a *great circle*. These circles can be found as the intersections of the sphere with an array (sheaf) of planes that has the line joining the input and output junctions as its axis.

S160   The two current-carrying wires produce a stationary, *planar* magnetic field.[70] Analogous planar fields occur in electrostatics when a charge density does not depend upon one of the space coordinates. An example in which this is the case is that of one or more 'infinitely' long, parallel, insulating rods that are uniformly charged.

Magnetostatic vector fields are divergence-free (i.e. solenoidal) and rotation-free (implying that a potential can be defined) in those regions of space where there are no electric currents. Electrostatic vector fields have the same properties if there are no charges in the region concerned. For both kinds of (planar) vector fields, we can assume that there are field lines and equipotential lines (the planar equivalent of equipotential surfaces) that form arrays of mutually perpendicular curves.

In what follows, we show that, subject only to the relevant region being charge- and current-free, planar steady-state electrical and magnetic vector fields can be related to each other in a one-to-one correspondence. In this so-called *dual connection*, the equipotential lines of one of the vector fields correspond to the field lines of the other, and vice versa.

Vector fields are not determined unequivocally just by being divergence- and rotation-free – they become unequivocal as the result of imposed boundary conditions. Electric field lines start at charges (sources) situated on the boundaries of the region, or (depending on the sign of the charge) they terminate there, when the charges are known as sinks. Magnetic field lines are always closed loops encircling the electric currents that produce them.

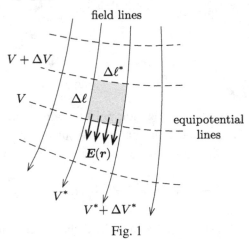

Fig. 1

---

[70] A vector field F is said to be planar if, when expressed in a system of orthogonal coordinates u_i (say, Cartesian), one of its components F_{u_3} (say, F_z) is zero everywhere, and the other two components do not depend upon the value of u_3 (i.e. F_x and F_y are independent of z).

Let $E(r)$ be a planar, electrostatic vector field with corresponding electrostatic potential $V(r)$, as illustrated in Fig. 1. The equipotential lines ($V =$ constant) are denoted by dashed lines, and the field lines by continuous lines with arrowheads; by convention, the direction of the electric field is that of decreasing potential. These lines form two arrays of curves, which everywhere intersect each other orthogonally.

Since the field lines do not cross each other, for each field line, the total *flux* that passes through between it and an arbitrarily chosen reference field line can be found. This is done by calculating at each point the product of the length of a small line segment and the strength of the field component perpendicular to it, and summing or integrating as necessary along a line that joins two points, one on the particular field line and the other on the reference line.

> *Note.* For three-dimensional vector fields, the flux is defined as the product of a small surface area and the strength of the field component perpendicular to that area. In planar (two-dimensional) vector fields, the flux can be defined as the product of the length of a small line segment and the strength of the field component perpendicular to it. The consistency between the two definitions is obvious, if the three-dimensional flux of a two-dimensional vector field is calculated for a 'rectangular area' that has equal small sides in two x–y planes and unit height in the z direction.

Let the total flux passing between the field line at position r and a reference field line be denoted by $V^*(r)$. The function $V^*(r)$ thus associates a scalar quantity with each point of the field, and the equation specifying an individual field line is of the form $V^*(r) =$ constant. Two such field lines are marked in Fig. 1.

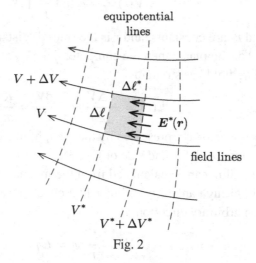

Fig. 2

Now consider two field lines (close to each other), and two equipotential lines (also close to each other) that cross them. These lines define a small region in the plane of the vector field (shaded in Fig. 1) that, to a good approximation, can be considered rectangular. The magnitude of the field strength can be related in two different ways to the lengths of the rectangle's sides. On the one hand, from the definition of a potential, we have

$$\Delta V = -|E|\,\Delta\ell,$$

where the negative sign reflects the fact that the field conventionally points in the direction of decreasing potential. But, on the other hand, we can express a connection between the flux and the magnitude of the field strength by

$$\Delta V^* = |E|\,\Delta\ell^*.$$

Eliminating the magnitude of the field strength, we get the following relationship:

$$\frac{\Delta V}{\Delta\ell} + \frac{\Delta V^*}{\Delta\ell^*} = 0. \tag{1}$$

The all-too-obvious symmetry of this relationship presents an opportunity to relate the vector field $E(r)$ to another 'dual' vector field, one with changed roles for the vector field and scalar potential. If the function $V^*(r)$ is considered as the potential function of a vector field $E^*(r)$, and $V(r)$ is considered as the quantity giving the flux for this field, then relation (1) will also hold for this new field (*see* Fig. 2); i.e. it is both divergence- and rotation-free. We also note that, in the dual field, because of their changed roles, we have to interchange $\Delta\ell$ and $\Delta\ell^*$.

These observations can be illustrated by a well-known example. If the potential is

$$V(r) = -K\ln\left(\frac{r}{r_0}\right), \tag{2}$$

where K and r_0 are constants, and r is the (radial) distance of a general point from the origin O of the plane, then the equipotential curves are concentric circles. The corresponding field has strength

$$E(r) = -\frac{\Delta V}{\Delta r} \approx -\frac{dV}{dr} = \frac{K}{r},$$

and its direction can be represented by the straight line connecting the origin to the particular point (*see* left-hand side of Fig. 3).

Further, the flux can be calculated from the magnitude of the field strength. The flux passing through an arc of radius r that covers a sector with opening angle φ, relative to an arbitrary direction, is

$$V^* = \frac{K}{r}r\varphi = K\varphi.$$

This field is the same as the electrostatic field associated with an 'infinitely' long rod carrying a uniform linear charge density of $\lambda = 2\pi\varepsilon_0 K$.

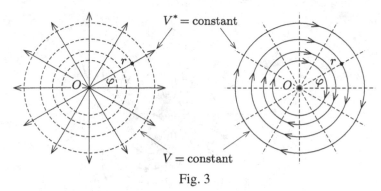

$V^* = \text{constant}$

$V = \text{constant}$

Fig. 3

Starting from a field whose *equipotential lines* (two-dimensional surfaces) are given by $V = -K \ln(r/r_0) = \text{constant}$, the *field lines* of the *dual* field $\boldsymbol{E}^*$ are *concentric circles*. The magnitude of that field can be calculated from the requirement that, if we move a little along an arbitrary field line, we must have

$$\boldsymbol{E}^* \cdot r\Delta\varphi = -\Delta V^* = -K\Delta\varphi,$$

from which we get $|\boldsymbol{E}^*| = -K/r$. If the value of the constant K is chosen to be $\mu_0 I/(2\pi)$, then the dual vector field represents the magnetic field of a long straight wire carrying an electric current I (*see* right-hand side of Fig. 3).

> *Notes.* 1. The magnetic field of a long straight current-carrying wire is locally rotation-free (as it can be derived from a potential function), but it is *not conservative*: its integral around a closed loop is not necessarily zero. This peculiarity is reflected mathematically in the fact that the potential function V^*, which is proportional to the angle φ, is a *multi-valued* function. When the point O (i.e. the current-carrying wire) is encircled by the loop, the value of the potential will not return to its initial value, even though the loop is closed. This situation cannot happen with electrostatic potentials, because, if it did, it would be possible to create *perpetuum mobile*. The 'magnetic potential' has no direct physical meaning, and is only a convenient mathematical device – no possibility of *perpetuum mobile* exists, alas!
>
> 2. A possible physical interpretation for the dual field associated with a planar electrostatic field can be given only with knowledge of the imposed boundary conditions. It is not true that the dual field of an electrostatic field always represents some magnetic field. The dual field of the homogeneous electrostatic field of a parallel-plate capacitor is also a homogeneous field, which could, for example, be the electrostatic field of another parallel-plate capacitor, one rotated by 90°. The dual field of a two-dimensional (planar) ideal electric dipole can also be interpreted as an electrostatic field, that of another dipole rotated by 90° with respect to the original (*see* Fig. 4).

Now, let us return to the original problem, the description of the magnetic field of two parallel current-carrying wires with oppositely directed currents. This arrangement is the dual companion of two very long, parallel, uniformly

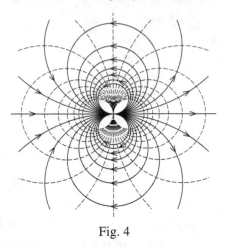

Fig. 4

charged, insulating rods with opposite charges. On the left-hand side of Fig. 5, the (planar) magnetic field, produced by the currents, is illustrated. The magnetic field lines are denoted by continuous lines, and the 'contour lines' of the magnetic potential are drawn using dashed lines. Because of the equal currents, the field is symmetrical with respect to the perpendicular bisector of the line joining the two wires.

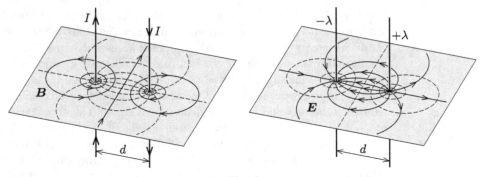

Fig. 5

The dual arrangement, shown on the right-hand side of Fig. 5, represents an electric field in which the roles have been reversed. There, the equipotential lines, denoted by dashed lines, encircle the infinitely long rods, and the array of curves, marked by continuous lines and perpendicular to the dashed ones, represent the system of electric field lines. The correspondence is a symmetrical one, provided the electric field is produced by two, very long, parallel, insulating rods carrying uniform linear charge densities $+\lambda$ and $-\lambda$.

The electrostatic field of a single, long, insulating rod with uniform, linear charge density λ has a potential given by expression (2), providing the substitution $K = \lambda/(2\pi\varepsilon_0)$ is made. The potential function for two 'infinitely' long rods, with linear charge densities of identical magnitude, but opposite signs, can be found by superposing those of two single rods:

$$V(r) = \frac{\lambda}{2\pi\varepsilon_0} \ln \frac{r_1}{r_0} - \frac{\lambda}{2\pi\varepsilon_0} \ln \frac{r_2}{r_0} = \frac{\lambda}{2\pi\varepsilon_0} \ln \frac{r_1}{r_2},$$

where r_1 and r_2 are the distances between the particular point r and the rods.

Equipotential curves in the electrostatic configuration are characterised by the requirement that $V = $ constant, i.e. the quotient r_1/r_2 is constant; the same condition provides the equations of the magnetic field lines in the dual problem. In a plane, the locus of all points whose distances from two fixed points are in a constant *ratio* is an *Apollonian circle*. Consequently, in the dual magnetic field, the field lines are circles.

> *Note.* It can be further proved that, for two parallel current-carrying wires, the equipotential curves associated with the magnetic field (and therefore, at the same time, the field lines in the electrostatic problem) are also *circles*. Both of them are determined by the condition $V^* = $ constant, and, using the 'flux function' for a single long rod found earlier, this is equivalent to the requirement that $\varphi = \varphi_2 - \varphi_1 = $ constant. Here φ is the angle subtended by points A and B, the positions of the two wires, at an arbitrary point on the corresponding electric field line, as shown in Fig. 6. Since the locus of points at which the two end-points of a given line segment subtend a constant angle is a circle, the electric field lines of parallel rods with opposite charges must be *circles* or arcs of circles. This result was obtained earlier – using other, more elementary, considerations – in the solution on page 338.

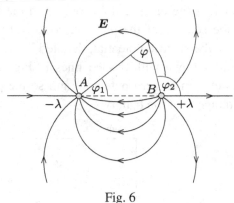

Fig. 6

S161   If all the squares on the chessboard were made from a homogeneous metal plate with conductivity σ_1, then the magnitudes of the electric field strength and current density would be $E = V/L$ and $j_1 = \sigma_1 E$, respectively, and the total current flowing through the board would be

$$I_1 = j_1 Lt = \sigma_1 ELt = V\sigma_1 t.$$

Similarly, in another homogeneous metal plate with conductivity σ_2, under the same voltage V, the total current I_2 would be $V\sigma_2 t$. We note in passing that the current *does not depend* on the length L of the board.

In the (somewhat lengthy) development that follows, we prove that the current flowing through the inhomogeneous chessboard is the geometric mean of I_1 and I_2, that is

$$I = Vt\sqrt{\sigma_1\sigma_2}. \tag{1}$$

This statement (about the geometric mean giving the required current) is valid for any set of square plates of constant thickness that satisfy the following requirement:

> That, for a pair of points, A and B, which can be transformed into each other by rotating the square through 90°, the product of the conductivities at those two points must have a value that is independent of how the point pair is chosen.

It is clear that, in our case, this condition will be satisfied, because a rotation by 90° will always transform a point on a light square into one on a dark square, and vice versa; i.e. points A and B will always be on different coloured squares and $\sigma_A\sigma_B$ will always have the value $\sigma_1\sigma_2$.

In the following (for the sake of simplicity and clarity), instead of a normal chessboard, a 2×2 version will be illustrated in our not-to-scale sketch figures.

We consider first the electric field formed in the board, and the currents that flow through the squares of the plate. A conservative electric field can be fully characterised by the electric potential, denoted by Φ, and represented graphically by equipotentials, drawn with dashed lines in Fig. 1. The 'bottom' edge of the square plate is chosen as zero potential, and so the potential of the 'top' of it is equal to the voltage V of the battery.

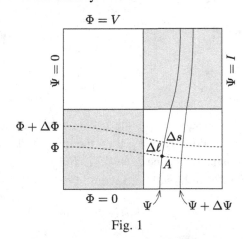

Fig. 1

The electric field, and consequently the current density vector, which is proportional to it, are both perpendicular to the equipotential curves. It follows that the current-streamlines, drawn in the figure with continuous lines, are also perpendicular to the same lines. Let us associate with each current-streamline a number Ψ that is equal to the total current that flows between it and the left-hand edge of the plate. It is obvious that the left-hand edge of the plate is a streamline, and corresponds to $\Psi = 0$, and that the right-hand edge is $\Psi = I$. From here on, let us call Φ the *voltage-potential* and Ψ the *current-potential*.

If, in the vicinity of an arbitrary point A of the plate, a distance $\Delta\ell$ between two equipotential lines corresponds to a small potential difference $\Delta\Phi$ between them, then the magnitude of the electric field at A is

$$E_A = \frac{\Delta\Phi}{\Delta\ell},$$

and the local electric current density is

$$j_A = \sigma_A E_A = \sigma_A \frac{\Delta\Phi}{\Delta\ell}.$$

Here, σ_A denotes the conductivity around point A, which could be σ_1 or σ_2, depending on the 'colour' of the particular square in which A is located.

Now, from the definition of the current-potential, the current element $j_A t \Delta s$ that flows through a cross-sectional area $t \Delta s$ of the plate must be equal to the change $\Delta\Psi$ in the current-potential over the small distance Δs:

$$j_A t \Delta s = \sigma_A \frac{\Delta\Phi}{\Delta\ell} t \Delta s = \Delta\Psi,$$

from which we get

$$t\sigma_A \frac{\Delta\Phi}{\Delta\ell} = \frac{\Delta\Psi}{\Delta s}. \tag{2}$$

Next, consider rotating Fig. 1 anticlockwise by 90° in the plane of the plate, and obtaining the arrangement shown in Fig. 2.

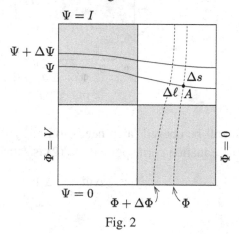

Fig. 2

The question now arises whether this arrangement, after interchanging the roles of the voltage- and current-potentials, and multiplying them by appropriately chosen factors, is a suitable description of the currents flowing through the original plate.

Let the renaming and scaling be

$$\Psi' = \frac{I}{V}\Phi \quad \text{and} \quad \Phi' = \frac{V}{I}\Psi,$$

and denote the separations of the neighbouring pairs of curves by

$$\Delta\ell' = \Delta s \quad \text{and} \quad \Delta s' = \Delta\ell.$$

The scaling factors are chosen so that the maximal value for Φ' (the new voltage-potential) is V, and that for Ψ' (the new current-potential) is I. If these new (primed) variables are used to label the original (unrotated) plate, then we get the arrangement shown in Fig. 3. Point B, shown in Fig. 3, is the place to which point A has been moved by the rotation. As noted earlier, B is bound to be on a different coloured square than A was, and the product of the corresponding conductivities is

$$\sigma_A\sigma_B = \sigma_1\sigma_2. \tag{3}$$

This always has the same value, wherever A, and hence B, are situated on the chessboard.

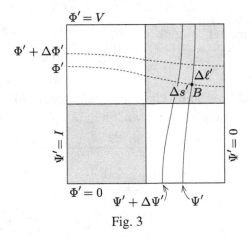

Fig. 3

Ohm's law will be obeyed (as it needs to be) by the new voltage- and current-potentials, if an equation corresponding to (2) is satisfied, that is, if

$$t\sigma_B\frac{\Delta\Phi'}{\Delta\ell'} = \frac{\Delta\Psi'}{\Delta s'}.$$

Using the connections between the primed and unprimed quantities and equation (2), this requirement can be written in the form:

$$t\sigma_B \frac{V}{I} \frac{\Delta\Psi}{\Delta s} = t\sigma_B \frac{V}{I} \cdot t\sigma_A \frac{\Delta\Phi}{\Delta\ell} = \frac{I}{V} \frac{\Delta\Phi}{\Delta\ell}.$$

This can be simplified and, using (3), put in the form

$$I = Vt\sqrt{\sigma_A\sigma_B} = Vt\sqrt{\sigma_1\sigma_2},$$

i.e. as stated in equation (1).

> *Notes.* 1. All of the arguments used in the solution are valid not only for the 2×2 board but also for the normal 8×8 chessboard. More generally, it can be stated that the final result is true for all boards with $n \times n$ squares if n is even, but not true if n is an odd number. In the latter case, a rotation by $90°$ does not change the 'colour' of the square on which any particular point is located, and further, since the numbers of light and dark squares are different, the result cannot be symmetric in σ_1 and σ_2.
>
> 2. Using numerical methods, the current-streamlines and the equipotential lines for any particular chessboard can be found and plotted. Such a computer-generated map, for a 2×2 'chessboard', is shown in Fig. 4; in it, the conductivity σ_2 of the dark squares is twice that, σ_1, of the light ones. Because of the boundary conditions, both the current-streamlines and the equipotential lines change direction at the interfaces – in much the same way that light is refracted at the boundary between two different media.

current-streamlines

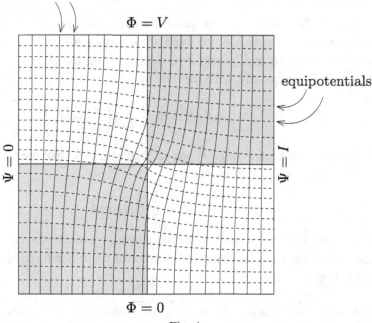

Fig. 4

S162 *Solution 1.* Applying the principle of superposition, it can be shown that the magnetic flux crossing the end of a long solenoid is exactly one-half of the flux passing through a single turn deep inside the solenoid.[71] Accordingly, in the arrangement described in the problem, the magnitude of the magnetic flux leaving the first of the solenoids, and penetrating into the second, is

$$\Phi = \frac{1}{2}\mu_0 I_1 nA.$$

Virtually all of this flux leaves that coil through its lateral surface, because, if the solenoids are sufficiently long, the part of it that escapes through the 'far end' of the second coil is negligibly small.

We now investigate the force produced by the magnetic field of one of the solenoids (say, 1) on the wire carrying the current in the other coil (2). The axial component of the magnetic field vector exerts no net force on the near-circular wire turns, as the magnetic forces it produces act radially. So we only need to deal with the magnetic field components that are perpendicular to the axis.

Denote the *radial* component of the magnetic field vector produced by solenoid 1, at the position of the ith turn from the end of coil 2, by B_i. Then the magnitude of the force, exerted by solenoid 1 on the turns of coil 2, each of radius R and length $2\pi R = 2\pi\sqrt{A/\pi} = 2\sqrt{\pi A}$, is $F_i = 2\sqrt{\pi A}\, I_2 B_i$, and the force acting on the whole of coil 2 is

$$F = \sum_i F_i = 2\sqrt{\pi A}\, I_2 \sum_i B_i. \tag{1}$$

However, we also know how much flux from solenoid 1 'escapes' through the lateral surface of coil 2. Since there are n turns per unit length, we may take the width of one turn as $1/n$, and its (small) area on the lateral surface as $2\pi R/n = 2\sqrt{\pi A}/n$. Consequently, we have

$$\frac{1}{2}\mu_0 I_1 nA = \Phi = \sum_i \left(\frac{2\sqrt{\pi A}}{n} B_i\right). \tag{2}$$

Comparison of equations (1) and (2) makes it clear that the force in question has magnitude

$$F = \frac{1}{2}\mu_0 I_1 I_2 n^2 A.$$

If the solenoids are wound in the same sense, and the current directions are the same, then the force is attractive; if either of these conditions is reversed, but not both, then it is repulsive.

---

[71] The proof of this statement can be found in, for instance, the predecessor of this book: see 'Problem 120' in P. Gnädig, G. Honyek & K. F. Riley, *200 Puzzling Physics Problems* (Cambridge University Press, 2001).

Solution 2. The force between the coils is proportional to both I_1 and I_2, since the magnetic field B, proportional to one of the currents, exerts a Lorentz force on the other coil, and this force is proportional to the current in that coil (as well as to B). So the 'wanted' force can be written in the form $F = KI_1I_2$, where K depends only on the specification of the solenoids, and is independent of the currents. If the force is calculated for the special case $I_1 = I_2 = I$, thus determining K, the required solution for arbitrary currents follows.

The magnitude of the force depends only on the magnitudes of the electric currents, but not on the details of how those currents are produced. If, for instance, with suitable cooling, the current-carrying coils are transformed into superconductors (with zero electrical resistance), and their terminals are connected so as to produce one double-length solenoid, then the current in it will be I everywhere. As the solenoid coil is superconducting, the current persists, even without any power supply. Inside the new solenoid the magnetic field strength is $B = \mu_0 nI$, and the stored magnetic energy is

$$W_{\text{magn}} = \frac{1}{2\mu_0} B^2 V,$$

where $V = \ell A$ is the volume of the double solenoid, which has cross-sectional area A and total length ℓ. For the effects of fringing fields, see Note 1.

What happens if the solenoids are moved apart by a small distance Δx ($\ll \ell$)? The magnetic fields inside them cannot change, as that would induce voltages, which would, because of the zero resistance of the wires, generate 'infinitely large' currents. But the energy of the magnetic field does change; the field in the gap is essentially the same as (the unchanged field) in the solenoids, but the volume 'filled' with magnetic flux has increased by $\Delta V = A \Delta x$. The corresponding energy change is

$$\Delta W_{\text{magn}} = \frac{1}{2\mu_0} B^2 \Delta V = \frac{1}{2} \mu_0 I^2 n^2 A \Delta x.$$

This change must be equal to the work done by F during the coil separation, $W = F\Delta x$, from which it can be seen that

$$F = \frac{1}{2} \mu_0 I^2 n^2 A = KI^2, \qquad \text{that is} \qquad K = \frac{1}{2} \mu_0 n^2 A.$$

From this, the attractive or repulsive force can be found, even when the electric currents are not the same:

$$F = KI_1I_2 = \frac{1}{2} \mu_0 I_1 I_2 n^2 A.$$

Notes. 1. The effects of fringing fields do not change the result, because, when the coils are moved apart a little, they 'carry' the fringing fields at the far ends of the coils with them. Consequently, the fringing effect does not appear in the energy change.

2. The same result can be found without resorting to the 'superconducting trick'. Connect the two coils at their adjacent ends, and provide a constant common electric current I with the help of an independent current source. If the two coils, connected in this way, are moved apart a little, then – despite the constant value of the current supplied – the magnetic field, as well as the magnetic flux, decrease a little, because the act of separating the two parts generates a current opposing the original, and consequently reduces the magnetic field.

It can be shown that the stored magnetic energy of the coils decreases, despite positive work having been done on them. This strange situation is explained by the induced voltage caused by the change in magnetic flux. The induced electromotive force and the current from the source together mean that electrical energy is transferred into the source; the amount involved is equal to exactly twice the energy decrease in the coils. So, overall, there is an increase in the energy of the coils plus current source, and the force F, calculated from this, is the same as the value found earlier.

S163   *a*) The wire is thin, and the field is strong, and so any effects due to gravity can be neglected; in particular, we do not need to know the orientation of the line P_1P_2. The wire takes up a shape that lies in a plane that is perpendicular to the magnetic induction. As the force acting on any small piece of the wire is everywhere perpendicular to that piece, the tension in the wire is the same throughout its length. The tension force in a small piece of the wire with radius of curvature r is $F = IBr$, as can be proved by investigating the forces acting on it (*see* Fig. 1).

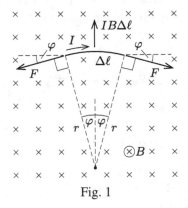

Fig. 1

The condition for equilibrium of a small piece of wire that has length $\Delta\ell = 2\varphi r$, and subtends an angle 2φ at the centre of the osculating circle, is

$$2F \sin \varphi = IB\Delta\ell.$$

Because the angles involved are (vanishingly) small, $\sin \varphi \approx \varphi$, which immediately yields the stated result, $F = IBr$.

It follows from the above that the shape of the wire is a planar curve with constant radius of curvature, and so it must be a *circular arc*.[72]

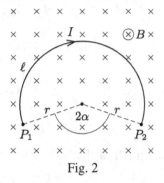

Fig. 2

The length of the circular arc and the distance between points P_1 and P_2 are both given, and so the following relations have to be satisfied by the arc radius r and the angle α shown in Fig. 2:

$$2r \sin \alpha = \ell/2, \qquad 2r(\pi - \alpha) = \ell.$$

From these, we get the transcendental equation $2 \sin \alpha = \pi - \alpha$, whose numerical solution is

$$\alpha \approx 1.246 \text{ rad} = 71.40°.$$

Inserting this back into the equations above, we find that the radius of the circle is $r \approx 0.264\ell$, and that the tension in the wire is $F \approx 0.264IB\ell$.

b) There can be no magnetic force acting on any piece of the wire that is parallel to the magnetic field, and so, even in this case, the component of the tension in the wire that is parallel to B is constant throughout the wire. It is also true that the force exerted by the magnetic field on any small piece of the wire is everywhere perpendicular to that piece; this is why the magnitude (but not the direction) of the force is the same at each point of the wire.

From these two facts, it follows that the component of the tension in the wire, perpendicular to the magnetic field, must be constant, i.e. if we look at the wire along the direction of the magnetic field, then we see a circle. In three dimensions, the shape of the wire is a one-turn helix with an axis parallel to the magnetic field, and uniform pitch (*see* Fig. 3). In principle, a multi-turn helix is also a possible equilibrium state, but, as in part *a*), that kind of shape is unstable.

---

[72] In principle, a 'circular coil' with at least one complete turn would also be in equilibrium, but its state would be unstable. Such a configuration would not be formed, for much the same reason that a pencil cannot stand on its point.

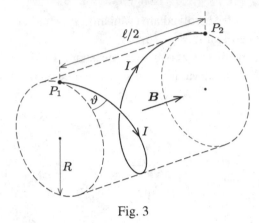

Fig. 3

The angle ϑ of the pitch of the helix (that is, the angle between the tangent to the helix at a particular point, and the plane perpendicular to the magnetic field B that goes through that same point) can be calculated from the given arc length ℓ:

$$\sin \vartheta = \frac{\ell/2}{\ell} \quad \text{and from this} \quad \vartheta = 30°.$$

The radius R of the imaginary cylinder that fits inside the helix is

$$R = \frac{\ell \cos \vartheta}{2\pi} = \frac{\sqrt{3}\,\ell}{4\pi} \approx 0.138\ell.$$

Finally, we come to calculating the force acting on P_1 (or P_2). Looking along the axis of the helix, we see a whole circle of radius R, which the magnetic Lorentz force is trying to enlarge. Opposing this force is the component of the tension in the wire that is perpendicular to the magnetic field, namely $F \cos \vartheta$:

$$IBR = F \cos \vartheta.$$

Using our previous expression for R, we get the tension force in the wire as

$$F = \frac{IB\ell}{2\pi} \approx 0.159 IB\ell.$$

It seems that the tension in the wire is independent of the distance d between the points P_1 and P_2 ($0 < d < \ell$).

S164 The electron's trajectory is confined to the mid-plane (plane of symmetry) between the circular current loops; and the magnitude of its velocity (its speed) is always v_0, as the magnetic field cannot do any work on it.

The direction of its velocity is much harder to determine, since it is continually changing because of the effect of the Lorentz force. The magnitude of the magnetic field strength (and consequently that of the Lorentz force) depends quite sophisticatedly on the electron's position, but, in spite of this, we can deduce something specific about the direction of its final velocity.

When the electron is at a distance r from point O, its angular momentum about that point is given by mrv_t, where v_t is the *tangential* component of its velocity. The angular momentum changes continuously as a result of the torque generated by the Lorentz force, but only the radial component v_r of the electron's velocity contributes to that force:

$$\frac{d(mrv_t)}{dt} = ev_rB(r)r = e\frac{dr}{dt}B(r)r, \tag{1}$$

where m is the mass of the electron and e is the elementary charge; in the figure in the problem, the positive direction has been chosen to be vertically upwards.

It follows that the change in the electron's angular momentum, as it increases its radial distance by a small amount Δr, is

$$\Delta(mrv_t) = eB(r)r\Delta r = \frac{e}{2\pi}\Delta\Phi, \tag{2}$$

where $\Delta\Phi = B(r)2\pi r\Delta r$ is the flux from the axially symmetric magnetic field that crosses a centrally placed annulus of average radius r and width Δr.

According to equation (2), the change in the angular momentum of the electron, as its distance from point O increases from r_1 to r_2, is proportional to the total magnetic flux that crosses a centred annulus that has inner and outer radii r_1 and r_2, respectively. As the initial angular momentum of the electron is zero, at $r_1 = 0$, the angular momentum of the electron at any time is proportional to the total flux $\Phi(r)$ crossing a centred disc whose radius is equal to the current value of r.

After the electron has moved 'sufficiently far' from the circular current loops ($r \to \infty$):

$$\Phi_{total} \to \Phi(\infty) = 0.$$

The final equality follows because the magnetic field is divergence-free, and the electron's trajectory plane, taken as a whole, is pierced by just as many magnetic field lines directed from top to bottom, as from bottom to top. The figure shows the structure of the magnetic field in a plane section that contains point O, and is perpendicular to the planes of the current loops; the horizontal line (with no arrowheads) is a section of the plane containing the electron's trajectory.

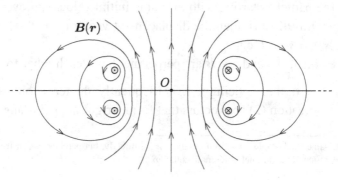

So, the angular momentum of the electron, when it is very far from the current loops, becomes zero again, and this implies that v_t has also returned to zero. Accordingly, the electron must ultimately be moving away from point O *in a radial direction*, with its unchanged speed of v_0.

S165   We first write the differential equation of motion for the charged particle in terms of its position and velocity vectors r and v. In the presence of both the magnetic field B and the braking force ($\propto v$) we have

$$m\frac{dv}{dt} = Q\frac{dr}{dt} \times B - k\frac{dr}{dt}. \tag{$*$}$$

The quantities m and Q (the mass and charge of the particle), B and k (the drag coefficient) are all constants, and so, in any small time interval Δt, the small changes in r and v satisfy

$$m\Delta v = Q\Delta r \times B - k\Delta r.$$

This equation can be summed over the whole motion:

$$m\sum \Delta v = Q\left(\sum \Delta r\right) \times B - k\sum \Delta r. \tag{1}$$

On the left-hand side of this equation, the sum of the velocity changes is just (-1) times the initial velocity of the particle, and each sum on the right-hand side is the total displacement vector of the particle (from the entry point to the final stopping place):

$$\sum \Delta v = -v_0, \qquad \sum \Delta r = s.$$

Using this notation, equation (1) can be written in the form:[73]

$$mv_0 = -Qs \times B + ks. \tag{2}$$

In the absence of any magnetic field, the particle stops after covering a path of length $|s_1| = s_1 = 10$ cm, and so, from (2), we have

$$mv_0 = ks_1. \tag{3}$$

If the particle, starting with the same initial velocity, moves in a magnetic field of strength $|B| = B$, then the displacement vector of its final stopping place is s_2, with $|s_2| = s_2 = 6$ cm.

We note that s and $s \times B$ are perpendicular to each other for any s. Then, either

a) by further noticing that, consequently, the terms on the right-hand side of equation (2) are vectors that form a right-angled triangle, or

---

[73] The same result could have been obtained more directly, but perhaps less instructively, by straightforward integration of the original differential equation ($*$).

b) from taking the squared modulus of both sides of (2), and using the orthogonality of s and $s \times B$,

$$(mv_0)^2 = Q^2[(s_2 \times B)^2] + [Q(s_2 \times B) \cdot s_2] + (ks_2)^2 = (Qs_2B)^2 + (ks_2)^2,$$

we find that

$$(mv_0)^2 = (Qs_2B)^2 + (ks_2)^2. \tag{4}$$

Finally, in the doubled magnetic field of strength $2B$, the magnitude of the particle's total displacement s_3 is given by

$$(mv_0)^2 = (2Qs_3B)^2 + (ks_3)^2. \tag{5}$$

By eliminating all the constant physical quantities from equations (3), (4) and (5), we get the required distance as

$$s_3 = \frac{s_1 s_2}{\sqrt{4s_1^2 - 3s_2^2}} = \frac{30}{\sqrt{73}} \text{ cm} = 3.51 \text{ cm}.$$

S166 The ball, at position r relative to its rest position, and moving with velocity v, is acted upon by the tension force in the string, the gravitational force and the Lorentz force caused by the magnetic field. Using the usual approximations for small-amplitude motions of pendulums, the horizontal projection of the tension in the string is $-m\omega^2 r$, where ω $(= \sqrt{g/\ell})$ is the angular frequency of the pendulum without any magnetic field. The ball's equation of motion, in the presence of the magnetic field, is

$$m\dot{v} = qv \times B - mr\omega^2. \tag{1}$$

This quite sophisticated (vector) differential equation can be solved without using calculus, if we notice that it is very similar to the equation of motion of a simple pendulum in a rotating frame of reference (a *Foucault pendulum* at the North Pole).

To describe the small-amplitude swinging of a planar pendulum with angular frequency ω_0, in a frame of reference rotating with angular velocity Ω relative to the inertial reference frame (Ω is a vertical vector), we need an equation of motion that contains the *Coriolis force* as well as the centrifugal one:

$$m\dot{v} = 2mv \times \Omega - mr\omega_0^2 + mr\Omega^2. \tag{2}$$

The solution to this equation is well known. Since the oscillation plane is fixed in the inertial frame of reference, in a reference frame rotating with angular velocity Ω, the plane rotates with angular velocity $-\Omega$; the period of a whole revolution is $T = 2\pi/\Omega$.

Comparing equations (1) and (2), it can be seen that the observed rotation rate is $\Omega = qB/(2m)$, and that the angular velocity vector of the rotating plane has a direction *opposed* to that of $\boldsymbol{B}$ (*see* figure). The actual angular frequency ω_0 of the pendulum can be expressed entirely in terms of the other parameters of the problem:

$$\omega_0 = \sqrt{\frac{g}{\ell} + \left(\frac{qB}{2m}\right)^2},$$

but this has no bearing on the rotation rate of the plane of oscillation.

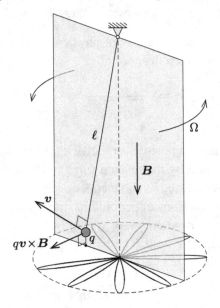

So, the time needed for the plane containing the pendulum's motion to complete one revolution is

$$T = 4\pi\frac{m}{qB}.$$

Notes. 1. It is interesting that the rotational period of the pendulum's plane depends only on the strength of the magnetic field and the data associated with the small ball, and that it does not depend on the natural period of the pendulum (i.e. it is independent of ℓ and g).

2. The quantity qB/m is called the *cyclotron angular frequency* associated with the given magnetic field; it is the angular velocity of a particle, with charge-to-mass ratio q/m, that will orbit in a homogeneous magnetic field of strength B. The plane of the pendulum in the problem rotates with one-half of the cyclotron angular frequency.

S167 *a)* For the distance between them not to change, the two electrons need to move on the same trajectory, around a circle of radius $d/2$; they need to be diametrically opposite each other, and move with the same speed (*see* Fig. 1). Each electron, moving in the magnetic field, needs to experience a Lorentz force that both compensates for the mutually repulsive electrostatic force and provides the centripetal force required for uniform circular motion.

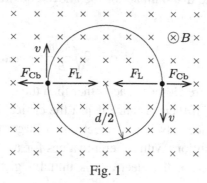

Fig. 1

We write the equation of motion for one of the electrons. The Lorentz force, acting on a particle with charge $-e$, mass m and speed v, has magnitude

$$F_L = evB,$$

and the Coulomb force is

$$F_{Cb} = k_e \frac{e^2}{d^2},$$

where k_e is the constant in Coulomb's law, $k_e = 1/(4\pi\varepsilon_0)$. The Coulomb force is always repulsive, but the use of Fleming's left-hand rule – remember that the electron is negatively charged – shows that the Lorentz force is always directed towards the other electron.

> *Note.* In principle, we should have taken into consideration the force effects arising from the magnetic field produced by the moving electrons (and calculated from the Biot–Savart law):
>
> $$F_{magn} = ev \frac{\mu_0}{4\pi} \frac{ev}{d^2} = \frac{v^2}{c^2} F_{Cb},$$
>
> where c is the speed of light. But, for the motion of classical (non-relativistic) charged particles, $v \ll c$, and this force is negligible compared to the Coulomb force. We have therefore ignored it.

The equation of motion of the electrons, taking account of the relevant directions, is

$$evB - k_e \frac{e^2}{d^2} = m \frac{v^2}{d/2}.$$

This is a quadratic equation for the required speed v, and its roots are

$$v = \frac{edB}{4m} \pm \sqrt{\left(\frac{edB}{4m}\right)^2 - \frac{k_e e^2}{2md}}.$$

The situation described in the problem can be realised if the value of v is real (and positive), i.e. the discriminant of the above equation is non-negative. So, we find that the condition on d is

$$d \geq 2\sqrt[3]{\frac{k_e m}{B^2}} = d_{\text{crit}}.$$

If the electrons are closer to each other than the critical distance, then the problem has no solution; if their distance is just the critical one, then there is one solution; and if the distance between the electrons is larger than the critical value, then we can find two solutions with different values for v.

b) If only one of the electrons is initially given a velocity, then the motion becomes more complex, even in the special case in which their separation – as in the given problem – remains constant.

We can get closer to the solution if we first seek the answer to an auxiliary question: 'What is the trajectory of the system's centre of mass (CM) in this case?' Write the (vector) equations for the electrons' motions using the usual notation:

$$m\boldsymbol{a}_1 = k_e \frac{e^2}{|\boldsymbol{r}_1 - \boldsymbol{r}_2|^3}(\boldsymbol{r}_1 - \boldsymbol{r}_2) - e\boldsymbol{v}_1 \times \boldsymbol{B}, \qquad (1)$$

$$m\boldsymbol{a}_2 = k_e \frac{e^2}{|\boldsymbol{r}_1 - \boldsymbol{r}_2|^3}(\boldsymbol{r}_2 - \boldsymbol{r}_1) - e\boldsymbol{v}_2 \times \boldsymbol{B}. \qquad (2)$$

For two particles with identical masses, we can write

$$\boldsymbol{r}_{\text{CM}} = \frac{\boldsymbol{r}_1 + \boldsymbol{r}_2}{2}, \qquad \boldsymbol{v}_{\text{CM}} = \frac{\boldsymbol{v}_1 + \boldsymbol{v}_2}{2}, \qquad \boldsymbol{a}_{\text{CM}} = \frac{\boldsymbol{a}_1 + \boldsymbol{a}_2}{2}.$$

These quantities appear in our formulae if the equations of motion of the two electrons are added,

$$m(\boldsymbol{a}_1 + \boldsymbol{a}_2) = 0 - e(\boldsymbol{v}_1 + \boldsymbol{v}_2) \times \boldsymbol{B},$$

from which it follows that

$$m\boldsymbol{a}_{\text{CM}} = -e\boldsymbol{v}_{\text{CM}} \times \boldsymbol{B}.$$

As would be expected, the (internal) Coulomb interaction has been eliminated from the equation of motion for the centre of mass.

This last equation enables us to state the important result that the centre of mass of a system consisting of two electrons moves in the same way in a magnetic field as would a single electron. If the field is homogeneous, and the motion is perpendicular to the magnetic field lines, the centre of mass moves in uniform circular motion!

We conclude that the centre of mass moves uniformly along a circular track, while the two electrons 'wobble' around it. The angular velocity of the centre of mass is

$$\omega_{CM} = \frac{a_{CM}}{v_{CM}} = \frac{e}{m}B = \omega_c,$$

which is just the *cyclotron angular frequency* of the electron (*see* Note 2 on page 408). In a given magnetic field – e.g. inside a cyclotron particle accelerator – the particles orbit with this angular velocity ω_c. The radius of the circular trajectory of the centre of mass is

$$R_{CM} = \frac{v_{CM}}{\omega_{CM}} = \frac{v_{CM}}{\omega_c}.$$

How do the electrons move around the centre of mass? It is obvious that the answer to this question leads to the solution of the final part of this problem. Write the position vector of electron 1 in the form $r_1 = r_{CM} + R$. It is clear that the position vector of electron 2 is then $r_{CM} - R$. Using the definition of centre of mass, it follows that

$$R = r_1 - r_{CM} = r_1 - \frac{r_1 + r_2}{2} = \frac{r_1 - r_2}{2}.$$

An equation giving the rate of change of this vector can be found if we take the difference between equations (1) and (2):

$$m(a_1 - a_2) = 2k_e \frac{e^2}{|r_1 - r_2|^3}(r_1 - r_2) - e[(v_1 - v_2) \times B]. \tag{3}$$

The difference between the position vectors is just twice the vector R given above; so the difference between the velocity vectors is just twice the vector V, which gives the rate of change of R; and similar relations hold for the accelerations:

$$r_1 - r_2 = 2R, \qquad v_1 - v_2 = 2V, \qquad a_1 - a_2 = 2A.$$

Using this notation, equation (3) can be transformed into

$$mA = k_e \frac{e^2}{|2R|^3}2R - e(V \times B). \tag{4}$$

This equation is intrinsically the same as the equation of motion in part *a*) (in which the centre of mass remains at rest), and so – given suitable initial velocities – this equation can also have a solution describing uniform circular motion.

Assuming this form of solution, if the magnitude of vector $R(t)$ is constant in time with value R, and its direction rotates with angular velocity ω, then according to the well-known formulae of uniform circular motion, we have $A = -\omega^2 R$ and $V \times B = R\omega B$. Then, from equation (4), we obtain the following quadratic equation for the angular velocity ω of the electron pair, as they orbit around the centre of mass:

$$\omega^2 - \frac{e}{m}B\omega + \frac{k_e}{m}\frac{e^2}{4R^3} = 0. \tag{5}$$

A real root for ω exists only if the discriminant is non-negative, i.e. if

$$R \geq \sqrt[3]{\frac{k_e m}{B^2}}.$$

The minimal electron separation for such a motion to exist is

$$d_{\min} = 2R_{\min} = 2\sqrt[3]{\frac{k_e m}{B^2}} = d_{\text{crit}}.$$

Our analysis is also valid when the centre of mass is at rest, so it is not surprising that this result for d_{crit} is the same as that obtained in part a).

Let us try to sketch the trajectory of the particle in the special case in which $d = d_{\min}$, that is,

$$R = R_{\min} = \sqrt[3]{\frac{k_e m}{B^2}}.$$

Inserting this value into (5), the angular velocity of the electrons as they orbit around the centre of mass is

$$\omega = \frac{1}{2}\frac{e}{m}B = \frac{1}{2}\omega_{\text{c}},$$

that is, just one-*half* of the angular velocity of the centre of mass.

Start the system, as prescribed in part b), so that only one of the electrons is given an initial velocity v_0. This has to be in a direction perpendicular to the line segment connecting the two particles, because, if it were not, the electron separation would change 'before the second electron could do anything about it'.

Initially, that second electron is at rest, and so the centre of mass starts off with velocity $v_0/2$, and each electron has a velocity relative to the centre of mass of the same magnitude (but oppositely directed).

For circular motion around the centre of mass, we must have

$$\frac{v_0}{2} = R\omega = R\frac{\omega_{\text{c}}}{2}.$$

Simultaneously, for the circular motion of the centre of mass, we need

$$\frac{v_0}{2} = R_{CM}\omega_c, \quad \text{that is} \quad R_{CM} = \frac{R}{2}.$$

Accordingly, the centre of mass orbits around a circular trajectory that has half the radius of the circle in which the electrons orbit the centre of mass; further, the period of the centre of mass's motion is one-half that of the electrons that go round it.

A sketch of the particle trajectories is shown in Fig. 2. To make things a little clearer, the trajectory of the centre of mass is also included (drawn with a dashed line). While the 'catapulted' electron's radius arm (of length R) rotates through an angle θ around the centre of mass, that of the centre of mass itself has moved through 2θ around its own circular trajectory, which has radius $R/2$.

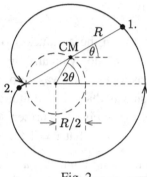

Fig. 2

During a period $T = 2\pi/\omega_c$, the centre of mass completes a whole circle, but the two electrons have each covered only a semicircle; they have just *swapped places with each other*. At that instant, the position and velocity of the centre of mass in space is exactly as it was at the start ($v_0/2$), the initially stationary electron has a velocity v_0, and the 'catapulted' one is at rest. It therefore first stops after a time

$$T = 2\pi \frac{m}{eB}.$$

Notes. 1. The trajectory curve of the two electrons (called a cardioid or 'heart curve', because of its shape) is simple and clear-cut only if their initial separation has the minimal (critical) value in the given magnetic field. If the constant distance is larger than the minimal one, then the trajectories and motions of the particles are more complicated, and generally the trajectories are not closed.

2. The situation is even more complicated if the initial velocity does not meet the conditions for a constant separation. But, even then, it can be shown that the particles cannot approach each other too closely, nor can they move very far apart – their separation always remains between two limiting values and oscillates between them.

S168   Instead of the original $\triangle$ circuit, we consider a $\curlyvee$ configuration (as shown in Fig. 1). The following simple equations can be written for the latter:

$$Y + Z = 1, \qquad X + Z = 2, \qquad X + Y = 3.$$

Their solution is $X = 2\ \Omega$, $Y = 1\ \Omega$ and $Z = 0\ \Omega$.

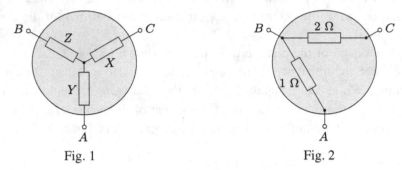

Fig. 1   Fig. 2

The connections between these values and those in the actual circuit take the form $X = yz/(x + y + z)$, and cyclically. In particular, $Z = xy/(x + y + z)$ and, since none of the measured values was zero, none of x, y and z can be zero. The only possible conclusion is that z is infinite, and that between terminals C and A there is an *open-circuit* (*see* Fig. 2). Further, $X = 2\ \Omega$ then implies that $y = 2\ \Omega$, and similarly $Y = 1\ \Omega$ implies that $x = 1\ \Omega$.

> *Notes.* 1. Infinity is *not* a number, and the usual rules of algebra are not valid for it; for this reason, our formal calculation has produced a contradiction. If the value of z is considered not infinitely large, but much larger (by many orders of magnitude) than $1\ \Omega$ (in physics, the meaning of 'infinite' is always of this kind), then the equations can be satisfied in a self-consistent way (and in accord with the accuracy of the given data). This procedure is equivalent to letting $1/(z + w)$, for any w, tend to zero; if this is done initially (with either foresight or hindsight!), then equations (1), (2) and (3) have a straightforward (and correct) solution.
>
> 2. The question can also be phrased in terms of the usual (non-transparent) black box containing a few ohmic resistors. In this case the solution (finding the minimal number and the values of the resistors inside the black box) could be 'heuristic'.

S169   Let every resistor of the actual chain have unit resistance. The one directly connected between points A and B can be notionally replaced by two resistors, one at each end of the strip, and each having 2 units of resistance. When the end-points of the chain are connected pairwise, these two resistors will be connected in parallel, with an equivalent resistance equal to the original 1 unit. This equivalent (for our purposes) circuit has symmetry properties that can be exploited.

If a given voltage V is connected across one end of this transformed chain, and, at the same time, a voltage $\pm V$ (i.e. with the same magnitude and the same or opposite polarity) is connected across the other, then we can dispense with the need to pairwise connect the two ends of the chain. The electric current distribution produced – as well as the equivalent resistance calculated from the currents – must be the same as when the chain is made into a closed strip, with case *a*) being simulated by the same polarity $(+V)$ connection, and case *b*) by the $-V$ arrangement.

The potential and current distributions in the (ordinary) strip connected to sources with the same polarity have *reflectional symmetry* about the mid-perpendicular (*ST* in Fig. 1), while in the Möbius strip connection (with sources of opposite polarity) the potential distribution has *two-fold rotational symmetry* with respect to the centre O of the chain. Points at the same potential can be connected together without affecting anything, and if we then find two 1 unit resistors are connected in parallel, they can be replaced by a single $\frac{1}{2}$ unit resistor. In the following figures, for both cases *a*) and *b*), 1 unit resistors are denoted by *empty* rectangles, and resistors with $\frac{1}{2}$ unit of resistance by *light grey* rectangles.

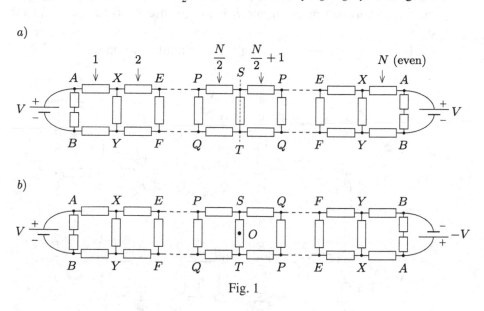

Fig. 1

If N is even, then Fig. 1 shows the two possible configurations, with points at the same potential denoted by the same letter; C and D have been replaced by A and B, as appropriate. These two circuits can be transformed into their half-length equivalents, as shown in Fig. 2.

Because of the rotational symmetry in case *b*), the potentials of points S and T are the same, and so the resistor between them, through which no current flows, can be replaced by a short-circuit. As the two networks differ only in the final loop on

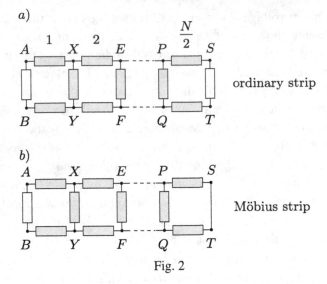

ordinary strip

Möbius strip

Fig. 2

the right-hand side, and that loop has a smaller resistance in case *b*), the equivalent resistance between points *A* and *B* is *larger* in case *a*) than it is for the Möbius strip.

Fig. 3 shows the corresponding pair of circuits for odd *N*.

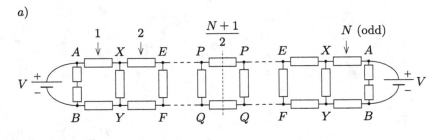

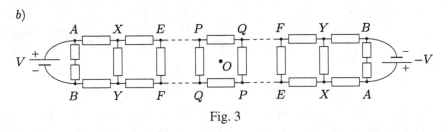

Fig. 3

The transformed networks, after equipotential points have been connected and parallel resistors have been substituted, can be seen in Fig. 4. Again, the equivalent resistance between points *A* and *B* is smaller in case *b*), as the two networks differ only in the last loop on the right-hand side, and its resistance is smaller in the case of the Möbius strip network.

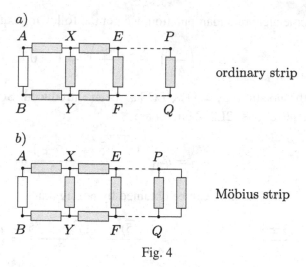

ordinary strip

Möbius strip

Fig. 4

S170 *Solution 1(a).* Imagine that any wire of the circuit (say, the segment AB in Fig. 1) is cut with scissors, and a tunable *voltage generator* is connected between the two loose ends. The circuit then undergoes a forced oscillation with the angular frequency ω of the generator.

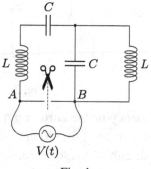

Fig. 1

When the latter approaches one of the natural frequencies of the system, the current strength between points A and B rises sharply because of the resonance. In this situation, the current flowing in the circuit is very large, even for a very small driving voltage, and, for ideal components, the system would oscillate without any voltage generator at all.

So, at the resonance frequency, the impedance Z_{AB} between points A and B approaches zero. Using complex impedances:

$$Z_{AB} = iL\omega + \frac{1}{iC\omega} + \frac{iL\omega \times 1/(iC\omega)}{iL\omega + 1/(iC\omega)} = 0.$$

After some algebraic manipulation, we get the following equation:

$$\left(\frac{\omega}{\omega_0}\right)^4 - 3\left(\frac{\omega}{\omega_0}\right)^2 + 1 = 0,$$

where the notation $\omega_0 = 1/\sqrt{LC}$ has been introduced. So the natural (angular) frequencies of this '2L2C' circuit are

$$\omega_{1,2} = \omega_0\sqrt{\frac{3 \pm \sqrt{5}}{2}} = \frac{\sqrt{5} \pm 1}{2}\,\omega_0.$$

The final simplification can be obtained by noting that

$$\frac{3 \pm \sqrt{5}}{2} = \frac{6 \pm 2\sqrt{5}}{4} = \frac{(\sqrt{5})^2 + (1)^2 \pm 2\sqrt{5}}{4} = \left(\frac{\sqrt{5} \pm 1}{2}\right)^2.$$

Solution 1(b). Choose two end-points of any circuit element, say points A' and B' in Fig. 2, and notionally connect a tunable a.c. *current generator* across them.

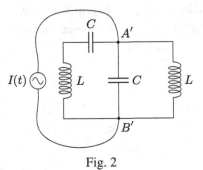

Fig. 2

If the angular frequency of the current generator is changed and it approaches one of the natural frequencies of the circuit, then the voltage between the arbitrarily chosen points rises sharply. At resonance, a very small input current produces a great voltage, i.e. the impedance $Z_{A'B'}$ between the points A' and B' approaches infinity:

$$Z_{A'B'} = \left(\frac{1}{iL\omega} + iC\omega + \frac{1}{iL\omega + 1/(iC\omega)}\right)^{-1} \longrightarrow \infty.$$

This happens if the value of the expression in parentheses is zero. From this, we get the same equation as in Solution 1(a), and, of course, the same final result.

Solution 2. Denote the charges on the two capacitors at a given moment by Q_1 and Q_2, and the current intensities flowing through them by I_1 and I_2 (*see* Fig. 3). Then the currents through the two inductors are I_1 and (according to the junction law) $I_1 - I_2$.

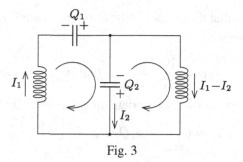

Fig. 3

Write Kirchhoff's second rule for the two clockwise loops shown in Fig. 3:

$$\frac{Q_1}{C} + \frac{Q_2}{C} - L\dot{I}_1 = 0,$$

$$-\frac{Q_2}{C} - L(\dot{I}_1 - \dot{I}_2) = 0.$$

Using the physical relationships $I_1 = -\dot{Q}_1$ and $I_2 = -\dot{Q}_2$, and the notation $\omega_0 = 1/\sqrt{LC}$, we find the following equations:

$$\ddot{Q}_1 = -\omega_0^2 Q_1 - \omega_0^2 Q_2, \tag{1}$$

$$\ddot{Q}_2 = -\omega_0^2 Q_1 - 2\omega_0^2 Q_2. \tag{2}$$

This system of (second-order) differential equations describes how the charges on the capacitors change with time. From the theory of coupled mechanical oscillations, we know that, in such a system, the quantities $Q_1(t)$ and $Q_2(t)$ can be written, quite generally, as the sum of two harmonic terms with different angular frequencies. But we can also find a linear combination of $Q_1(t)$ and $Q_2(t)$ that is a pure sinusoidal function of time. Let it be

$$q(t) = Q_1(t) + \alpha Q_2(t).$$

To find the constant α, add equation (1) to α times equation (2):

$$\ddot{Q}_1 + \alpha \ddot{Q}_2 = -\omega_0^2[(1+\alpha)Q_1 + (1+2\alpha)Q_2].$$

Factorise out the coefficient of Q_1 from the square brackets (so that its contents have the same form as that assumed for $q(t)$):

$$\underbrace{\ddot{Q}_1 + \alpha \ddot{Q}_2}_{\ddot{q}(t)} = -\underbrace{\omega_0^2(1+\alpha)}_{\omega^2}\Big[\underbrace{Q_1 + \underbrace{\frac{1+2\alpha}{1+\alpha}}_{\alpha} Q_2}_{q(t)}\Big]. \tag{3}$$

It can be seen that the linear combination $q(t)$ varies harmonically according to the equation $\ddot{q} = -\omega^2 q$, provided

$$\alpha = \frac{1 + 2\alpha}{1 + \alpha}, \qquad \text{that is} \qquad \alpha_{1,2} = \frac{1 \pm \sqrt{5}}{2}.$$

So, we have found two appropriate linear combinations:

$$q_1(t) = Q_1(t) + \alpha_1 Q_2(t), \qquad q_1(t) = q_{0,1} \sin(\omega_1 t + \varphi_1), \qquad (4)$$

$$q_2(t) = Q_1(t) + \alpha_2 Q_2(t), \qquad q_2(t) = q_{0,2} \sin(\omega_2 t + \varphi_2), \qquad (5)$$

and the corresponding natural frequencies, according to equation (3), are

$$\omega_{1,2} = \omega_0\sqrt{1 + \alpha_{1,2}} = \omega_0\sqrt{\frac{3 \pm \sqrt{5}}{2}} = \frac{\sqrt{5} \pm 1}{2}\,\omega_0.$$

The amplitudes $q_{0,1}$ and $q_{0,2}$ in equations (4) and (5), and the initial phase angles φ_1 and φ_2, can be determined from the initial conditions, that is, from the initial charges on the capacitors, and the initial currents in the inductors, so the functions $q_1(t)$ and $q_2(t)$ are known.

With arbitrary initial conditions, the charges on the capacitors can be expressed as functions of time using equations (4) and (5):

$$Q_1(t) = \frac{\alpha_2}{\alpha_2 - \alpha_1} q_1(t) - \frac{\alpha_1}{\alpha_2 - \alpha_1} q_2(t)$$

$$Q_2(t) = \frac{1}{\alpha_1 - \alpha_2} q_1(t) - \frac{1}{\alpha_1 - \alpha_2} q_2(t).$$

These expressions show that the charges on the capacitors (as functions of time) are each superpositions of two sinusoidal oscillations. By choosing proper initial conditions, it can be arranged that one of $q_1(t)$ and $q_2(t)$ is zero, and so produce one of the two natural frequencies, either ω_1 or ω_2, of the circuit.

S171 The connection between the voltage across the terminals of a capacitor with capacitance C and the alternating current (with angular frequency ω) flowing through it is[74]

$$\frac{V_C}{I_C} = \frac{1}{\omega C},$$

and so the electrical impedance (reactance) of the capacitor is $1/(\omega C)$; furthermore, the phase of the sinusoidal alternating current *leads* that of the capacitor voltage by 90°. For a coil, the same ratio is

---

[74] Here V_C and I_C can denote either peak values or r.m.s. values of the current and voltage.

$$\frac{V_L}{I_L} = \omega L,$$

i.e. the electrical impedance (reactance) of a coil with inductance L is ωL, and the current flowing through it *lags* the coil voltage by 90°.

As the phase shift between voltage and current in an inductor has the opposite sign to that in a capacitor, a coil can be considered as a capacitor with negative capacitance C_L (< 0):

$$\omega L = -\frac{1}{\omega C_L}, \quad \text{that is} \quad C_L = -\frac{1}{\omega^2 L}.$$

Being able to make such a substitution will allow us to work with an equivalent network consisting *entirely of capacitors*.

In what follows, it will be convenient to use the following notation:

$$k^2 = \frac{1}{\omega^2 LC} = \left(\frac{\omega_0}{\omega}\right)^2.$$

The dimensionless number k shows how many times larger than the applied frequency ω is the natural (resonance) frequency ω_0 of a simple LC oscillator, consisting of only one coil with inductance L and one capacitor with capacitance C. Using this notation, the 'effective capacitance' of each coil in the chain can be written in the form $C_L = -k^2 C$.

In a physics problem, the term 'infinite chain' means that the chain is 'very long' – the number of its elements is very large, but it is finite. Consider, first, a chain of length n (consisting of n coils and n capacitors). As the chain does not have any ohmic resistors, the phase shift between the current flowing through it and the voltage is either $+90°$ or $-90°$. In one of these cases, the chain can be replaced by a capacitor with a particular capacitance C_n, and in the other by a coil that has a suitable inductance. We will suppose that the first situation holds, and determine the value of C_n for the first few n. If it turns out that C_n is negative, then the chain behaves as a coil.

In the case of $n = 1$, we have to determine the equivalent impedance of a capacitor with capacitance C and a coil with inductance L (treated as another capacitor with capacitance C_L) connected in series:

$$\frac{1}{C_1} = \frac{1}{C} + \frac{1}{C_L},$$

from which

$$C_1 = \frac{CC_L}{C + C_L} = \frac{C \times (-k^2 C)}{C + (-k^2 C)} = \frac{k^2}{k^2 - 1} C.$$

For $n = 2$, a capacitor with capacitance C is connected in series with the parallel resultant of a coil and C_1:

$$\frac{1}{C_2} = \frac{1}{C} + \frac{1}{C_1 + (-k^2 C)}.$$

In general,

$$\frac{1}{C_{n+1}} = \frac{1}{C} + \frac{1}{C_n - k^2 C},$$

from which

$$C_{n+1} = \frac{C(C_n - k^2 C)}{C_n + (1 - k^2)C}.$$

It is convenient to express all of the capacitances in the form $C_n = x_n C$ (i.e. all capacitances are measured in units of C), because then the recursion formula assumes a clearer form:

$$x_{n+1} = \frac{x_n - k^2}{x_n + 1 - k^2}. \tag{1}$$

Question. Does the sequence C_n tend to a limit as $n \to \infty$? May we say that 'if $n \gg 1$, then $x_n \approx x_{n+1}$'? If so, by equating the approximately equal numerical values x_n to a common value x, the recursive relationship (1) can be transformed into a quadratic equation. Let us try it:

$$x = \frac{x - k^2}{x + 1 - k^2}, \qquad \text{that is} \qquad x^2 - k^2 x + k^2 = 0.$$

In the case of $k > 2$ (corresponding to low input frequencies), this equation has two real roots:

$$x_\pm = \frac{k^2}{2} \pm \frac{k}{2}\sqrt{k^2 - 4}. \tag{2}$$

Which one is the 'correct' root? Or perhaps both of them have physical meanings! You might suspect that, of the roots in expression (2), the smaller one x_- is the 'right' one. This is because, if coils for which L is very small were used (i.e. $\omega_0 \gg \omega$), then they would behave as virtual short-circuits, and the reactance between points A and B would be approximately the same as that of a single capacitor of capacitance C, i.e. $x \approx 1$. Now, as can be easily verified, it is the case that, if $k \gg 1$, then $x_- \approx 1$, corresponding to the impedance between points A and B being the expected $1/\omega C$. By contrast, for large values of k, $x_+ \approx k^2 \gg 1$, which is not in line with the physical situation. The conclusion is that the intuitive choice is the correct one.

These considerations show that the equivalent impedance of the 'infinite chain', at an angular frequency $\omega < 1/(2\sqrt{LC})$ – independently of the termination at its far end, i.e. independently of the numerical value of x_1 – can be only one well-defined value:

$$Z_{chain} = \frac{1}{\omega x_- C} = \frac{1}{\omega C} \frac{2}{k(k - \sqrt{k^2 - 4})} = \frac{1 + \sqrt{1 - 4\omega^2 LC}}{2\omega C}.$$

But what about the case of $k < 2$, i.e.

$$\omega > \frac{1}{2\sqrt{LC}} = \frac{\omega_0}{2}?$$

Then expression (2) is complex, with both its real and imaginary parts non-zero, showing that the impedance of the chain is not a real multiple of C. In other words, the recurrence relation, although correctly calculating x_n for each n, does not produces a sequence that tends to a limit. Interpreted physically, this means that the equivalent impedance of the long (but not infinitely long!) chain really does depend on the actual size of the chain – it depends on how long 'very long' actually is. This (higher) range of frequencies is discussed more quantitatively in the Note that follows.

Note. The problem can also be tackled using complex impedances: $i\omega L$ for an inductance; $1/(i\omega C)$ for a capacitance. If Z is the impedance of the 'infinite' chain, then it is also the impedance of the same chain with an additional capacitor–inductor unit attached to its front end, i.e. a capacitor in series with a parallel arrangement consisting of an inductor and the chain:

$$Z = \frac{1}{i\omega C} + \frac{i\omega L Z}{i\omega L + Z}.$$

Using the previous notation, this can be rearranged as

$$2Z\sqrt{\frac{C}{L}} = -ik \pm \sqrt{4 - k^2}$$

This shows that, if $k > 2$, then Z is purely imaginary, and since the second term is always less than the first in magnitude, Z has a negative imaginary value, i.e. it is capacitive in nature. The resolution of the $\pm$ dichotomy is as previously discussed.

For $k < 2$ (i.e. $\omega > 1/(2\sqrt{LC})$), the second term is real and the equivalent impedance of the *infinite chain* contains an ohmic part. This is surprising, because an arbitrarily long chain, but one necessarily containing a finite number of (ideal) coils and capacitors in practice, cannot have any ohmic resistance, it must have a purely capacitive or inductive equivalent impedance.

The appearance of an ohmic resistance for the 'infinite' (in reality finite, but very long) chain when $\omega > \omega_0/2$ is connected with the propagation of a wave along the chain. This wave transports energy from the power supply along the chain, and in the case of a very long chain this energy reflects back after a very long time. If the chain is really 'infinitely long', the wave never comes back, and so the inductor–capacitor chain behaves like a circuit containing ohmic elements as well.

S172   Positive charges accumulate on the surface of one of the wires, and negative ones on the surface of the other; denote the corresponding surface charge densities by $+\sigma$ and $-\sigma$. The electric fields created by these charges will be calculated separately for each conductor, and then the two fields will be superimposed.

If we had only a single charged wire, then its electric field would have axial symmetry, and the magnitude of the electric field strength, calculated at a distance r from the axis of the cylindrical conductor, would be[75]

$$E(r) = \frac{\sigma d}{2\varepsilon_0} \frac{1}{r}.$$

The electric field of the charge on a single wire must produce a potential difference between the conductors of $V/2$, because the net field of the two wires has to give a p.d. of V, to match the e.m.f. of the battery. The connection between the voltage and the electric field is the following:

$$\frac{V}{2} = \int_{r=d/2}^{D-(d/2)} E(r)\,dr \approx \frac{\sigma d}{2\varepsilon_0} \int_{r=d/2}^{D} \frac{1}{r}\,dr = \frac{\sigma d}{2\varepsilon_0} \ln \frac{2D}{d},$$

from which it follows that the surface charge density on the wires is

$$\sigma = \frac{\varepsilon_0\, V}{d \ln 100}.$$

The attractive electric force between the wires can be found as the product of the (average) electric field produced by one of the wires at the location of the other, and the electric charge on the latter:

$$F_{\text{electric}} = E(r=D)Q = \frac{\sigma d}{2\varepsilon_0} \frac{1}{D} \sigma \pi d L = \frac{\varepsilon_0 \pi L V^2}{2D(\ln 100)^2}.$$

The repulsive magnetic force is the Lorentz force produced by the magnetic field associated with one of the currents I as it acts upon the other. The magnetic induction $\boldsymbol{B}$, calculated from Ampère's rule, has magnitude $\mu_0 I/(2\pi r)$ at a radial distance r from the wire. So, the magnetic force is

$$F_{\text{magnetic}} = BIL = \frac{\mu_0 I}{2\pi D} IL = \frac{\mu_0}{2\pi} \frac{L}{D} \frac{V^2}{R^2}.$$

In the final step, Ohm's law, $I = V/R$, has been used.

To attain the given condition, $F_{\text{electric}} = F_{\text{magnetic}}$, we must have

$$\frac{\varepsilon_0 \pi L V^2}{2D(\ln 100)^2} = \frac{\mu_0}{2\pi} \frac{L}{D} \frac{V^2}{R^2},$$

---

[75] If not already known, this can be proved by applying Gauss's law to a cylinder, of radius r and length L, that is coaxial with the wire: $E \cdot 2\pi r L = (1/\varepsilon_0)\pi d L \sigma$. From this the expression for $E(r)$ follows immediately.

from which

$$R = \frac{1}{\pi}\sqrt{\frac{\mu_0}{\varepsilon_0}} \ln 100 = 553 \ \Omega.$$

With this load resistor, the attractive electric force between the unlike charges just balances the repulsive magnetic force between the opposing currents.

> *Note.* It is interesting to note that the resistance R *does not depend* on the length of the wires, nor on the battery voltage – even the (large) ratio D/d (treated as $\gg 1$) appears only in a logarithmic way (i.e. only as the argument of a slowly changing function). The order of magnitude of R is effectively determined by the universal quantity $\sqrt{\mu_0/\varepsilon_0} \approx 377 \ \Omega$, which is called the *wave impedance of free space*.

S173 Consider first a single ring, and concentrate on a small part of it, introducing a reference frame $\mathcal{K}$ in which that part is at rest. As the ring is moving with constant angular acceleration α, this reference system is not an inertial one and a certain linear acceleration is generated within it.

As the ring is very thin, the radial component of this acceleration produces no significant effects. This is not the case, however, for the tangential component of the acceleration, which acts along the part of the ring under consideration and has magnitude $r\alpha$. In $\mathcal{K}$ the positive ions forming the metal crystal lattice are at rest, but a certain inertial force acts on the electrons, which have (effective) mass m. This inertial force has magnitude $mr\alpha$ and its direction is opposite to that of the tangential acceleration.

The interaction between the electrons and the crystal lattice prevents the former from increasing their speed indefinitely; according to Ohm's law, the interaction increases as the velocity of the electrons, relative to the crystal lattice, increases. After some time, equilibrium between the inertial force and the lattice-induced braking force is reached. Consequently, the positive ions and the negative electrons move (tangentially) with different speeds, and as measured in $\mathcal{K}$ an electric current flows.

The magnitude of the inertial force is the same everywhere, and, since it acts tangentially at all points on the ring, it has the same effect as that of a fictitious electric field similarly directed. The magnitude of this fictitious electric field can be found by equating the force due to it to the inertial force:

$$eE = mr\alpha \qquad \text{and so} \qquad E = \frac{mr\alpha}{e}.$$

In a single ring of resistance R (at rest), the above electric field generates a current:

$$I = \frac{2\pi r E}{R} = \frac{2\pi m r^2 \alpha}{eR}.$$

The field E is a fictitious electric field, but the interaction between this field and the electrons is real, and so is the electric current!

The system described in the problem can be considered as a very long solenoid with n loops per unit length, in which a current I is flowing. It is well known that the magnetic field B inside a long solenoid (well away from its ends) is homogeneous and has magnitude $B = \mu_0 n I$, where μ_0 is the permeability of free space. Since the points along the axis are not rotating, but are at rest in both $\mathcal{K}$ and the laboratory frame, the axial magnetic field at the centre of the cylinder, measured in the laboratory frame, is

$$B = \frac{2\pi\mu_0 nmr^2\alpha}{eR}.$$

Notes. 1. The Stewart–Tolman effect is very instructive, because, despite the fact that the rings are electrically neutral, in this system – because of the finite inertial mass of the charge carriers – there is, *perhaps unexpectedly*, a current flowing and a magnetic field produced.

2. Experimentally, it is easier to demonstrate the Stewart–Tolman effect in electrolytes than in metals because the effect is directly proportional to the mass of the charge carriers. A similar phenomenon – also related to the inertia of the charge carriers – can be observed during the linear acceleration along its axis of a metal rod; in this case, charge separation occurs.

3. Of the three fictitious forces appearing in rotating frames (the centrifugal force, Coriolis force and Euler force), the first two, which act radially, result in a finite but negligible charge separation in the thin rings, and only the third one ($-m\alpha \times r$) gives a tangential force relevant to the solution.

4. The Stewart–Tolman effect can also be interpreted in an inertial reference frame. The free, or quasi-free, conduction electrons in the metal must be accelerated by something (they can hardly stand still in the laboratory frame), and this something cannot be anything else other than the interaction between the conduction electrons and the crystal lattice of positive ions; this interaction might even be called an *electromotive force*.

5. If the rings are constantly accelerated, say, in the positive sense (anticlockwise), then the electrons always lag a little bit behind the crystal lattice. Consequently, the current of positive ions is larger than that of the electrons, and the net current is also anticlockwise. The resulting magnetic field vector (determined using the right-hand rule for solenoids) points in the same direction as the positive axis for the cylinder's rotation (to the left in the figure in the problem).

S174 Our overall aim is to find the connection between the voltage V of the voltmeter and the current I flowing into and out of the plate. If we knew the (position-dependent) current density $j(r)$ of the two-dimensional current distribution, then, using the continuum form of Ohm's law $E(r) = \varrho j(r)$, we could calculate the electric field inside the plate. From that, the voltage difference between points C and D could be found by using the integral calculus. This is a rather daunting mathematical task – but, fortunately, there is a much easier way.

At an edge of the plate, the current density vector $j(r)$ must have no component perpendicular to that edge. This unusual boundary condition can be managed relatively easily if we follow a method similar to that of image charges, as used in electrostatics (*see* figure).

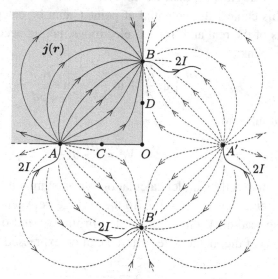

Denote by A' and B' the mirror image points of, respectively, A in side edge BD and B in AC. The current distribution in the actual *finite* plate is exactly the same as it would be in one quarter of an *infinite* metal plate, into which currents $2I$ were injected at both A and A', and currents $2I$ left it at each of points B and B'. Note that, although the current entering at A is $2I$, that flowing through the actual plate is only I.

Now consider a situation in which current enters/leaves the infinite metal plate at a single electrode and leaves/enters it at infinity. If a current $2I$ enters the plate, then the magnitude of the current density at a distance r from the input terminal would be

$$j(r) = \frac{2I}{2\pi rt}.$$

The corresponding electric field is

$$E(r) = \frac{\varrho I}{\pi rt},$$

and the potential (relative to an arbitrarily chosen point that is a distance r_0 from the electrode) is given by

$$V(r) = -\int_{r_0}^{r} E(r)\,dr = \frac{\varrho I}{\pi t}\ln\frac{r_0}{r}.$$

In the following, in order to find the potentials at C and D, we will superimpose four partial potentials. Although, the results cannot depend on the arbitrary position of the zero of potential, the calculation is much shorter if it is chosen as the centre O of the square $ABA'B'$; then we have $AO = BO = A'O = B'O = r_0$.

With this choice, the potential at point C can be calculated by superimposing the effects of the real and 'image' electrodes, taking account of whether each is a source or a sink:

$$V_C = \frac{\varrho I}{\pi t}\left(\ln\frac{r_0}{d} + \ln\frac{r_0}{3d} - \ln\frac{r_0}{\sqrt{5}d} - \ln\frac{r_0}{\sqrt{5}d}\right),$$

which simplifies to

$$V_C = \frac{\varrho I}{\pi t}\ln\frac{5}{3}.$$

Either by an exactly parallel calculation, or on symmetry grounds, the potential at point D is given by $V_D = -V_C$. So the voltage between C and D is $2V_C$, and this will be the reading V on the voltmeter. From this and our expression for V_C, the resistivity ϱ of the metal plate material can be expressed as

$$\varrho = \frac{\pi t}{2\ln(5/3)}\frac{V}{I}.$$

It is interesting to note that the result does not depend on the choice of d, providing it is much smaller than the length of a plate edge, but, at the same time, much larger than the plate thickness.

S175   If a capacitor of capacitance C and initial charge Q_0 discharges through an ohmic resistor R, then the charge $Q(t)$ on the capacitor decreases with time according to

$$Q(t) = Q_0\,e^{-t/(RC)},$$

and it reaches the value of $Q_0/2$ in a time[76]

$$T_{1/2} = RC\ln 2. \tag{1}$$

In the analogous situation given in the problem, C is the capacitance of the 'isolated' metal sphere, and R is the 'effective resistance' of the poorly conducting air. The sphere's capacitance is straightforward to determine. The potential, relative

---

[76] This time interval is similar to the parameter used to describe radioactive decays, and can also be called a *half-life*.

to a reference point at infinity, of a sphere of radius r_0 and carrying a charge Q is $Q/(4\pi\varepsilon_0 r_0)$, and so its capacitance is

$$C = 4\pi\varepsilon_0 r_0. \tag{2}$$

The determination of the effective resistance of the air around the sphere is a somewhat more sophisticated task. The air, with resistivity $1/\sigma$, can be notionally divided into thin spherical shells concentric with the sphere. Let the thickness of a shell at radius r be Δr. Its surface area is $4\pi r^2$, and so its electrical resistance (for radial currents) is

$$\Delta R = \frac{1}{\sigma}\frac{\Delta r}{4\pi r^2},$$

and the net resistance of all of the spherical shells, connected in series, is

$$R = \sum \Delta R = \frac{1}{4\pi\sigma}\sum \frac{\Delta r}{r^2}. \tag{3}$$

This sum can be evaluated (without approximation) by letting $\Delta r \to 0$, yielding the integral:

$$R = \frac{1}{4\pi\sigma}\int_{r_0}^{\infty}\frac{dr}{r^2} = \frac{1}{4\pi\sigma r_0}. \tag{4}$$

The same result can also be found without integration, by using the fact that the potential of a point charge, relative to a reference point at infinity, can be calculated from its electric field strength; this procedure produces a very similar sum to that in (3).

Substituting results (2) and (4) into expression (1), we get that the half-life of the charge on the metal sphere is

$$T_{1/2} = RC\ln 2 = \frac{1}{4\pi\sigma r_0}4\pi\varepsilon_0 r_0 \ln 2 = \frac{\varepsilon_0}{\sigma}\ln 2.$$

It is interesting to note that the half-life does not depend upon the size of the metal sphere. It is determined solely by the conductivity of air and a universal physical constant.

S176 Let us, hypothetically, enclose the figure within a closed convex surface. Now, as shown in the figure, we consider the surface to be divided into small segments and denote the area of the ith segment by ΔA_i, the electric field produced by the aluminium-clad Santa Claus at the position of this segment by E_i, and the current density there by j_i.

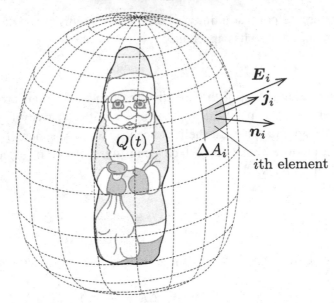

The total electric current passing through the closed surface gives the rate of decrease at time t of Santa Claus's charge, which is denoted by $Q(t)$:

$$\frac{dQ(t)}{dt} = -\sum_i j_i \cdot n_i \, \Delta A_i,$$

where n_i is the outward normal of the ith area segment. According to Ohm's law, the current density j_i is proportional to the electric field E_i, and the proportionality factor is the conductivity of air, σ, thus $j_i = \sigma E_i$.

Using this to eliminate the current, the rate of decrease of Santa Claus's charge is given by

$$\frac{dQ(t)}{dt} = -\sigma \sum_i E_i \cdot n_i \, \Delta A_i.$$

However, the sum on the right-hand side of this equation is the total flux of the figure's electric field *out of* the closed surface, and according to Gauss's law this is the product of ε_0^{-1} and the total charge *inside* the closed surface:

$$\frac{dQ(t)}{dt} = -\frac{\sigma}{\varepsilon_0} Q(t).$$

The solution to this differential equation, which is analogous to that describing the decay of radioactive nuclei, is

$$Q(t) = Q(0) \exp\left(-\frac{\sigma}{\varepsilon_0} t\right).$$

It follows that Santa's charge is halved in a time $T_{1/2}$ given by

$$\frac{1}{2} = \exp\left(-\frac{\sigma}{\varepsilon_0} T_{1/2}\right) \qquad \longrightarrow \qquad T_{1/2} = \frac{\varepsilon_0 \ln 2}{\sigma}.$$

We have obtained a rather surprising result – that the half-life is *independent* of both the size and shape of the chocolate figure, and depends only on the conductivity of air. But given this, it is not so surprising that the result is exactly the same as that for a charged sphere (see page 428).

S177 At a time t, let the rod be a distance $x(t)$ from A. At that moment, the area of the triangle formed by the rod and the arms of the V-shaped wire is

$$T(x) = x^2 \tan\frac{\alpha}{2},$$

and so the magnetic flux it encloses is

$$\Phi(t) = BT(x).$$

The magnitude of the induced electromotive force (voltage) is given by the rate of change of this flux:

$$V = \frac{d\Phi}{dt} = B\frac{dT}{dt}.$$

The area T depends on time because it contains the function $x(t)$.

The instantaneous resistance of the (triangular) circuit is $R = r\ell$, where $\ell(x) = 2x\tan(\alpha/2)$ is the distance between the points at which the rod and the wire make contact. Because of the induced voltage, a current flows in the closed circuit, and has magnitude

$$I = \frac{V}{R} = \frac{B}{r\ell}\frac{dT}{dt},$$

and so the Lorentz force acting on the rod is

$$F = IB\ell = \frac{B^2}{r}\frac{dT}{dt}.$$

This force decreases the speed of the rod, whatever the direction of the vertical field; if the field were reversed, then the induced e.m.f. and the current direction would also be reversed, but the Lorentz force would not be.

In accord with Newton's second law,

$$m\frac{dv}{dt} = F = -\frac{B^2}{r}\frac{dT}{dt},$$

from which the following 'conservation law' follows:

$$\frac{B^2}{r}T(x) + mv = \text{constant},$$

that is, independent of time. Using the formula for $T(x)$, and the initial conditions, we can write

$$\frac{B^2}{r}x_0^2 \tan\frac{\alpha}{2} + mv_0 = \frac{B^2}{r}x^2 \tan\frac{\alpha}{2} + mv.$$

It can be seen that, as x increases, v decreases, and, at the position

$$x = x_{\text{max}} = \sqrt{x_0^2 + \frac{mv_0 r}{B^2 \tan(\alpha/2)}},$$

the rod stops.

The time interval until the rod stops is, in principle, infinitely long, although in reality, because of some unavoidable friction and air drag, it is only 'very long'. The time dependence of the velocity, and the position function $x(t)$, can be determined using integral calculus or numerical methods.

S178 Almost immediately after release, following a short transient process, the magnet moves uniformly, with its weight compensated by the magnetic forces produced by the eddy currents induced in the wall(s) of the non-ferromagnetic tube(s); it covers the distance between the markers at a constant speed.

We can assume that the magnetic dipole moment of the strong cylindrical magnet is vertical, and that the eddy currents form horizontal loops in the neighbourhood of the magnet, with those above and below it having opposite senses of rotation. Note that there are no eddy currents in the walls of the tubes in the horizontal plane containing the midpoint of the magnet. This is because, in this plane, the magnetic field lines are vertical, and so, in its vicinity, the field does not change during the slow motion of the magnet. The magnetic field lines are mirror-symmetric about this mid-plane, and have opposite slopes above and below it. This is the very reason why the eddy currents have opposite senses of rotation above and below the magnet.

The magnetic braking force is proportional to the 'average' magnitude of the eddy currents, which is proportional to the induced electromotive force (voltage), which, in turn, is proportional to the speed of the magnet (Faraday's law). Consequently, we can write for the two cases in which the magnet falls through a single tube:

$$mg = A_1 v_1 = A_1\frac{\ell}{t_1} \quad \text{and} \quad mg = A_2 v_2 = A_2\frac{\ell}{t_2}.$$

Here m is the mass of the magnet, v_1 and v_2 are the terminal speeds of the magnet in the two different tubes, and ℓ is the distance between the markers. The proportionality factors A_1 and A_2 are constants that depend on the physical and geometrical properties of the tubes.

Because of the special geometry of the arrangement, with the two thin-walled tubes fitting tightly into each other with negligible mutual inductance, we assume that in the double-tube case we can use the same proportionality factors:

$$mg = (A_1 + A_2)v = (A_1 + A_2)\frac{\ell}{t}.$$

Here v is the terminal speed of the magnet in the double tube, and t is the unknown time to be determined. This procedure is equivalent to assuming that the eddy currents produced in the two tubes by the moving magnet are independent of each other.

After substituting for A_1 and A_2 and some simple algebraic manipulation, we find that the transit time through the double tube is $t = t_1 + t_2$.

> Note. The final result is so simple that it hardly seems believable. However, quite surprisingly, measurements we have made agree with the simplified theory used in the solution, within an error margin of 10 %.

S179 The induced voltage is determined by the mutual inductance of the small circular wire loop with respect to the larger one. It is not easy to calculate this quantity, as the magnetic field of the small loop is strongly inhomogeneous inside the larger circular loop, changing significantly (but continuously, of course) from point to point.

It is much easier to determine the mutual inductance of the large circular wire loop with respect to the smaller one. This is because a current I flowing in the larger loop produces a magnetic field at the position of the smaller one that can, to a good approximation, be considered homogeneous, and its magnitude is simple to calculate:

$$B \approx \frac{\mu_0 I}{2R}.$$

The magnetic flux of this field through the small circular loop of area πr^2 is

$$\Phi(t) = \frac{\mu_0 \pi r^2}{2R} I(t),$$

and so, from Faraday's law, the induced voltage is

$$V^*(t) = -\frac{d\Phi(t)}{dt} = -\frac{\mu_0 \pi r^2}{2R}\frac{dI(t)}{dt}. \tag{1}$$

According to the definition of the mutual inductance, we have

$$V^*(t) = -M \frac{dI(t)}{dt},$$

and comparing this with expression (1), the mutual inductance of the large circular wire loop, relative to the smaller one, can be found:

$$M = \frac{\mu_0 \pi r^2}{2R}. \tag{2}$$

Because of the symmetry property of mutual inductance, formula (2) also represents the mutual inductance of the small circular wire loop relative to the larger one. It follows that the magnitude of the induced voltage in the large circular loop is

$$V^* = \frac{\mu_0 \pi r^2}{2R} \frac{dI_1}{dt} = \frac{\mu_0 \pi r^2}{2R} \frac{I_0}{t_0},$$

with its polarity determined by Lenz's law, i.e. in the figure in the problem, it would cause a current to flow anticlockwise.

S180   The magnetic flux through the circular loop, produced by the small bar magnet, changes as the latter falls, and so an induced voltage V^* appears in the loop, in accordance with Faraday's law:

$$V^* = -\frac{d\Phi}{dt}. \tag{*}$$

During the motion of the magnet, the polarity of the induced voltage can be of either sign, depending on the momentary sign of the flux change; accordingly, the diode either blocks the current (behaves like an open-circuit), or it conducts the current (becomes a short-circuit with no resistance).

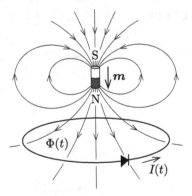

The magnitude of the magnetic flux through the loop, starting from an almost-zero value, increases as the magnet approaches the loop (*see* figure); later, with the magnet at the centre of the loop, it reaches its maximal value Φ_{max}; finally, as the magnet leaves the loop, the flux decreases. Applying Lenz's law, it can be seen that, in the first part of the motion, the current 'would like' to flow through the diode in the conducting direction (forward bias), and, because nothing prevents this, some charge flows through the diode. As the magnet moves away from the centre of the loop, the polarity of the induced voltage reverses and the diode no longer conducts (reverse bias); no current flows during this part of the magnet's motion.

For the forward-biased diode, the charge flow rate (the current) through it is given by Ohm's law:

$$\frac{d\Phi}{dt} = R\frac{dQ}{dt}.$$

The negative sign in (∗) has been dropped, because it refers only to the direction of the charge flow. From this, we get

$$\frac{d(\Phi - RQ)}{dt} = 0 \qquad \longrightarrow \qquad \Phi(t) - RQ(t) = \text{constant}.$$

Initially, the magnetic flux through the loop is negligible, and the charge flowing through the loop is zero, so the value of the constant is 0. All the charge that is allowed to flow has done so by the time the flux reaches its maximum – there is no more after that. So, the total amount of charge that flows through the diode is

$$Q = \frac{\Phi_{\text{max}}}{R}.$$

Our only remaining task is the determination of the magnetic flux through the loop at the moment when the bar magnet is at the centre of it. To do this, we notionally replace the magnet with a small circular current loop of radius r_0 that carries a current I_0, with r_0 chosen to make the loop's magnetic moment equal to m, i.e. such that

$$|m| = \pi r_0^2 I_0.$$

In the solution on page 433, it was proved that the mutual inductance of two concentric circular wire loops, in the same plane, and with radii r and $r_0 \ll r$, is

$$M = \frac{\mu_0 \pi r_0^2}{2r}.$$

It follows that the maximum flux through the circular loop, produced when the small bar magnet bar is at its centre, is given by

$$\Phi_{\text{max}} = MI_0 = \frac{\mu_0 \pi r_0^2}{2r} I_0 = \frac{\mu_0 |m|}{2r}.$$

Finally, the total amount of charge flowing through the diode is therefore

$$Q = \frac{\mu_0 |m|}{2rR}.$$

S181   Number the wire loops by size, i.e. let the smallest one be 1, and the largest be 3. Further, denote the mutual inductance of the ith loop with respect to the jth loop by M_{ij}, noting that, perfectly generally, $M_{ij} = M_{ji}$.

If the rate of change of the current in the middle loop is dI/dt, then in the other two loops the induced voltages are

$$V_1 = M_{21}\frac{dI}{dt} \quad \text{and} \quad V_3 = M_{23}\frac{dI}{dt}.$$

At the precise moment that $V_1 = V_0$, the induced voltage between the terminals of the largest loop is

$$V_3 = \frac{M_{23}}{M_{21}}V_0 = \frac{M_{32}}{M_{21}}V_0.$$

Although the individual determinations of the coefficients M_{21} and M_{32} need sophisticated calculations, their quotient can be found quite easily. We note that the system consisting of loops 2 and 3 is simply that consisting of loops 1 and 2, scaled up by a factor of 2. A two-fold magnification of the smaller loops has two effects: (i) it reduces the magnitude of the magnetic field that is produced by the larger loop and passes through the smaller one by a factor of $\frac{1}{2}$; (ii) it increases the surface area through which the linking flux passes by a factor of 4. Taking both effects into account, we see that the mutual inductance will be doubled: $M_{32} = 2M_{21}$.

Accordingly, the answer to the question in the problem is that voltage $V_3 = 2V_0$. As this result holds for any V_0, we conclude, more generally, that, at any moment, the induced voltage in the largest loop is twice that in the smallest one.

S182   The rings are identical, and so their self-inductances are the same (equal to L, say). Denote the (necessarily) common mutual inductance by M. This value is not fixed, as it depends on the relative orientation, and the distance apart, of the rings. When they are very far from each other, $M = 0$, but as they get closer, M increases.

In each of the two superconducting rings (which have zero resistance), the value of the magnetic flux cannot change; if it could, the induced voltage would produce an infinitely large current. Denoting the currents by I_A and I_B, the magnetic fluxes through the individual rings, and their constant values, are described by the equations below. Each contribution to a flux can be expressed as the product of a current and the appropriate inductance:

$$\Phi_A = LI_A + MI_B = LI_0, \tag{1}$$

$$\Phi_B = MI_A + LI_B = 0. \tag{2}$$

Eliminating M, we get

$$I_A^2 - I_0I_A - I_B^2 = 0, \tag{3}$$

and its solution for I_A (when $I_B = I_1$) is

$$I_A = \frac{I_0 \pm \sqrt{I_0^2 + 4I_1^2}}{2}.$$

Of the two roots, the physically correct one is that with a positive sign in front of the square root, because we must have $I_A = I_0$ when $I_1 = 0$. In summary, the current in ring A, under the conditions stated in the question, is

$$I_A = \frac{I_0 + \sqrt{I_0^2 + 4I_1^2}}{2}.$$

Note. Of the two roots of the quadratic equation, the physically correct one can be determined in a different way. We rewrite (3) as an equation for the ratio of the currents, i.e. for $x = I_B/I_A$:

$$x^2 + ax - 1 = 0,$$

where $a = I_0/I_A$. The product of the two roots of this equation is -1, so the absolute value of one of the roots must be greater than 1. However, this cannot be the physically correct root, because according to (2)

$$|x| = \left|\frac{I_B}{I_A}\right| = \left|\frac{M}{L}\right|, \qquad (*)$$

and this ratio cannot be greater 1 because of the physical inequality $M \le \sqrt{L_1 L_2} = L$.

So, the correct root for x is the one with a modulus less than unity; as the choices are $\frac{1}{2}(-a \pm \sqrt{a^2 + 4})$, this is the one with the plus sign before the square root. This root lies in the range $0 \le x \le 1$, showing that $|I_B| \le |I_A|$, in agreement with $(*)$.

S183 If a current I flows in the wire forming the square in Fig. 1a), then its magnetic field produces a magnetic flux $\Phi_{\text{itself}} = IL_1$ through the square face. Taking the positive directions as pointing outwards from the cube, this flux is positive. But the current also produces some (negative) magnetic flux $\Phi_{\text{neighbour}}$ through each of the neighbouring faces of the cube.

a)

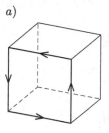

b)

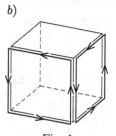

c)

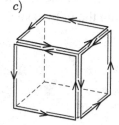

Fig. 1

We can find a connection between this flux and the self-inductance measured in the arrangement in Fig. 1*b*). As noted in the hint and illustrated in Fig. 1, the arrangement in Fig. 1*b*) is equivalent to squares on two neighbouring faces of the cube (with the opposing currents in the two sides that are thereby superimposed cancelling each other). Since each square has the other square as a neighbour, the total flux through the configured wire circuit is related to L_2 by

$$IL_2 = \Phi^{(b)} = 2\Phi_{\text{itself}} + 2\Phi_{\text{neighbour}} = 2IL_1 + 2\Phi_{\text{neighbour}}.$$

This can be rearranged to give

$$\Phi_{\text{neighbour}} = I(\tfrac{1}{2}L_2 - L_1).$$

As shown in Fig. 1*c*), the third wire arrangement can be replaced by three squares. Each of the three squares has two neighbouring squares, and, with current I flowing in each, the total magnetic flux is

$$\Phi^{(c)} = 3\Phi_{\text{itself}} + 3 \times 2\Phi_{\text{neighbour}}.$$

This must be equal to IL_3 and so

$$IL_3 = 3\Phi_{\text{itself}} + 6\Phi_{\text{neighbour}} = 3IL_1 + 6I(\tfrac{1}{2}L_2 - L_1) = 3I(L_2 - L_1).$$

So, finally, the coefficient of self-inductance in the arrangement in Fig. 1*c*) is

$$L_3 = 3(L_2 - L_1).$$

S184   We use the numbering shown in the figure of the problem. As the coils are identical, their self-inductances L are all equal, and, because of their symmetrical arrangement around the toroid's circumference, their mutual inductances are also all equal. We denote the common value by M.

Before the switch is closed, $I_2 = 0$ and $I_3 \approx 0$, because the internal resistance of the voltmeter is very large; consequently, only the changing current I_1 produces induced voltages. The induced voltages in coils 1 and 3 are

$$V_1 = L\frac{dI_1}{dt} \qquad \text{and} \qquad V_3 = M\frac{dI_1}{dt},$$

respectively. However, we are told that $V_3 = V_1/2$, implying that $M = L/2$.

After the switch is closed, current also flows in coil 2. Now the induced voltages in the individual coils can be written as

$$V_1 = L\frac{dI_1}{dt} + M\frac{dI_2}{dt},$$

$$V_2 = M\frac{dI_1}{dt} + L\frac{dI_2}{dt},$$

$$V_3 = M\frac{dI_1}{dt} + M\frac{dI_2}{dt}.$$

Because the second coil is short-circuited, we must have that $V_2 = 0$, and from this we can find a simple connection between the rates of change of the two currents:

$$\frac{dI_2}{dt} = -\frac{M}{L}\frac{dI_1}{dt}.$$

From this and our previous result that $M = L/2$, we can find the ratio V_3/V_1:

$$\frac{V_3}{V_1} = \frac{M - (M^2/L)}{L - (M^2/L)} = \frac{1}{3}.$$

So, after the switch is closed, the reading on the voltmeter is *one-third* of the r.m.s. voltage of the source.

> *Notes.* 1. We used the constancy of the magnetic permeability of the iron core when we assumed that the self- and mutual inductances were unchanged when the switch was closed, and, for example, that $M = L/2$ was still the case.
>
> 2. It is quite unusual, in very transformer-like problems, that voltage ratios are markedly different from the ratios of the numbers of turns in the coils. The well-known formula $V_{primary}/V_{secondary} = N_{primary}/N_{secondary}$ is valid only for ideal transformers, and real transformers only approximately obey this formula. In practice, some flux always follows a path that takes it outside the windings, though normally this 'fringing field' is not very significant. The unusual feature of this problem, in which the coils are described as 'wide' and the iron core as 'narrow', is the strong spreading of the magnetic field away from the coils, with the result that, despite the presence of the iron core, a significant part of the flux through any one coil does not pass through the others.

S185 The induced voltage V_2 in coil 2 produced by the changing current in coil 1 is given in terms of their mutual inductance M by

$$V_2(t) = M\frac{dI_1}{dt}.$$

Since, as a result of the voltmeter's 'infinite' internal resistance, the current in coil 2 is zero, we can write the loop rule for the coil 1 circuit in the form:

$$V_1(t) = L_1\frac{dI_1}{dt},$$

where L_1 is the self-inductance of coil 1. From these two expressions, we have directly that, at any time,

$$V_2(t) = \frac{M}{L_1}V_1(t). \tag{1}$$

A similar connection holds between the r.m.s. values of the voltages. So basically our task comes down to finding the mutual inductance of the two coils.

The alternating current $I_1(t)$ flowing in coil 1 produces a time-varying magnetic field inside it, whose magnitude is given by Ampère's circuit law as

$$B_1(t) = \frac{\mu_0 N_1}{\ell} I_1(t),$$

where ℓ is the circumference of (either of) the axial circles of the toroids. If the coils have cross-sectional areas A, this magnetic field produces, through the windings of coil 1, a total magnetic flux of $\Phi_1(t) = N_1 B_1(t) A$. It follows that the self-inductance of the toroidal coil 1 is

$$L_1 = \frac{\Phi_1(t)}{I_1(t)} = \frac{\mu_0 N_1^2 A}{\ell}.$$

Coil 2 encircles the magnetic field formed *inside* coil 1 only once (and even then, in a very 'devious' way). For this reason, we might 'at first sight' think that the (time-dependent) magnetic flux through coil 2 produced by the magnetic field of coil 1 is simply[77]

$$\Phi_{12}^I(t) = B_1(t) A = \frac{\mu_0 N_1 A}{\ell} I_1(t).$$

If this were the case, then the mutual inductance of coil 1, with respect to coil 2, would be

$$M = \frac{\Phi_{12}^I(t)}{I_1(t)} = \frac{\mu_0 N_1 A}{\ell}.$$

But this is *not symmetrical* with respect to interchanging N_1 and N_2 – and it should be! It can be proved that, if the mutual inductance were not symmetric, the law of conservation of energy would be broken. Clearly, something important has been left out of our considerations.

To resolve this dilemma, we have to notice that coil 1 produces a magnetic field not only *inside* its windings (this part of the field is denoted 'I' in the figure), but also *outside* the coil (denoted 'II'). Usually, for toroidal coils, this weak outside magnetic field (equivalent to that of a single circular current loop) is neglected, but here the magnetic field lines outside the windings of coil 1 flow through all N_2 windings of coil 2, and so they can produce a significant flux!

At the position of the ith turn of coil 2, denote the component of the local magnetic field that is perpendicular to that turn by $B_i(t)$; the field in question is the one in region II of coil 1. As the average distance between the windings of coil 2 is ℓ/N_2, this is the element of path length, parallel to B_i (and the same size for all i), needed to make an Ampère's circuit law calculation. We can now write the

---

[77] The reason for the presence of the superscript 'I' will become clear later.

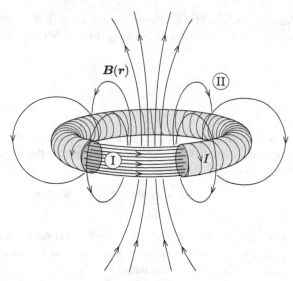

calculation for a path taken around the second coil enclosing, as noted above, the equivalent of a single circular current loop carrying a current I_1:

$$\sum_i B_i(t)\frac{\ell}{N_2} = \mu_0 I_1(t).$$

Multiplying through by $N_2 A/\ell$, on the left-hand side we get just the magnetic flux Φ_{12}^{II} that is produced by the *outside* magnetic field of coil 1 and passes through the windings of coil 2:

$$\Phi_{12}^{\text{II}}(t) = \sum_i B_i(t)A = \frac{\mu_0 N_2 A}{\ell}I_1(t).$$

The total magnetic flux through coil 2 produced by the magnetic field of coil 1 is the sum of Φ_{12}^{I} and Φ_{12}^{II}, and so we have

$$\Phi_{12}(t) = \Phi_{12}^{\text{I}}(t) + \Phi_{12}^{\text{II}}(t) = \frac{\mu_0(N_1 + N_2)A}{\ell}I_1(t),$$

and the corresponding implied mutual inductance of

$$M = \frac{\Phi_{12}(t)}{I_1(t)} = \frac{\mu_0(N_1 + N_2)A}{\ell}.$$

Happily, this M is symmetrical with respect to N_1 and N_2!

Finally, the voltmeter reading $V_{\text{voltmeter}}$ can be found from equation (1), and our expressions for L_1 and M:

$$V_{\text{voltmeter}} = \frac{M}{L_1}V_0 = \frac{N_1 + N_2}{N_1^2}V_0.$$

Notes. 1. We can also find a much shorter, but quite heuristic, solution. Because of the necessary symmetry of the mutual inductance, we can write it in two different ways:

$$\frac{\mu_0 N_1 A_1}{\ell_1} + f(2) = M = \frac{\mu_0 N_2 A_2}{\ell_2} + f(1).$$

In both expressions, the first terms correspond to the contributions from fluxes due to magnetic fields I, and the second terms arise from the flux contributions of fringing fields II. Rearranging the equation, we obtain

$$\frac{\mu_0 N_1 A_1}{\ell_1} - f(1) = \frac{\mu_0 N_2 A_2}{\ell_2} - f(2).$$

Now, since we have a free choice for all coil parameters, the two sides can only be equal if both sides of this equation are zero; this determines both $f(1)$ and $f(2)$ – and also the mutual inductance:

$$M = \frac{\mu_0 N_1 A_1}{\ell_1} + \frac{\mu_0 N_2 A_2}{\ell_2} = \frac{\mu_0 (N_1 + N_2) A}{\ell}.$$

2. Our result can also be demonstrated experimentally. But if the experiment is carried out at room temperature with air-core coils, then their ohmic resistances are negligible compared to the inductive impedances only if a high-frequency power supply is used. The results of the experiments are very convincing.

3. When summing the fluxes Φ_{12}^{I} and Φ_{12}^{II}, we assumed – tacitly – that the turns of both coils were wound in the same sense (as the toroidal coils were said to differ *only* in their numbers of turns). If, accidentally, this is not the case (i.e. one of the coils is wound clockwise, and the other anticlockwise), then we have to take the difference of the two flux contributions. In this case, the reading on the voltmeter is

$$V_{\mathrm{voltmeter}} = \frac{|N_1 - N_2|}{N_1^2} V_0.$$

S186   Inside the long solenoid, which has radius R, say, a homogeneous, time-dependent magnetic field $B(t)$ is formed. The changing magnetic field induces an axially symmetric azimuthal electric field around the solenoid. This field, in turn, causes the charged pearl to accelerate along a tangential field line – so our first result is that it does not stay still!

The strength E of the induced electric field, at a distance r from P, can be determined if we write Faraday's law for a circle with centre P and radius r $(r > R)$, whose plane is perpendicular to the solenoid's axis:

$$-\frac{\mathrm{d}\Phi}{\mathrm{d}t} = E \cdot 2\pi r.$$

Here $\Phi(t) = B(t)\pi R^2$ is the magnetic flux through the circle. From this, the magnitude of the electric field, as a function of distance r, measured from the centre of the solenoid, is

$$E(r) = -\frac{R^2}{2r}\frac{\mathrm{d}B}{\mathrm{d}t}.$$

The electric force QE that acts on the pearl carrying a charge Q has a non-zero torque about point P, and this must be equal to the rate of change of the pearl's angular momentum, J, relative to P:

$$\frac{\mathrm{d}J}{\mathrm{d}t} = rQE.$$

Using the expression for the electric field given earlier, we get[78]

$$\frac{\mathrm{d}J}{\mathrm{d}t} + \frac{QR^2}{2}\frac{\mathrm{d}B}{\mathrm{d}t} = 0.$$

When the sum of the rates of change of two quantities is zero, then the sum of these quantities is constant in time, and so

$$J + \frac{QR^2}{2}B = \text{constant}.$$

With the help of this 'modified' conservation law, which is similar to the law of conservation of angular momentum, we can find the answer to the question in the problem. Initially, the pearl is at rest, and the electric current in the solenoid is zero, so the value of the constant on the right-hand side of our 'conservation law' is zero. At the end of the process, the magnetic field strength is again zero, and so the angular momentum of the pearl must also be zero. This can happen only if the pearl is finally moving either directly towards, or directly away from, point P.

Which of the two possibilities actually happens can be determined as follows. The pearl always accelerates in a direction perpendicular to its position vector relative to point P, and so its acceleration *never* has any non-zero component directed *towards* P. For this reason, the pearl moves continuously *away* from P and continues to do so at the end of the process.

S187 According to Faraday's law of induction, the induced voltage around an arbitrary closed loop (which could be described as its 'electric rotation') is

$$\sum E(r)\Delta r = -\frac{\mathrm{d}\Phi}{\mathrm{d}t}. \tag{1}$$

Ampère's circuit law produces a similarly structured statement for the magnetic rotation (which, in turn, could be described as the 'magnetic loop voltage'):

---

[78] It can be checked that, whatever the sign of Q and the direction of the current, the signs in this equation are in agreement with the left-hand rule specifying the direction of the induced electric field.

$$\sum B(r)\Delta r = \mu_0 I. \tag{2}$$

These two equations, together with Gauss's law for a charge- and dielectric-free region, and Maxwell's equation expressing the non-existence of free magnetic monopoles,[79] *uniquely* determine the electromagnetic field produced by given electric currents and magnetic flux changes.

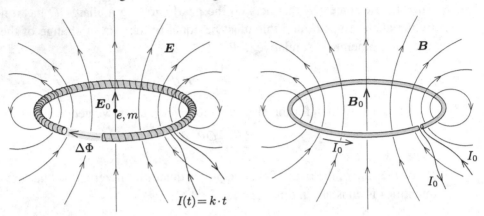

Formally, equations (1) and (2) are similar, and they can be transformed into each other if the following interchanges are made:

$$-\frac{d\Phi}{dt} \quad\longleftrightarrow\quad \mu_0 I, \tag{3}$$
$$E \quad\longleftrightarrow\quad B.$$

Using this symmetry feature, if we know the solution to some particular problem involving one of the two phenomena, then we 'automatically' have the solution to the geometrically 'analogous' situation in the context of the other.

In our case, the (uniformly) changing magnetic flux in the thin toroidal coil produces an induced electric field E with just the same geometrical structure as that of the magnetic field B around a circular loop carrying a steady current I_0 (compare the two situations shown in the figure).

From the Biot–Savart law, we know that the magnetic field at the centre of a circular wire loop of radius r that has an electric current I flowing through it has magnitude

$$B_0 = \frac{1}{2r}(\mu_0 I),$$

and is perpendicular to the plane of the loop. From this and relationships (3), it follows that the changing magnetic flux in the toroidal coil induces an electric field at the centre of the toroid of magnitude

---

[79] More mathematically, both the electric and magnetic fields have zero divergence: $\nabla \cdot E = 0$ and $\nabla \cdot B = 0$.

$$E_0 = -\frac{1}{2r}\frac{d\Phi}{dt}.$$

The negative sign reflects the fact that the direction of E_0, relative to the direction of the magnetic field inside the toroidal coil, follows the left-hand rule.

We know that, inside an air-core toroidal coil with N turns, major radius r and with an electric current $I(t)$ flowing through it, the magnitude of the magnetic field is

$$B = \frac{\mu_0 NI}{2\pi r},$$

and, consequently, the varying magnetic flux due to this field induces an electric field with magnitude

$$E_0 = \frac{\mu_0}{4\pi}\frac{NA_c}{r^2}\frac{dI}{dt} = 10^{-7}\,\frac{V\,s}{A\,m} \times \frac{200 \times 2 \times 10^{-4}\,m^2}{(0.1\,m)^2} \times 10\,\frac{A}{s}$$
$$= 4 \times 10^{-6}\,V\,m^{-1},$$

where A_c is the cross-sectional area of the coil. This electric field causes a proton, which carries an elementary charge e and has mass m_p, to have an initial acceleration of

$$a_0 = \frac{eE_0}{m_p} \approx 380\;m\;s^{-2}.$$

S188   The capacitance of the capacitor can be calculated (in the approximation of $d \ll R$) in the same way as for a parallel-plate capacitor:

$$C = \varepsilon_0 \frac{2\pi R\ell}{d},$$

and so the magnitude of the electric charge on each of its plates is

$$Q = CV = \varepsilon_0 \frac{2\pi R\ell V}{d}. \tag{1}$$

a) If the cylinders forming the capacitor become electrically connected, then a rapidly changing current $I(t)$ is formed in the wire. The magnetic field exerts a Lorentz force IBd on the wire, and that means a torque $IBd \times R$ acts on the capacitor as a whole (*see* Fig. 1). This torque causes rotation of the system, which has moment of inertia MR^2. We note that the Lorentz force would, if it could, also cause the capacitor to swing as a pendulum, but, in the given problem, this is not possible.

In accord with the dynamical equation of rotation, we can write

$$MR^2 \frac{d\omega}{dt} = IBRd = BRd\frac{dQ}{dt}.$$

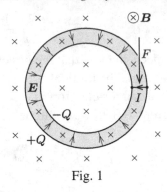

Fig. 1

It follows that, during any short time interval Δt, the charge transferred between the plates and the corresponding increase in the capacitor's angular velocity are related by

$$MR^2 \Delta\omega = BRd\,\Delta Q. \tag{2}$$

We now sum both sides of (2) for the whole discharge process. As the initial angular velocity is zero, and the sum of the 'charge quanta' ΔQ flowing through the wire is just the initial charge Q on the capacitor, given by (1), the final (maximal) angular velocity of the rotating capacitor is given by

$$\omega_{\text{max}} = \sum \Delta\omega = \frac{BRd}{MR^2} \sum \Delta Q = \frac{BdQ}{MR} = 2\pi\varepsilon_0 \frac{VB\ell}{M}.$$

It is interesting that the value of the maximal angular velocity does not depend on the lengths R and d, and, furthermore, it is independent of the time dependence of the discharge current.

b) When the magnetic field is switched off, an electric field that has a non-zero azimuthal component is induced; this exerts a force on the capacitor's charge and a torque on the capacitor. As a general rule, the induced electric field does *not* have axial symmetry, and its accurate values can be found only with knowledge of the whole magnetic field.

However, the loop voltage (characterising the azimuthal component) can be calculated for *any* closed loop. Consider, for instance, the closed loop shown in Fig. 2, a loop that includes the cross-section of the capacitor's inner cylinder, and for which the loop voltage, according to Faraday's law of induction, is

$$\sum E_\| \Delta s = -\frac{d\Phi}{dt} = -2\pi Rd\frac{dB}{dt}, \tag{3}$$

where Φ denotes the magnetic flux through the darker area in Fig. 2, Δs is the length of a small piece of the curve, and $E_\|$ denotes the component of the induced electric field that is parallel to it.

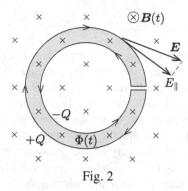

Fig. 2

The amount of charge on a strip of the cylinder's lateral area that has width Δs is $\Delta Q = Q\Delta s/(2\pi R)$, and the induced electric field exerts a torque $\Delta Q \times E_\parallel \times R = QRE_\parallel \Delta s/(2\pi R)$ on it. Accordingly, the sum on the left-hand side of (3),[80] multiplied by $Q/(2\pi)$, is just the torque acting on the whole capacitor.

The dynamical equation of rotation can be written as

$$MR^2\frac{d\omega}{dt} = \frac{Q}{2\pi}\sum E_\parallel \Delta s = -\frac{Q}{2\pi}\frac{d\Phi}{dt} = -\frac{Q}{2\pi}2\pi Rd\frac{dB}{dt}.$$

That is, the connection between small changes in B and ω is

$$\Delta\omega = -\frac{Qd}{MR}\Delta B.$$

Carrying out a summation for the whole process, and also using (1), we get the final angular velocity of the capacitor:

$$\omega_{max} = \sum\Delta\omega = -\frac{Qd}{MR}\sum\Delta B = \frac{BQd}{MR} = 2\pi\varepsilon_0\frac{VB\ell}{M},$$

which is (perhaps surprisingly) just the same as in case a).

> *Note.* The calculations show that the final angular velocity of the charged cylindrical capacitor, located in the magnetic field, will be the same, whether it is the electric field or the magnetic field that is 'disconnected'. We may wonder whether there is any deeper physical reason for this.
>
> The cylindrical capacitor itself does not form a closed system, but together with the electromagnetic field around it, it does – and angular momentum must be conserved in this closed system. It is a plausible interpretation that somehow the electromagnetic field has angular momentum. It would not be related to a geometrical point, but 'smeared' out in space, in much the same way that energy and linear momentum are 'smeared' in an electromagnetic field.
>
> The angular momentum of the electromagnetic field can be written in terms of the Poynting vector:

---

[80] We can neglect the contributions of the two small radial segments, which, in any case, cancel each other as the gap between them becomes arbitrarily small.

$$S = \frac{1}{\mu_0} E \times B.$$

This vector is an expression for the density of energy flow, i.e. it gives the energy that passes through unit area, perpendicular to the direction of energy propagation, in unit time. If the Poynting vector is divided by the speed of light, we obtain the linear-momentum flow density. We can get the 'angular-momentum flow density', at any particular point, if we take the vector product (cross-product) of the position vector of that point and the linear-momentum flow density.

The special geometry of the arrangement in the problem greatly eases this calculation, as, in the cylindrical capacitor, the energy of the electromagnetic field only propagates between and 'parallel to' the walls of the two cylinders. Further, at all points in this volume, the electric and magnetic field vectors are perpendicular to each other.[81] The 'angular-momentum flow density' of the system, relative to its symmetry axis, is thus

$$\frac{S}{c}R = \frac{EBR}{\mu_0 c}.$$

If this is multiplied by the cross-sectional area of the volume, ℓd, and the time the flow takes to travel the length of the volume, $2\pi R/c$, then we get the total angular momentum (circulating in the cylindrical capacitor) of the (stationary) electromagnetic field:

$$|J| = \frac{2\pi EBR^2 \ell d}{\mu_0 c^2}.$$

Since $E = V/d$, and the speed of light is given by $1/c^2 = \varepsilon_0 \mu_0$, we can express the angular momentum of the electromagnetic field as

$$|J| = 2\pi\varepsilon_0 VB\ell R^2,$$

which is in line with the mechanical angular momentum $MR^2\omega_{max}$, calculated in the solution.

In the problem, this fixed amount of electromagnetic angular momentum is transformed into mechanical angular momentum, in *a*) because the electric field disappears, and in *b*) because the magnetic field is switched off. In view of this, we can hardly be surprised that the final mechanical angular momentum (or velocity) of the capacitor is the same in both cases.

What is the origin of the initial electromagnetic angular momentum of the system? It is found experimentally that some torque acts on the capacitor, either if the capacitor is charged in the presence of a homogeneous magnetic field, or if the magnetic field is built up around a previously charged capacitor. This torque rotates the initially stationary capacitor, and the rotation needs to be stopped (by hand) in order to establish the starting conditions given in the problem.

---

[81] When this phenomenon is first met, it might be difficult to understand that, in a stationary combined electrostatic and magnetostatic field, the energy is moving around with the speed of light!

S189 Let a moving observer inside a charged parallel-plate capacitor have a speed v_0 parallel to its rectangular plates. For simplicity, we assume that the space between the plates is evacuated.

Let us denote the lengths of the sides of the plates by a and b, and the plate separation by d. If $d \ll a, b$, the charges will be nearly uniformly distributed over the inner surfaces of the plates and produce a homogeneous electric field E_0 between them. The total charge Q on one plate is given numerically by Gauss's law as

$$Q = \varepsilon_0 E_0 ab.$$

The moving observer would 'see' that the charged plates are moving with velocity $-v_0$ parallel, let us say, to the edge of length a. The full length of the plates would pass by the observer in a time $T = a/v_0$ and so be equivalent to currents

$$I = \pm \frac{Q}{T} = \pm \frac{Qv_0}{a}.$$

This pair of currents produce magnetic fields, which reinforce each other inside the capacitor, and give a net magnetic field, assumed homogeneous, inside the plates, and a negligible field outside. This line of reasoning is similar to that used in the determination of the electric field inside a parallel-plate capacitor or the magnetic one inside a long solenoid: it involves consideration of symmetry and the application of integral laws.

The magnetic field vector B_0 is parallel to the side of length b, and so is perpendicular to both v_0 and E_0. Its magnitude is determined by applying Ampère's rule to a circuit surrounding one of the plates: $B_0 \cdot b + 0 \cdot b = \mu_0 I$. Thus

$$B_0 = \frac{\mu_0 I}{b} = \frac{\mu_0 Q v_0}{ab} = \mu_0 \varepsilon_0 v_0 E_0 = \frac{1}{c^2} v_0 E_0.$$

Taking into account the directions of the vectors and the right-hand rule governing a vector product, the observed magnetic field can be written as

$$B_0 = -\frac{1}{c^2} v_0 \times E_0.$$

Our conclusion is: 'Yes, the observer *does* experience a magnetic field.'

> *Note.* According to special relativity, the transformation formulae for the electric and magnetic fields perpendicular to the direction of motion are
>
> $$B'_\perp = \frac{B_\perp - c^{-2}(v_0 \times E_0)}{\sqrt{1 - v_0^2/c^2}},$$

$$E'_\perp = \frac{E_\perp + v_0 \times B_0}{\sqrt{1 - v_0^2/c^2}},$$

while the components parallel to the velocity remain unchanged.

For the case of non-relativistic motions ($v_0 \ll c$), these formulae reproduce the results of our more elementary considerations.

S190 The magnetic field in the region of the compass can be found with the help of the Biot–Savart law. This law is an equation describing the magnetic field generated by a real electric current, and relates the magnetic field to the magnitude, direction, length and proximity of the current. The law is valid in the magnetostatic approximation, and is consistent with both Ampère's circuit law and Gauss's law for magnetism.

In the Biot–Savart law we have to take into consideration the contributions from real currents, but not those from *displacement currents*. This is because, if you do consider the contributions of the displacement currents in any (quasi-)steady circuit, their net magnetic effect at all points will turn out to be zero. Given this, it is perhaps surprising that, generally, the contributions of displacement currents to Ampère's circuit law are not zero.

Therefore, considering only real currents, when determining the magnetic field at the position of the compass (using the Biot–Savart law), we have three contributions to consider: first, that from the essentially vertically upward current in the left-hand plate; second, the contribution of the horizontal current from left to right through the rod; and finally, the effect of the essentially vertically downward current in the right-hand plate. Using the right-hand rule, we can say that, between the plates and below the rod, all three reinforce each other and produce an approximately horizontal magnetic induction, parallel to the plates of the capacitor, and pointing backwards in the figure of the problem. This would cause the compass to rotate clockwise, which is the opposite direction to that in the 'official solution', which is therefore wrong!

> *Note.* Through the conducting rod, the electric discharge current is homogeneous (assuming that the conductivity of the rod is not too large), but the discharge currents in the plates are not. The metal plates can be considered to be equipotential surfaces, and so, in practical terms, the charge densities on them are always homogeneous, though decreasing with time. This means that the currents at the lower edges of the plates are zero, and that they vary linearly with height along any particular plate, with the absolute value of the current reaching that in the rod only at the tops of the plates. This means that, near the compass, the weak current elements cannot make strong contributions to the magnetic field, and where the current elements are larger, the distances to the compass are greater. As a summary, we can say that the effect on the compass of the discharge of the capacitor is much smaller than might be expected.

S191 Around Santa's charged figure there is an electrostatic field, which acts on the charged particles in the air and so causes electric currents to flow to 'infinity' (or, more precisely, to distant grounded conductors). We need to investigate whether or not these 'current-streamlines' produce a magnetic field around the figure.

We first note that any magnetic field lines associated with the chocolate figure must form closed loops; this follows from the fact that the magnetic field contains no sources (stated more mathematically, its divergence is zero – or, more prosaically, magnetic monopoles do not exist).

Now consider an arbitrary magnetic field-line loop L in the space surrounding Santa and a surface area bounded by it (say, the grey area shown in the figure). Consider, further, a line integral $\oint \boldsymbol{B} \, d\boldsymbol{\ell}$ taken around the closed loop (taking the positive direction to be in the direction of $\boldsymbol{B}$). Since the angle between the direction of the magnetic field and the tangent to L is always zero, this integral cannot give a negative value. Further, the integral can only be zero if the value of the magnetic field is zero at every point of the loop.

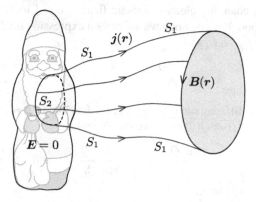

According to Ampère's law, the value of the same integral is proportional to the net current passing through any surface area bounded by L; this net current is the sum of the conduction current due to moving charges and the displacement current produced by the changing electric field. Although displacement currents can never produce magnetic fields around any circuit, they do make contributions to the current flowing through an area bounded by any closed curves in the context of Ampère's circuit law (*see* solution on page 450). The direct calculation of the real and displacement currents through the grey area is practically hopeless, but its value can be found with the help of a 'cunning plan'.

Rather than the grey area, let us choose another, but more complicated, surface S, one that bulges so that its 'cap' penetrates inside the chocolate figure. Outside the aluminium foil, the surface S_1 is defined as being bounded by those electric current-streamlines that pass through L. Inside the foil, the cap, S_2, can be any open surface whose boundary coincides with the closed loop defined by where these streamlines meet the foil. The two parts, S_1 and S_2, of the new surface S $(= S_1 \cup S_2)$ are indicated in the figure.

Inside the chocolate figure (with its aluminium coating), both the electric field and the current density are zero. Outside it, because of the way the S_1 part of the new surface is defined, the conduction current can have no component perpendicular to it; the same is true for the electric field vector, since the two are parallel. Consequently, we can conclude that no current or electric field crosses any part of S, and it follows that the integral $\oint B \, d\ell$ is zero for the loop L. As we have already shown, this implies that the magnetic field is zero at all points on this *arbitrarily* chosen loop.

In summary, there are no magnetic field lines in the vicinity of the chocolate figure during the discharge process, and the value of the magnetic field is zero throughout the space.

Note. An alternative approach to this problem is to use some of the results appearing in the solution that starts on page 429. A general displacement current j^d in a changing electromagnetic field is equal to $[1/(\mu_0 c^2)] \, dE/dt = \varepsilon_0 \, dE/dt$. Although we do not have an explicit expression for E, from that solution we know that $j_i = \sigma E_i$ and

$$\frac{dQ(t)}{dt} = -\sigma \sum_i E_i \cdot n_i \, \Delta A_i = -\sum_i j_i \cdot n_i \, \Delta A_i.$$

If we differentiate both sides of the first equality with respect to time and, recalling that $Q(t) = Q(0) \exp(-(\sigma/\varepsilon_0)t)$, replace

$$\frac{d^2 Q}{dt^2} \quad \text{by} \quad -\frac{\sigma}{\varepsilon_0} \frac{dQ}{dt} \quad \text{and} \quad \frac{\partial E_i}{\partial t} \quad \text{by} \quad \frac{j_i^d}{\varepsilon_0},$$

then we obtain

$$\frac{\sigma}{\varepsilon_0} \sum_i j_i \cdot n_i \, \Delta A_i = -\frac{\sigma}{\varepsilon_0} \frac{dQ(t)}{dt} = -\sigma \sum_i \frac{j_i^d}{\varepsilon_0} \cdot n_i \, \Delta A_i.$$

Since this equality must hold however the surface elements ΔA_i are chosen, we must have that $j_i^d = -j_i$ for all i, i.e. the displacement current at any point is equal and opposite to the real current, and so, in Ampère's law, their net contribution is zero, and there is no magnetic field.

S192   *a)* The linear momentum of the electron parallel to the x-axis is conserved, as no force acts on the particle in this direction:

$$\frac{mv_x}{\sqrt{1 - v^2/c^2}} = \text{constant}. \tag{1}$$

Here m is the (rest) mass of the electron, v is its speed at any particular moment and v_x is the x component of that speed. The fact that the x-directed linear momentum is constant does not mean that v_x also remains constant – in fact, according to (1), v_x must decrease as the total speed v increases!

It is convenient to compare all speeds to that of light, so let us use the notation $\beta_0 = v_0/c = 0.6$ and $\beta_1 = v_1/c$, as is usual in relativity theory. Conservation law (1) can now be applied to the electron as it enters, and later leaves, the electric field, in the form

$$\frac{\beta_0 mc}{\sqrt{1 - \beta_0^2}} = \frac{\beta_1 mc}{\sqrt{1 - \beta_1^2}} \cos 45°,$$

from which

$$\beta_1 = \sqrt{2} \frac{\beta_0}{\sqrt{1 + \beta_0^2}} = \frac{3}{\sqrt{17}} \approx 0.728.$$

So, the electron leaves the electrostatic field with 72.8 % of the speed of light.

b) The work done on the electron by the electric field, while it is being displaced sideways by a distance d in a direction opposed to that of the field, is eEd, where e denotes the elementary charge. This work must equal the change in the electron's total energy:

$$eEd = \frac{mc^2}{\sqrt{1 - \beta_1^2}} - \frac{mc^2}{\sqrt{1 - \beta_0^2}} \approx 0.208 mc^2.$$

Noting that, by definition, 1 keV $= e \times 1$ kV, the required distance d is

$$d = \frac{0.208 mc^2}{eE} = 0.208 \times \frac{510 \text{ keV}}{e \times 510 \text{ kV m}^{-1}} = 0.208 \text{ m} = 20.8 \text{ cm}.$$

S193 The equation of motion of a particle with mass m and electric charge $\pm e$ moving along a circular trajectory of radius R in a homogeneous magnetic field B is

$$\frac{d\boldsymbol{p}}{dt} = \pm e \, \boldsymbol{v} \times \boldsymbol{B},$$

where $\boldsymbol{p}$ is the linear momentum of the particle and $\boldsymbol{v}$ is its velocity. This equation also applies to particles moving with relativistic speeds.

Since, in the current problem, $\boldsymbol{v}$ and $\boldsymbol{B}$ are orthogonal, when we take the absolute values on both sides of the equation we get

$$\left| \frac{d\boldsymbol{p}}{dt} \right| = evB. \tag{1}$$

Mechanically, the magnitude of the required rate of change of a vector $\boldsymbol{p}$ that has constant absolute value and rotates with angular velocity $\omega = v/R$ is $p\omega = pv/R$.

Inserting this into the equation of motion (1), we get the linear momentum of the particle moving along a circular trajectory:

$$p = eBR.$$

Using the well-known value of the elementary charge, and the relevant data about the Earth (equatorial radius, $R = 6.378 \times 10^6$ m; average surface magnetic field strength at the Equator, $B \approx 3.5 \times 10^{-5}$ T, directed northwards), we find that the required linear momentum of the proton (or electron) is $p \approx 3.6 \times 10^{-17}$ kg m s^{-1}.

If we calculated using the non-relativistic formula for momentum, $p = mv$, then we would find that the 'required speeds' for the proton and electron need to be 71 times and 130 000 times greater than the speed of light, respectively. This is clearly impossible, and a relativistic calculation is required.

We need the relativistic formula

$$p = \frac{mv}{\sqrt{1 - v^2/c^2}},$$

where m is the 'rest' mass of the relevant particle. It is convenient to define the following dimensionless quantities:

$$\beta \equiv \frac{v}{c}, \qquad K \equiv \frac{p}{mc} = \frac{\beta}{\sqrt{1 - \beta^2}} = \frac{eBR}{mc}.$$

Note that K essentially expresses how many times larger than the speed of light the speed of the 'classical' particle has to be to give momentum p; it reproduces the previously calculated values,

$$K_{\text{electron}} = 1.3 \times 10^5 \qquad \text{and} \qquad K_{\text{proton}} = 71.$$

a) With both values of $K \gg 1$, the second equality above shows that both particles must move with practically the speed of light, $\beta \approx 1$. The precise value of the required speed is a little smaller for a proton – it is about 30 km s^{-1} less than the speed of light. But for an electron this difference is only about 1 cm s^{-1}. So the electron needs to move at virtually the speed of light! Taking into consideration the direction of the Lorentz force, in order to generate the required centripetal force, the negative electrons must move *to the east* and the protons *to the west*.

b) The energy of one of the particles is

$$E = \frac{mc^2}{\sqrt{1 - \beta^2}} = \frac{K}{\beta}mc^2 \approx Kmc^2 = eBRc = BRc \text{ [in eV]} = 67 \text{ GeV}.$$

It is interesting that this energy does not depend on the mass of the particle, and is therefore the same for protons and electrons.

> *Note.* Along the Earth's magnetic equator, the magnetic field strength is approximately constant. But the magnetic equator is not a circle – it is not even a planar

curve. Furthermore, a homogeneous magnetic field only provides for particle trajectories around a circular orbit; it does not prevent the beam from diverging, and so, for the realisation of around-the-Earth orbits, some focusing magnets (and many other things!) are needed.

S194   A charged particle moving, without slowing down, in a plane perpendicular to a homogeneous magnetic field performs uniform circular motion, on top of which it may also have a uniform rectilinear motion parallel to the magnetic field.

The sketch in the figure of the problem (based on a photograph, which may or may not exist) has been made looking along the direction of the magnetic field (as the trajectories are circular arcs); any motion along the magnetic field – not seen in the photo – can be ignored. We assume that the magnetic field direction is perpendicularly *outwards* from the plane of the figure.

Number the trajectories of the particles from inside to outside, and denote the radii of the trajectories by R_1, R_2 and R_3. Accordingly, it can be seen in the figure of the problem that

$$R_1 < R_2 < R_3. \tag{1}$$

Using the same order, denote the magnitudes of the linear momenta of the particles by p_1, p_2 and p_3, and the magnitudes of their electric charges by q_1, q_2 and q_3. All of the values of R_i, p_i and q_i ($i = 1, 2, 3$) are positive, as they denote the magnitudes of the particular quantities. If the linear momentum or charge of one of the particles were opposite to that of another, that would be allowed for by using negative signs in the relevant formulae.

The (non-relativistic) equation of motion of a particle with mass m moving with speed v along a circular trajectory of radius R in a homogeneous magnetic field B is

$$\frac{mv^2}{R} = qvB, \qquad \text{that is} \qquad mv = qBR,$$

which can be expressed in terms of the linear momentum $p = mv$ of the particle:

$$p = qBR. \tag{2}$$

> *Note.* As stated in the solution on page 453, equation (2) is obeyed even if the motion of the particle is described by the laws of relativistic dynamics. So the following arguments are valid also for particles that move relativistically.

In a particle decay, the total electric charge of the particles (written with the correct signs) and the signed sum of their linear momenta are conserved – so these quantities must remain unchanged by the process. The figure of the problem does not tell us which of the three particles decays, and so we have to investigate all three possibilities, separately.

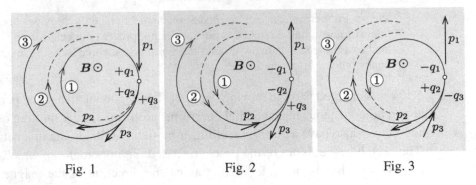

Fig. 1   Fig. 2   Fig. 3

(i) If particle 1 decays (*see* Fig. 1), then all of the three particles are 'turning to the right', and so all of them have positive charge. According to the laws of conservation of linear momentum and electric charge, we can write

$$p_1 = p_2 + p_3 \quad \text{and} \quad q_1 = q_2 + q_3,$$

from which, using (2), we get

$$R_1 = \frac{q_2 R_2 + q_3 R_3}{q_2 + q_3}. \tag{3}$$

But this is *impossible*, because the arithmetic mean of R_2 and R_3, weighted by positive factors, cannot be equal to a radius that is smaller than both of them.

(ii) If particle 2 decays (*see* Fig. 2), then only particle 3 has a positive charge, the other two ('turning to the left') must be negatively charged. The equations for the conservation of linear momentum and electric charge are now

$$p_2 = p_1 - p_3 \quad \text{and} \quad -q_2 = -q_1 + q_3,$$

from which, using (2), we again get the unacceptable expression (3).

(iii) Finally, if particle 3 decays (*see* Fig. 3), then only the charge of particle 2 is positive and the other two are negatively charged. Then, according to the conservation laws:

$$p_3 = p_1 - p_2 \quad \text{and} \quad -q_3 = -q_1 + q_2,$$

which yet again yields the unphysical, and therefore unacceptable, expression (3).

In summary, we can state that the particle 'trails', shown in the figure of the problem, *cannot have appeared* in any cloud chamber photo of a real decay process!

S195 Chose a system of units in which the speed of light is unity, i.e. $c = 1$. The vectorial form of the linear-momentum conservation law for the decay is

$$\boldsymbol{p}_\pi = \boldsymbol{p}_e + \boldsymbol{p}_\nu.$$

Taking its square, and using the fact that p_e and p_v are perpendicular to each other, we get

$$p_\pi^2 = p_e^2 + p_v^2. \tag{1}$$

In the chosen system of units, the connection between the total (relativistic) energy of a particle with rest mass m and linear momentum p can be written in the following form:

$$E = \sqrt{p^2 + m^2}, \tag{2}$$

which is called the *mass-shell* condition.

In collisions and decays of subatomic particles, conservation of energy is valid for the sum of the total energies; in the given case it can be written as

$$E_\pi = E_e + E_v. \tag{3}$$

Applying the mass-shell condition (2) to the squares of the total energies, we get

$$m_\pi^2 + p_\pi^2 = m_e^2 + p_e^2 + p_v^2 + 2E_vE_e.$$

Using connection (1) between the linear momenta, this reduces to

$$E_vE_e = \frac{m_\pi^2 - m_e^2}{2}. \tag{4}$$

That is, the product of the energies of the two decay products is constant (irrespective of the actual electron and neutrino directions, so long as they are orthogonal).

Our aim is to minimise the possible speed, and so the possible energy, of the pion. This energy equals the sum of the energies of the two decay products, and we are seeking its minimal value. Apply the general inequality between the arithmetic and geometric means to the particular case of these two energies:

$$\sqrt{E_vE_e} \le \frac{E_v + E_e}{2}.$$

Using equations (3) and (4), we get the possible minimal energy of the pion as

$$E_\pi = E_v + E_e \ge 2\sqrt{E_vE_e} = \sqrt{2(m_\pi^2 - m_e^2)}.$$

The total energy of the pion can be written in terms of its mass and speed as

$$\frac{m_\pi}{\sqrt{1 - v^2}} = E_\pi \ge \sqrt{2(m_\pi^2 - m_e^2)},$$

from which we get the following lower limit:

$$v \ge \sqrt{1 - \frac{m_\pi^2}{2(m_\pi^2 - m_e^2)}} = \sqrt{\frac{m_\pi^2 - 2m_e^2}{2m_\pi^2 - 2m_e^2}} = 0.707.$$

Because of the system of units chosen, this means that the speed of the pion must be at least 70.7 % of the speed of light.

S196   The linear momenta of an isolated system of particles are always unchanged by collisions, and, if the collisions are elastic, so is its total energy. When the particles are moving at close to the speed of light, these quantities must be calculated according to the rules of relativistic mechanics. The (total) energy and linear momentum of a particle with (rest) mass m and velocity v are

$$E = \frac{mc^2}{\sqrt{1 - v^2/c^2}} \quad \text{and} \quad p = \frac{mv}{\sqrt{1 - v^2/c^2}}, \tag{1}$$

and from these it follows that

$$p = \frac{E}{c^2}v.$$

In the ultra-relativistic limit (i.e. when $|v| = v \approx c$), the connection between the energy and linear momentum becomes very simple:

$$E \approx pc.$$

Note. This formula can be found directly from the mass-shell condition, which itself can be obtained by eliminating v from the two equations in (1),

$$E^2 = (pc)^2 + (mc^2)^2.$$

In the ultra-relativistic case, the rest-energy term $(mc^2)^2$ is neglected, and the expression for the total energy becomes simply $E = pc$.

Denote the final linear momenta of the particles in our problem by q_1 and q_2. Assuming that these are also ultra-relativistic linear momenta, the conservation laws can be written in the form:

$$p_1 + p_2 = q_1 + q_2, \tag{2}$$

and $p_1c + p_2c = q_1c + q_2c$, that is,

$$p_1 + p_2 = q_1 + q_2. \tag{3}$$

These equations are illustrated graphically in Fig. 1. For the given linear momenta p_1 and p_2 of the particles, the total linear momentum of the system P can be constructed; this also has to be the sum of the linear momenta after the collision. Using a point F_1 as an arbitrary origin, the value of $P = p_1 + p_2$ determines a second point F_2. If the vector $q_1 + q_2$ is similarly constructed, it follows from equation (2) that it too must lead to F_2.

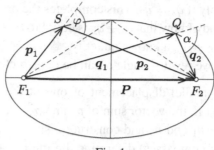

Fig. 1

If Q is the point that has vector position q_1 relative to F_1, then according to formula (3) for the conservation of energy, the sum of the distances of the (as-yet) unknown point Q to the end-points of vector P is fixed:

$$q_1 + q_2 = F_1Q + F_2Q = \text{constant} \, (= p_1 + p_2). \qquad (4)$$

Equation (4) shows that point Q lies on an *ellipse* with foci F_1 and F_2, and its major axis is the given (and constructible) distance $p_1 + p_2$.

How can we find the point on this ellipse that yields the minimum value for α, which is marked in Fig. 1, and gives the angle between the two final particle directions? We will prove that α is minimal if Q is one of the end-points of the ellipse's minor axis, i.e. if $q_1 = q_2$.

Square both sides of equation (2) (or write the cosine rule for the triangles F_1SF_2 and F_1QF_2):

$$p_1^2 + p_2^2 + 2p_1p_2 \cos \varphi = P^2 = q_1^2 + q_2^2 + 2q_1q_2 \cos \alpha, \qquad (5)$$

Also take the square of both sides of equation (3):

$$p_1^2 + p_2^2 + 2p_1p_2 = q_1^2 + q_2^2 + 2q_1q_2. \qquad (6)$$

From the difference between equations (6) and (5), we get

$$2p_1p_2(1 - \cos \varphi) = 2q_1q_2(1 - \cos \alpha),$$

from which the cosine of the angle in question can be expressed as

$$\cos \alpha = 1 - \frac{p_1p_2(1 - \cos \varphi)}{q_1q_2}.$$

It can be seen that the angle α is smallest when the product q_1q_2 is largest, as the other factors are independent of the position of Q. From the general inequality between arithmetic and geometric means, we can write

$$q_1q_2 \le \left(\frac{q_1 + q_2}{2}\right)^2 = \left(\frac{p_1 + p_2}{2}\right)^2,$$

with equality only if $q_1 = q_2$. This completes the proof of when α is minimal. But, further, we already know that $q_1 + q_2$ is equal to $p_1 + p_2$, and so the maximal value of $q_1 q_2$ is $\frac{1}{4}(p_1 + p_2)^2$, a known quantity.

Using the above results, angle α can be constructed as follows:

1. With a parallel displacement of one of the vectors (as shown in Fig. 1), construct $\boldsymbol{P}$, the vector sum of $\boldsymbol{p}_1$ and $\boldsymbol{p}_2$.
2. Measure p_1 and p_2 and construct a straight line of length $p_1 + p_2$; (using a pair of compasses) determine the midpoint of $p_1 + p_2$ by constructing a perpendicular there, as shown in Fig. 2.
3. Draw two circles, each of radius $(p_1 + p_2)/2$, with their centres (F_1 and F_2) at the end-points of vector $\boldsymbol{P}$. The points of intersection of the circles determine the directions of $\boldsymbol{q}_1$ and $\boldsymbol{q}_2$, and the angle between those directions gives the magnitude of the minimal separation angle (*see* Fig. 3).

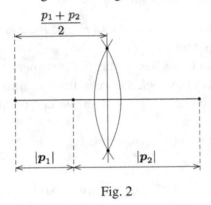
Fig. 2

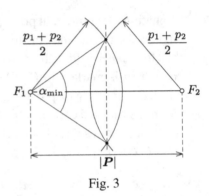
Fig. 3

What still remains to be proved is that, after the collision of the ultra-relativistic particles, they were still moving with speeds near the speed of light, and that using the ultra-relativistic energy formula in our solution was a justifiable approximation.

Assume the opposite, i.e. that one of the particles moves very slowly after the collision – both of them cannot decelerate, since their total energy is much larger than the rest energies of either. If that happened, then the equations for the conservation of energy and momentum would read as

$$p_1 c + p_2 c \approx q_1 c,$$
$$\boldsymbol{p}_1 + \boldsymbol{p}_2 \approx \boldsymbol{q}_1,$$

because both the energy and the momentum of the 'slow' (second) particle are negligible compared to the corresponding quantities for the 'fast' one. But this is not possible, since the two equations contradict each other. According to the triangle inequality

$$q_1 = |\boldsymbol{p}_1 + \boldsymbol{p}_2| \le |\boldsymbol{p}_1| + |\boldsymbol{p}_2|,$$

so

$$q_1c \leq p_1c + p_2c \approx q_1c.$$

Equality is possible, but only if p_1 and p_2 are parallel, and the figure in the problem shows that this is not the case. So we have a contradiction, and it was wrong to assume that one of the particles could move very slowly after the collision.

> *Notes.* 1. In a solution that uses the ultra-relativistic approximation, the rest masses of the particles do not appear in the formulae. For this reason, the solution remains valid even for collisions in which the final state consists of a different pair of particles from the original ones; they may have larger (or smaller) rest masses than those in the initial state. Such processes can be observed in Nature. One example of this is the following collision:
>
> $$e^+ + e^- \longrightarrow \mu^+ + \mu^-.$$
>
> Here μ denotes the elementary particle known as a *muon*, which has a mass about 207 times larger than that of an electron; e^+ denotes the *positron*, the antiparticle of the electron. The two muons involved, $\mu^\pm$, also form a particle–antiparticle pair.
>
> 2. The largest value of the angle between the momenta of the final particles can be as much as $180°$, and so they can even move in totally opposite directions.

S197 The magnitude of the magnetic induction at the centre of a circular current loop of radius R that carries a steady current I is

$$B_\odot = \frac{\mu_0 I}{2R}.$$

If the electron, with electric charge e, orbits with speed v, then the electric current is

$$I = \frac{ev}{2\pi R}.$$

The electron, of mass m, is kept in its circular orbit by the Lorentz force provided by the homogeneous magnetic induction field of strength B_0, and so the electron's equation of motion can be written as

$$evB_0 = \frac{mv^2}{R}.$$

Combining these equations, we get

$$\frac{B_\odot}{B_0} = \frac{\mu_0}{4\pi} \frac{e^2}{mR},$$

which can be transformed, using the connection $\mu_0 \varepsilon_0 = 1/c^2$, into

$$\frac{B_\odot}{B_0} = \left(\frac{e^2}{4\pi\varepsilon_0} \frac{1}{mc^2} \right) \frac{1}{R},$$

where c is the speed of light in vacuum.

It can be seen that the magnitude of the magnetic field produced by the electron's uniform circular motion would be larger than the homogeneous magnetic field B_0 if the radius of the motion, R, were smaller than the quantity in parentheses, which is called the *classical electron radius*. The numerical value of the classical electron radius is about 2.8×10^{-15} m, which is of the order of the size of an atomic *nucleus*; it makes no sense to consider an electron orbiting around such a trajectory!

So far as the laws of classical mechanics and electrodynamics are concerned, it would be possible to imagine the situation described in the problem, but in practice it is unrealistic. So, the answer to the question is 'no', the electron cannot produce a larger magnetic field than the homogeneous field that keeps it on a circular track.

Note. A formal 'derivation' of the classical electron radius is given by the following 'argument'. Equate the rest energy, mc^2, of the electron to the total electrostatic energy it would have if it had total charge e and radius r_0:

$$mc^2 = \frac{1}{4\pi\varepsilon_0}\frac{e^2}{r_0}.$$

From this we can express the radius r_0 as

$$r_0 = \frac{1}{4\pi\varepsilon_0}\frac{e^2}{mc^2}.$$

By the laws of classical physics, the electron's size cannot be smaller than r_0, because, if it were, its electrostatic energy would be larger than its total rest energy. We might add that the laws of classical electrodynamics fail if we try to apply them to events or processes involving lengths smaller than r_0. Further, and for different reasons (e.g. the wave nature of the electron), classical theory cannot even be applied to systems with sizes considerably larger than r_0.

S198   *Solution 1.* A particle of mass m enclosed in a 'box' of finite size cannot be at rest because of quantum effects, and it must have linear momentum of some average magnitude p. According to Heisenberg's uncertainty relationship

$$\Delta x \Delta p \geq \frac{h}{4\pi}, \tag{1}$$

where $\Delta x \approx d$ is the uncertainty in the neutron's position and $\Delta p \approx p$ is the uncertainty in its linear momentum.

Notes. 1. We might have written h, $h/2$ or some other similar expression on the right-hand side of inequality (1), since we need only the *order of magnitude* of the uncertainty. More precise values would make sense if the uncertainties in the linear momentum and position were defined more exactly, but we are not seeking this level of accuracy in an *estimation*.

2. It is obvious that the average value of the (vector) linear momentum must be zero, because the neutron cannot leave the box. This is why the uncertainty

(the mean deviation from the average value) of the linear momentum can be estimated as being equal to its (scalar) magnitude.

Considering, for the sake of definiteness, the limiting case of (1) in which equality holds, the magnitude of the linear momentum of the neutron is

$$p \approx \pm \frac{h}{4\pi} \frac{1}{d},$$

and its speed is

$$v \approx \pm \frac{h}{4\pi} \frac{1}{md}.$$

This result can be interpreted (following *classical physics*, for want of a better principle) as the neutron bouncing back and forth between the opposite walls of the box. So, at one of the walls, a change of linear momentum of $2mv$ occurs at regular intervals of $\Delta t = 2d/v$, and, as a result, the particle exerts a force of

$$F = \frac{2mv}{\Delta t} = \frac{mv^2}{d} = \frac{h^2}{16\pi^2 m} \frac{1}{d^3}$$

on the wall. The corresponding pressure is

$$P = \frac{F}{d^2} = \frac{h^2}{16\pi^2 m} \frac{1}{d^5}. \tag{2}$$

Numerically, this pressure is (using SI units)

$$P \approx \frac{1.6 \times 10^{-42} \ [\text{N m}^3]}{d^5},$$

which is a very low value (unless the dimension d of the box is really small).

Solution 2. According to quantum theory, the ground state of a particle enclosed in a box with edges of length d is described by a wave whose half-wavelength is equal to d (i.e. $\lambda/2 = d$). The de Broglie relations then give the components of the particle's linear momentum as

$$p_x = p_y = p_z = \frac{h}{\lambda} = \frac{h}{2d}. \tag{3}$$

The energy (all kinetic) of such a 'quantum particle' is

$$E = \frac{1}{2}m(v_x^2 + v_y^2 + v_z^2) = \frac{1}{2m}(p_x^2 + p_y^2 + p_z^2).$$

Using (3), this becomes

$$E(d) = \frac{3h^2}{8m} \frac{1}{d^2}.$$

Now suppose that the edges of the box are slowly reduced by an amount $\Delta d \ll d$; its volume consequently decreases by

$$\Delta V = d^3 - (d - \Delta d)^3 \approx 3d^2 \Delta d.$$

The work required to bring this about is

$$W = P\Delta V,$$

where P is the quantum pressure to be determined. The corresponding increase in the particle's energy is

$$\Delta E = E(d - \Delta d) - E(d).$$

The work–kinetic energy theorem states that $W = \Delta E$, i.e.

$$P \cdot 3d^2 \Delta d = \frac{3h^2}{8m} \Delta \left(-\frac{1}{d^2} \right) \approx \frac{3h^2}{8m} \frac{2}{d^3} \Delta d = \frac{3h^2}{4md^3} \Delta d. \qquad (4)$$

From equation (4) we see that the required pressure is

$$P = \frac{h^2}{4m} \frac{1}{d^5}, \qquad (5)$$

which is of the same order of magnitude as the expression given in formula (2) of Solution 1. So the two solutions are compatible estimates of the pressure exerted by the neutron.

> *Notes.* 1. The energy levels of a particle restricted to a finite volume are discrete, and the energy differences between successive levels increase as the size of the volume decreases. If the environment surrounding the box has temperature T, thermal excitation, involving energy transfers that are of order kT in magnitude, need to be considered. However, provided the length d is sufficiently small, the energy transfers possible are not large enough to raise the neutron from its ground state. In exceptional cases, where d is not very small, or T is relatively high, contributions to the pressure from the neutron in excited states (corresponding to smaller wavelengths) also need to be considered.
>
> 2. In the classical limit, when the average energy of particles, each of mass m, is roughly $\frac{1}{2}kT$, and so the average square of their velocity is kT/m, the pressure exerted on the walls can be calculated from
>
> $$P = kT \frac{1}{d^3}.$$
>
> This pressure, for thermal motion at room temperature, and for a box with dimensions of the order of, say, 1 nm, is approximately 100 times larger than the pressure of quantum origin. However, if the neutron is enclosed in a box that is much smaller than this, then the latter dominates.
>
> 3. A formula for quantum pressure can be found using dimensional analysis – it gives the correct order of magnitude. If we assume that the pressure depends only

on the size of the box, the mass of the particle and the Planck constant, the only way of combining these quantities to produce an expression with the dimensions of pressure is

$$P = \text{constant} \times \frac{h^2}{md^5},$$

where the constant is a dimensionless number with an expected magnitude of order unity.

The method of *dimensional analysis* can only be applied with confidence if there are strong arguments to say that the 'target quantity' in question does *not* depend on some excluded variables. In our case, the pressure might, in principle, have depended on Boltzmann's constant, or the speed of light, or maybe Newton's universal gravitational constant; but, in fact, these constants have no part to play in the determination of quantum pressure.

S199 The electron, with mass m and charge $-e$, and the positron, with the same mass but opposite charge, both orbit with velocity v around the same circular trajectory, of radius r and centred on their common centre of mass. Their common equation of motion is

$$\frac{mv^2}{r} = k_e \frac{e^2}{(2r)^2},$$

where $k_e = 1/(4\pi\varepsilon_0)$ is the constant in Coulomb's law, and the generalised Bohr's quantum condition states that

$$2mrv = n\hbar \qquad (n = 1, 2, \ldots).$$

Using these two equations, the orbital radius, corresponding to the nth quantum state, can be calculated as

$$r_n = \left(\frac{\hbar^2}{mk_e e^2} \right) n^2,$$

and the speed in that quantum state is

$$v_n = \left(\frac{k_e e^2}{2\hbar} \right) \frac{1}{n}.$$

Finally, the total energy of the bound state is

$$E_n = 2 \times \frac{mv_n^2}{2} - k_e \frac{e^2}{2r_n} = -\frac{m(k_e e^2)^2}{4\hbar^2} \frac{1}{n^2} = -\frac{6.8 \text{ eV}}{n^2}.$$

Notes. 1. If two particles with masses m and M orbit around their common centre of mass, then their Bohr-model energy levels differ from those of Bohr's original model – with its 'infinitely heavy' attractive centre – only in that the mass of the particle needs to be replaced by the so-called *reduced mass* $\mu = mM/(m + M)$. This relationship is in line with the results of exact quantum-mechanical calculations. In the hydrogen atom, the finite mass of the proton results in only a very small correction of about 0.05 %, but in positronium $\mu = m/2$, and so the magnitudes of the energy levels are exactly one-half of the corresponding values in the original Bohr formula.

2. For two-particle systems, the generalised Bohr condition is not the same as requiring *both* of the particles to have angular momenta that are integral multiples of $\hbar = h/(2\pi)$. This is why, in these systems, the qualitative 'explanation' of the quantisation of the angular momentum, namely that the perimeter of a particle's trajectory must be an integral multiple of the de Broglie wavelength, is no longer tenable.

3. Another exotic 'atom', similar to positronium, is the *muon–hydrogen* atom, which is the bound state of a negative muon and a proton. This observable system (also, like positronium, very unstable) is, because of the relatively large mass of the muon, about 200 times smaller in size than the hydrogen atom. For this reason, the Coulomb field of the proton is largely shielded (screened) by the negatively charged muon, and so (at least in principle) it is possible for another proton, with relatively low energy, to get close to the nucleus of the muon–hydrogen atom. Earlier, physicists were very optimistic about developing a so-called *muon-catalysed fusion reactor*, which could be operated at significantly lower temperatures than those required for the fusion of normal hydrogen. Unfortunately – because of other difficulties – these expectations have not yet been realised.

S200 The SI unit of volume is metre cubed (m^3) and the SI unit of distance is metre (m). Accordingly, the conversion is as follows:

$$\frac{4 \text{ litre}}{100 \text{ km}} = \frac{4 \times 10^{-3} \text{ m}^3}{10^5 \text{ m}} = 4 \times 10^{-8} \text{ m}^2 = 0.04 \text{ mm}^2.$$

It will be noticed that the fuel consumption is expressed as an area.

The SI value for the consumption is a very small area, equal to that of a square with sides of 0.2 mm – roughly the cross-sectional area of a fairly strong thread. So, an imaginative interpretation of the fuel consumption could be as a long fuel 'thread' in the road that the car sucks up as it is driven. This kind of fuel thread, with a cross-section of 0.04 mm^2 and a length of 100 km, has a volume of just 4 litres. If the car is driven in hilly terrain, then the thread has to be thicker when driving uphill and thinner when the car is on a downslope – or maybe occasionally disappears altogether, when the electronics of the car switches off the fuel intake!

Appendix

Useful mathematical formulae

Vectors and vector operations

The magnitude of a vector a represented by the Cartesian components (a_x, a_y, a_z):

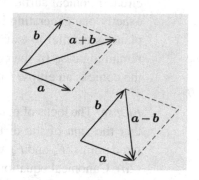

$$a = |a| = \sqrt{a_x^2 + a_y^2 + a_z^2}.$$

Addition:

$$a + b = b + a \qquad \text{(commutative)}$$
$$a + (b + c) = (a + b) + c \qquad \text{(associative)}$$
$$|a| + |b| \geq |a + b|$$

Scalar multiplication:

$$\lambda(a + b) = \lambda a + \lambda b, \qquad (\lambda + \mu)a = \lambda a + \mu a, \qquad (\lambda\mu)a = \lambda(\mu a)$$

Scalar product (or dot product): The scalar product of the vectors a and b (denoted by $a \cdot b$ or ab) returns a single number:

$$a \cdot b = |a||b| \cos\theta = a_x b_x + a_y b_y + a_z b_z,$$

where θ is the angle between a and b.

Vector product (or cross-product): The vector product of the vectors a and b (denoted by $a \times b$) returns a vector that is perpendicular to both and its magnitude is given by $|a||b| \sin\theta$. The direction of the vector $a \times b$ can be determined using the right-hand rule (see the figure).

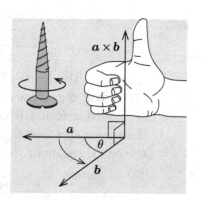

The components of $a \times b$ can be expressed with the components of a and b:

$$a \times b = \begin{pmatrix} a_y b_z - a_z b_y \\ a_z b_x - a_x b_z \\ a_x b_y - a_y b_x \end{pmatrix}.$$

Scalar triple product (or mixed product): The scalar triple product of vectors a, b and c returns a single number, which is the (signed) volume of the parallelepiped defined by the three vectors given:

$$abc \equiv (a \times b)c = a(b \times c).$$

Vector triple product expansion (Lagrange's formula):

$$a \times (b \times c) = b(ac) - c(ab)$$

Conic sections, Apollonian circle

Conic sections (or conics) are planar curves obtained as the intersection of a right-circular conical surface with a plane. If the plane that cuts the cone is parallel to exactly one generating line of the cone, then the conic is called a parabola. If the plane is parallel to exactly two generating lines, the conic is a hyperbola. In the remaining case (if the cutting plane is not parallel to any of the generating lines), the conic is an ellipse (or, in one special case, a circle).

Ellipse: The locus of points P in the plane such that the *sum* of the distances from P to two fixed points (F_1 and F_2, called foci) is constant ($2a$). Canonical equation:

$$\frac{x^2}{a^2} + \frac{y^2}{b^2} = 1.$$

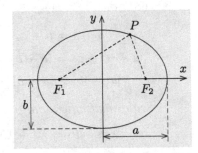

Hyperbola: The locus of points P in the plane such that the *difference* of the distances from P to two fixed points (F_1 and F_2, called foci) is constant ($2a$). Canonical equation:

$$\frac{x^2}{a^2} - \frac{y^2}{b^2} = 1.$$

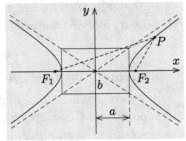

Parabola: The locus of points in the plane such that the distance to a fixed point (F, called a focus) is equal to the distance from a straight line (e, called the directrix). Canonical equation:

$$x^2 = 2py.$$

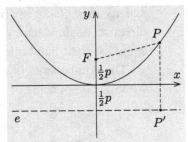

Apollonian circle: The locus of points P in the plane such that the *ratio* of the distances from P to two fixed points (A and B) is constant.
If the coordinates of the two fixed points are $(d/2, 0)$ and $(-d/2, 0)$, and the ratio of the distances is $\lambda > 1$, then the parameters of the circle are:

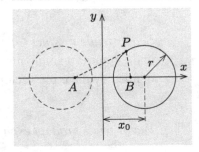

$$x_0 = \frac{d}{2} + \frac{d}{\lambda^2 - 1}, \qquad r = \frac{\lambda}{\lambda^2 - 1}d.$$

Trigonometric identities

Pythagorean trigonometric identity:

$$\sin^2\alpha + \cos^2\alpha = 1$$

Addition formulae:

$$\sin(\alpha \pm \beta) = \sin\alpha\cos\beta \pm \cos\alpha\sin\beta \qquad \tan(\alpha \pm \beta) = \frac{\tan\alpha \pm \tan\beta}{1 \mp \tan\alpha\tan\beta}$$

$$\cos(\alpha \pm \beta) = \cos\alpha\cos\beta \mp \sin\alpha\sin\beta \qquad \cot(\alpha \pm \beta) = \frac{\cot\alpha\cot\beta \mp 1}{\cot\beta \pm \cot\alpha}$$

Double-angle and half-angle formulae:

$$\sin 2\alpha = 2\sin\alpha\cos\alpha \qquad\qquad \sin^2\frac{\alpha}{2} = \frac{1 - \cos\alpha}{2}$$

$$\cos 2\alpha = \cos^2\alpha - \sin^2\alpha \qquad\qquad \cos^2\frac{\alpha}{2} = \frac{1 + \cos\alpha}{2}$$

$$\tan 2\alpha = \frac{2\tan\alpha}{1 - \tan^2\alpha} \qquad\qquad \tan\frac{\alpha}{2} = \frac{1 - \cos\alpha}{\sin\alpha}$$

Sum-to-product identities:

$$\sin\alpha + \sin\beta = 2\sin\frac{\alpha+\beta}{2}\cos\frac{\alpha-\beta}{2}$$

$$\sin\alpha - \sin\beta = 2\sin\frac{\alpha-\beta}{2}\cos\frac{\alpha+\beta}{2}$$

$$\cos\alpha + \cos\beta = 2\cos\frac{\alpha+\beta}{2}\cos\frac{\alpha-\beta}{2}$$

$$\cos\alpha - \cos\beta = -2\sin\frac{\alpha+\beta}{2}\sin\frac{\alpha-\beta}{2}$$

Approximation formulae ($|x| \ll 1$)

$$(1+x)^n \approx 1 + nx \qquad\qquad \frac{1}{1+x} \approx 1 - x$$

$$\sqrt{1+x} \approx 1 + \frac{x}{2} \qquad\qquad \tan x \approx x + \frac{x^3}{3}$$

$$\sin x \approx x - \frac{x^3}{6} \qquad\qquad \cos x \approx 1 - \frac{x^2}{2}$$

$$\ln(1+x) = x - \frac{x^2}{2} + \dots, \qquad\qquad e^x = 1 + x + \frac{x^2}{2} + \dots$$

Small differences

If the quantity x changes by a small amount Δx, then the small change in the quantity $f(x)$ (dependent on x) is

$$\Delta f \equiv f(x + \Delta x) - f(x) \approx f'(x)\Delta x.$$

In some common cases:

$$\Delta(x^2) \approx 2x\Delta x \qquad\qquad \Delta\left(\frac{1}{x}\right) \approx -\frac{\Delta x}{x^2}$$

$$\Delta(x^n) \approx nx^{n-1}\Delta x \qquad\qquad \Delta(ab) \approx a\Delta b + b\Delta a$$

$$\Delta(\sqrt{x}) \approx \frac{1}{2}\frac{\Delta x}{\sqrt{x}} \qquad\qquad \Delta\left(\frac{a}{b}\right) \approx \frac{b\Delta a - a\Delta b}{b^2}$$

$$\Delta(\sin x) \approx \Delta x \cos x \qquad\qquad \Delta(\cos x) \approx -\Delta x \sin x$$

Some useful limits

$$\lim_{x\to 0}\frac{\sin x}{x} = 1 \qquad\qquad \lim_{n\to\infty}\left(1 + \frac{x}{n}\right)^n = e^x$$

Some important sums

Sum of arithmetic sequence:

$$\sum_{k=1}^{n} k = \frac{n(n+1)}{2}$$

Sum of geometric sequence:

$$\sum_{k=0}^{n} q^k = \frac{q^n - 1}{q - 1} \qquad (q \neq 1)$$

Sum of squares of consecutive integers:

$$\sum_{k=1}^{n} k^2 = \frac{n(n+1)(2n+1)}{6}$$

Definition of the Riemann zeta function:

$$\zeta(s) = \sum_{n=1}^{\infty} \frac{1}{n^s} \qquad (s > 1),$$

and its values for selected points: $\zeta(2) = \pi^2/6$, $\quad \zeta(3) \approx 1.2021$, $\quad \zeta(4) = \pi^4/90$.

Differentiation rules, derivatives of elementary functions

$f(x)$	$f'(x)$	$f(x)$	$f'(x)$
$a \cdot f(x) + b \cdot g(x)$	$a \cdot f' + b \cdot g'$	$\ln x$	$\dfrac{1}{x}$
$f(x) \cdot g(x)$	$g \cdot f' + f \cdot g'$	$\sin x$	$\cos x$
$\dfrac{f(x)}{g(x)}$	$\dfrac{f' \cdot g - f \cdot g'}{g^2}$	$\cos x$	$-\sin x$
$f(k \cdot x)$	$k \cdot f'(k \cdot x)$	$\tan x$	$\dfrac{1}{\cos^2 x}$
x^n	$n \cdot x^{n-1}$	e^x	e^x

Integration rules, some important indefinite integrals

$f(x)$	$\int f(x)\,dx$	$f(x)$	$\int f(x)\,dx$
$x^n \quad (n \neq -1)$	$\dfrac{x^{n+1}}{n+1}$	x	$\dfrac{x^2}{2}$
x^2	$\dfrac{x^3}{3}$	$\dfrac{1}{x}$	$\ln |x|$
$\dfrac{1}{x^2}$	$-\dfrac{1}{x}$	$\sqrt{x}$	$\frac{2}{3}\sqrt{x^3}$
$\sin x$	$-\cos x$	$\cos x$	$\sin x$
$\tan x$	$-\ln |\cos x|$	e^x	e^x

Fundamental theorem of calculus (Newton–Leibniz formula):

$$\int_a^b f(x)\,dx = F(b) - F(a), \quad \text{where} \quad F'(x) = f(x)$$

Integration by parts:

$$\int_a^b f'(x)g(x)\,dx = \Big[f(x)g(x)\Big]_a^b - \int_a^b f(x)g'(x)\,dx$$

Important definite integrals:

$$\int_0^{\pi/2} \sin x\,dx = \int_0^{\pi/2} \cos x\,dx = 1, \qquad \int_0^{\pi/2} \sin^2 x\,dx = \int_0^{\pi/2} \cos^2 x\,dx = \frac{\pi}{4}$$

Some differential equations (with solutions) appearing in the book

$$y'(x) = 0 \qquad \longrightarrow \qquad y(x) = \text{arbitrary constant}$$

$$y'(x) = a \qquad \longrightarrow \qquad y(x) = ax + b$$

$$y''(x) = a \qquad \longrightarrow \qquad y(x) = a\frac{x^2}{2} + bx + c$$

$$y''(x) = -k^2 y(x) \qquad \longrightarrow \qquad y(x) = A_1 \cos(kx) + A_2 \sin(kx)$$

$$y'(x) = ky(x) \qquad \longrightarrow \qquad y(x) = y_0\, e^{kx}$$

(where b, c, A_1, A_2, y_0 are constants determined by the boundary conditions.)

Surface area (A) and volume (V) of some objects

sphere: $A = 4R^2\pi$

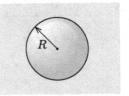

$$V = \tfrac{4}{3}R^3\pi$$

right-circular cone: $A = \pi r(r + a)$

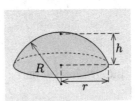

$$V = \tfrac{\pi}{3}r^2 h$$

spherical cap: $A = 2\pi Rh + r^2\pi$

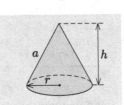

$$V = \tfrac{\pi}{3}h^2(3R - h)$$

cylindrical segment: $V = \tfrac{1}{2}hr^2(\alpha - \sin\alpha)$

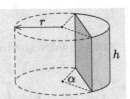

Position of the centre of mass of uniform bodies

triangular plate: $\qquad s = \dfrac{h}{3}$

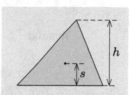

arc of a circle: $\qquad s = r\dfrac{\sin\alpha}{\alpha}$

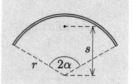

circle sector: $\qquad s = \dfrac{2}{3}r\dfrac{\sin\alpha}{\alpha}$

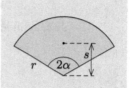

solid spherical cap: $\qquad s = \dfrac{R^2 - hR + (h^2/4)}{R - (h/3)}$

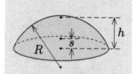

cap of spherical shell: $\qquad s = \dfrac{h}{2}$

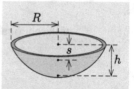

cone, pyramid: $\qquad s = \dfrac{h}{4}$

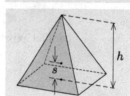

Moments of inertia of uniform bodies

thin rod: $\dfrac{1}{12}mL^2$

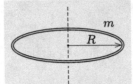

thin ring: mR^2

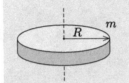

disc, cylinder: $\dfrac{1}{2}mR^2$

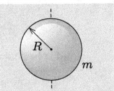

solid sphere: $\dfrac{2}{5}mR^2$

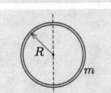

thin spherical shell: $\frac{2}{3}mR^2$

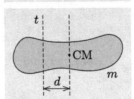

parallel axis theorem: $\Theta_t = \Theta_{CM} + md^2$

Physics reference tables

Fundamental physical constants

	notation	value	units
gravitational constant	G	6.673×10^{-11}	$N\,m^2\,kg^{-2}$
speed of light in vacuum	c	2.998×10^{8}	$m\,s^{-1}$
universal gas constant	R	8.314	$J\,mol^{-1}\,K^{-1}$
Avogadro constant	N_A	6.022×10^{23}	mol^{-1}
Boltzmann constant	k	1.381×10^{-23}	$J\,K^{-1}$
vacuum permittivity	ε_0	8.854×10^{-12}	$A\,s\,V^{-1}\,m^{-1}$
vacuum permeability	μ_0	$4\pi \times 10^{-7}$	$V\,s\,A^{-1}\,m^{-1}$
Coulomb's constant	k	8.988×10^{9}	$N\,m^2\,C^{-2}$
Stefan–Boltzmann constant	σ	5.670×10^{-8}	$W\,m^{-2}\,K^{-4}$
elementary charge	e	1.602×10^{-19}	C
Planck constant	h	6.626×10^{-34}	$J\,s$

Astronomical data of the Earth

mean radius	$6\,371$	km
radius at the Equator	$6\,378$	km
radius at the poles	$6\,357$	km
mass	5.973×10^{24}	kg
average density	$5\,514$	$kg\,m^{-3}$
moment of inertia (about the axis of rotation)	8.04×10^{37}	$kg\,m^2$
average distance from the Sun (= 1 AU)	1.496×10^{11}	m
smallest distance from the Sun	1.471×10^{11}	m
largest distance from the Sun	1.521×10^{11}	m
mean orbital speed around the Sun	29.78	$km\,s^{-1}$
largest orbital speed	30.29	$km\,s^{-1}$
smallest orbital speed	29.29	$km\,s^{-1}$
equatorial speed	465	$m\,s^{-1}$
gravitational acceleration in London	9.81	$m\,s^{-2}$

Data of the Sun and the Moon

the Sun

mean angular diameter	31′ 59″	
diameter	1.392×10^9	m
mass	1.989×10^{30}	kg
surface temperature	5 780	K
core temperature	*ca.* 1.5×10^7	K

the Moon

mean angular diameter	31′ 5″	
mean radius	1 737	km
mass	7.347×10^{22}	kg
average density	3 340	kg m^{-3}
moment of inertia (about the axis of rotation)	8.04×10^{37}	kg m^2
mean distance from Earth	3.844×10^8	m
orbital period around the Earth	27.32	days
gravitational acceleration on its surface	1.62	m s^{-2}
surface temperature at daytime	*ca.* 130	°C
surface temperature at night	*ca.* −150	°C

Planets and their average distance from the Sun

	AU	10^6 km
Mercury	0.387	57.9
Venus	0.723	108.2
Earth	1.000	149.6
Mars	1.524	227.9
Jupiter	5.203	778.3
Saturn	9.555	1 429
Uranus	19.22	2 875
Neptune	30.11	4 504
(Pluto)	(39.55)	(5 916)

The (rest) mass of some particles

	MeV/c^2	10^{-30} kg
photon	$0\ (< 10^{-24})$	
neutrino	$\leq 2 \times 10^{-6}$	
electron	0.511	0.9109
muon	105.6	
pion(charged)	139.6	
pion(neutral)	135.0	
proton	938.3	1 672.6
neutron	939.6	1 674.9
Higgs boson	1.26×10^5	

Densities of some materials, in normal state (kg m^{-3})

hydrogen	0.089	wood (pine)	480–620
helium	0.178	aluminium	2 700
air	1.293	silicon carbide (SiC)	3 210
water (at 4 °C)	1 000	titanium	4 510
ice (at 0 °C)	920	iron	7 860
ethanol	790	copper	8 960
dry sand	1 300–1 600	mercury	13 550
porcelain	2 200–2 500	tungsten	19 250
quartz	2 650	platinum	21 450

Thermodynamical properties of some materials

specific latent heat of fusion of ice	334	kJ kg^{-1}
heat of vaporisation of water	2 256	kJ kg^{-1}
molar heat of vaporisation of water	40.6	kJ mol^{-1}
specific heat of water	4 180	J kg^{-1} K^{-1}
specific heat of ice (at 0 °C)	2 090	J kg^{-1} K^{-1}
thermal conductivity of ice	2.3	W m^{-1} K^{-1}
thermal conductivity of water	0.56	W m^{-1} K^{-1}
boiling point of ether (at normal pressure)	34.6	°C
boiling point of oxygen (at normal pressure)	90.2	K
boiling point of nitrogen (at normal pressure)	77.4	K
specific latent heat of fusion of iron	272	kJ kg^{-1}
specific latent heat of fusion of nickel	292	kJ kg^{-1}
specific latent heat of fusion of aluminium	361	kJ kg^{-1}
specific heat of porcelain	800–900	J kg^{-1} K^{-1}
specific heat of quartz glass	700	J kg^{-1} K^{-1}
specific heat of quartz sand	830	J kg^{-1} K^{-1}
specific heat of steel	470	J kg^{-1} K^{-1}
linear thermal expansion coefficient of iron	1.18×10^{-5}	K^{-1}
linear thermal expansion coefficient of steel	$1.1\text{–}1.7 \times 10^{-5}$	K^{-1}

Density and pressure of saturated water vapour and density of water in vapour–liquid equilibrium as a function of temperature

T	p_{vap}	ϱ_{vap}	ϱ_{water}	T	p_{vap}	ϱ_{vap}	ϱ_{water}
(°C)	(kPa)	(kg m^{-3})		(°C)	(kPa)	(kg m^{-3})	
0.01	0.612	0.005	999.8	125	232	1.30	939.1
5	0.873	0.0068	999.9	150	476	2.55	917.1
10	1.23	0.0094	999.7	175	892	4.61	892.3
20	2.34	0.0173	998.2	200	1 554	7.86	864.7
30	4.24	0.0304	995.6	225	2 548	12.7	833.9
40	7.38	0.0511	992.2	250	3 974	20.0	799.1
50	12.3	0.0831	988.0	275	5 943	30.5	759.2
60	19.9	0.130	983.2	300	8 584	46.1	712.4
70	31.2	0.198	977.8	325	12 050	70.5	654.6
80	47.4	0.293	971.8	350	16 520	113.5	574.7
90	70.1	0.423	965.3	370	21 030	200.3	453.1
100	101.3	0.597	958.4	374.2*	22 060	326.2	326.2

*$T = 374.2\,°C$ is the critical temperature of water; the corresponding pressure and density are the critical pressure and critical density, respectively.

Ultimate tensile strength of some materials (MPa)

aluminium alloy	300–700
silicon carbide (SiC)	3 440
titanium alloy	540–1 300
iron	200
steel	400–1 800
copper	210–240
tungsten	400–1 200
platinum	120–220
glass fibres	3 100–4 800
Fe$_{80}$B$_{20}$ glassy metal	3600
spider silk	*ca.* 1 000

Surface tensions (in air environment)

surface tensions	(N m^{-1})
water (at 20 °C)	0.073
mercury	0.472
ethanol	0.227
glycerol	0.064

contact angles	
water–(clean) glass	0°
water–silver	90°
mercury–glass	140°
water–teflon	110°

Printed in the United States
by Baker & Taylor Publisher Services